AF646316

A Book of Physics
- In Perspective

[for the students of ICSE/CBSE, State Boards of Secondary and Intermediate Studies]

A Book of Physics
- In Perspective

[for the students of ICSE/CBSE, State Boards of Secondary and Intermediate Studies]

Dr. S. C. BHARGAVA
Retired Specialist (Electromagnetic Phenomena),
BHEL, Corporate R & D, Hyderabad
and variously Visiting Professor of Electrical Engineering
in
Engineering Institutes of Andhra Pradesh

BS Publications

BSP A unit of **BSP Books Pvt. Ltd.**
4-4-309/316, Giriraj Lane, Sultan Bazar,
Hyderabad - 500 095
Phone : 040 - 23445688, 23445600

A Book of Physics - *In Perspective by S.C. Bhargava*

© *Copyright* 2018, *by the Author*

All rights reserved. No part of this book or parts thereof may be reproduced, stored in a retrieved system or transmitted in any language or by any means electronic, mechanical, photocopying, recording or otherwise without prior written permission of the author.

Published and distributed by:

BS Publications

A unit of **BSP Books Pvt. Ltd.**

4-4-309/316, Giriraj Lane, Sultan Bazar,
Hyderabad – 500 095, India
Phone : 040 – 23445688, 23445600
e-mail : info@bspbooks.net
www.bspbooks.net

Printed at

Reprographic and Printing Services
13/3, Rasoolpura,
Begumpet,
Secunderabad 500 003. India

ISBN: 978-93-86819-96-3

Price: BSP BOOKS ₹695

Cover design: B.S. Publications Pvt. Ltd.

Science is all pervasive and physics is the essence of it

- anonymous

to

the 'students'

of

physics

Contents*

*Largely in accordance with the ICSE syllabus for 10th class physics

UNIT 4
Electricity and Magnetism

UNIT 5
Heat

UNIT 6
Modern Physics

Preface

Most physics books based on syllabi for class X whilst designed in a rather unfriendly manner, style and get-up for the students, are cluttered up with details not quite in sync with the subject matter and focus largely on a huge number of solved problems here and there and multitude of questions clubbed under "exercises" at the end of each topic, not just at the end of a chapter. The end result is that the students whilst not really learning the subject as it should be are bogged down by having to concentrate on solving countless problems; this being emphasised by the teachers, too, in the class. This is, in fact, counterproductive to learning the important 'science of physics' and a disservice to pupils who are discouraged to learn real facets of various topics of physics. The students' precious time and energy is mostly used up in solving nearly *all* of exercise questions with the aim of preparing for the board exam and "scoring good marks", almost 100 %! Some of the topics such as "Electricity and Magnetism" – so important in day-to-day life - whilst poorly documented in text books, with numerous errors and misleading figures and diagrams, fail to make it to the students in real sense.

In light of above, the need for a 'proper' book of physics judiciously dealing with most important topics of physics was long felt by this author who, after a painstaking task of reviewing the contents of some of the available books and lacunae therein, has developed the present script.

The salient feature of the book in hand is the presentation of all the relevant topics of physics ***in perspective***, thus providing an insight into the importance and prevalence of physics in the day-to-day phenomena of life. Whilst largely conforming to the revised ICSE syllabus of X class physics for board exams, the contents of the book are comprehensive and useful for students of CBSE and other boards of secondary education as well.

The presentation of subject matter in perspective naturally led to including matter related to numerous topics in addition to just following the syllabus and this pace has been maintained throughout the book. A specific example of this aspect being 'touching' upon "stereophonic sound" in Unit 3 on Sound and significance of this sub-topic in practice where nearly all sounds are stereophonic in nature as perceived by humans, gifted with TWO ears!

The contents of the book and the scope of presentation is especially directed to those students, and teachers, who have an innovative mindset and would be interested in learning the various concepts of topics, *not just limited to details listed in the syllabus and preparing for exams, but aiming for much more!*

The book whilst being objectively different from most other books of physics, would prove immensely helpful to the teachers and parents in understanding physics conceptually, and thus help them guide the students in learning the subject in a more simplified and meaningful manner and yet preparing them well for the board exams.

The book is enriched with numerous solved numerical problems, with *step-by-step* solution of each, at appropriate stage following discussion of a given topic and several questions in the exercises at the end of each chapter for practice by the students. Most questions, solved or unsolved, are adopted from the set of questions of ICSE board exams across ten years or so, to let the students have a feel of questions that are asked in the board exams.

The style of presentation of matter in the book is aimed at being *student friendly*, in an easy to follow language, and interspersed with helpful comments and footnotes for good grasping of the topics in each chapter.

An important asset of the book is inclusion of an extensive, comprehensive index, as also author/scientist index, that can be used by students as a "search engine" to help them locate a subject matter of interest or physical term in the book with much ease.

~ **The author**

Acknowledgements

At the outset the author would like to thank his granddaughter, Misha, the then student of ICSE, 10th class, as being largely instrumental in the initiation of writing of this book. She has also been a constant help during the course of the writing by providing several suggestions and comment, and checking of proofs at later stage.

The author is grateful to various colleagues, esp. (late) Shri A. H. Devadas, the author's colleague in BHEL, Corporate R & D, for his 'invention' of unique electromagnetic devices, some of these being discussed in the book, and resource persons for providing numerous numerical problems with solutions and other details relevant to different topics covered in the book.

His special thanks are due to his daughter, Anshu, for her untiring help in preparation of the final computer-script of the book, esp. in organising a number of equations etc. variously and "formatting" the matter involving photographic figures and illustrations.

The author would also like to put on record the painstaking and untiring efforts of Mr. Babu Rao of Reprographic and Printing Services in the design and formatting of the text till its final stage.

~ The author

About the Author

Dr S C Bhargava, having obtained his Ph D from the University of Aston in Birmingham, UK, served as Specialist, Electromagnetic Phenomena, at Corp. R & D, B H E L, Hyderabad, from 1974 to 1999, and, post retirement, has been a visiting faculty at several engineering institutes in Hyderabad, teaching graduate and postgraduate courses. He has also been variously associated, directly and indirectly, with a large number of ICSE schools in the twin cities for past several decades.

He is the Fellow or Senior Member of several Professional Societies such Institution of Engineers (India), Institution of Electrical Engineers (UK), Institution of Electronics and Telecommunication Engineers (India), Institute of Electrical and Electronics Engineers (USA) and Computer Society of India.

His fields of interest include "Biological effects of electromagnetic fields on living beings" and "Solar electricity".

He has published over 40 technical papers in national and international journals and conferences.

He is the author of the book "Electrical Measuring Instruments and Measurements", published in 2013 by B S Publications in Hyderabad and CRC Press in the UK. The book has been approved by the JNTU as one of the text books on electrical measurements, a subject taught to EEE students of the University in III semester.

His other book, titled "Finite Element Analyses of Eddy Current Effects in Turbogenerators", dealing with special problems of turbogenerators, and based on his vast experience related to electromagnetic fields and turbogenerators, has been recently published by CRC Press, Taylor & Francis Group, UK.

General List of Symbols, Abbreviations and Notation*

a	acceleration
	area
A	ammeter
	area
	atomic mass number
AU	angstrom unit (of wavelength of light)
B	flux density
BM	bending moment
c	critical angle
	specific heat
	velocity of light
C	angle of incidence
	centigrade (MKS unit of temperature)
	centre of curvature
	current
C’	heat capacity
CG	centre of gravity
d	distance
dB	decibel (unit associated with sound)
D	diopter (unit of power of lens)
	distance/displacement
DPST	double-pole single-throw (switch)
e	instantaneous induced voltage
eV	electron volt
E	effort
	energy
EA	effort arm
EMF	electromotive force
f	focal length
	frequency
F	Fahrenheit (MKS unit of temperature)
	force
	moment
g	acceleration due to gravity

*It may be noted that the same quantity is denoted differently variously in the book. This is followed inadvertently to maintain flow of presentation in a given chapter.

G	giga
GPE	gravitational potential energy
h	height
hp	horse power
H	heat (energy)
HP	horse power
i	angle of incidence
	instantaneous current
I	current
	intensity (of sound)
IF	infrared
J	joule (MKS unit of energy)
K	kelvin (MKS unit of absolute temperature)
	kinetic energy
l	length
L	latent heat
	length
	load
	loudness
LA	load arm
m	mass
	milli
M	magnification
	mass
	mega
	atomic mass number
MA	mechanical advantage
MCB	miniature circuit breaker
n	nano
	number of turns
	refractive index
N	number of turns
p	pico
	momentum
	power

P	number of poles
	power
PE	potential energy
PVC	photo voltaic cell
Q	charge
r	angle of refraction
	radius
	resistance
R	resistance
s	length of arc
S	displacement
SPDT	single-pole double-throw (switch)
SPST	single-pole single-throw (switch)
Sv	sievert (unit of radiation)
t	temperature
	time
T	temperature
	tension
	tera
	torque
TC	thermocouple
u	atomic mass unit
	distance of object from optical centre
U	gravitational potential energy
US	ultrasound
UV	ultra-violet
v	distance of image from optical centre
	EMF
	velocity
V	electrical potential
	velocity
	voltage
VR	velocity ratio
w	weight
W	watt (MKS unit of electric power)
	work
WE	water equivalent of calorimeter
Wh	watt-hour (MKS unit of electrical energy)
Z	atomic number

Non-English symbols

α	acceleration
	alpha particles
	temperature coefficient
β	beta particles
γ	gamma radiation
δ	angle of deviation
Δ	increment
η	efficiency
θ	angle
	angle of incidence
λ	wave length
μ	micro
	refractive index
π	constant [3.14157 . . .]
ρ	resistivity
τ	time period
	torque
ϕ	angle
	angular displacement
	magnetic flux
ω	angular velocity
Ω	ohm (MKS unit of resistance)
℧	mho (unit of conductance)
Σ	summation
α, ∝	proportional to

1. Force, Work, Power and Energy

Unit 1: Topics covered

Force, Work, Power and Energy

(i) Turning forces concept; moment of a force; forces in equilibrium; centre of gravity; [discussions using simple examples and simple numerical problems].

Elementary introduction of translational and rotational motions; moment (turning effect) of a force, also called torque and its cgs and SI units; common examples - door, steering wheel, bicycle pedal, etc,; clockwise and anti-clockwise moments; conditions for a body to be in equilibrium (translational and rotational); principle of moment and its verification using a metre rule suspended by two spring balances with slotted weights hanging from it; simple numerical problems; Centre of gravity (qualitative only) with examples of some regular bodies and irregular lamina.

(ii) Uniform circular motion.
As an example of constant speed, though acceleration (force) is present. Differences between centrifugal and centripetal force.

(iii) Work, energy, power and their relation with force.

Definition of work. $W=FS\cos\theta$; special cases $\theta=0^0$, 90^0. $W=mgh$ Definition of energy, energy as work done. Various units of works and energy and their relation with SI units. [erg, calorie, kWh and eV]. Definition of Power, $P=W/t$; SI and cgs units; other units, kilowatt (kW), megawatt (MW) and gigawatt (GW); and horse power (1hp=746W) [Simple numerical problems on work, power and energy].

(iv) Different types of energy (e.g. chemical energy, mechanical energy, heat energy, electrical energy, nuclear energy, sound energy, light energy).

Mechanical energy: potential energy $U=mgh$ (derivation included) gravitational PE, examples; kinetic energy $K=\frac{1}{2}mv^2$ (derivation included); forms of kinetic energy: translational rotational and vibrational - only simple examples. [Numerical problems on K and U only in case of translational motion]; qualitative discussions of electrical, chemical, heat, nuclear, light and sound energy, conversion from one form to another; common examples.

(v) Machines as force multipliers; load, effort, mechanical advantage, velocity ratio and efficiency; simple treatment of levers, pulley systems showing the utility of each type of machine.
Functions and uses of simple machines: Terms- effort E, load L, mechanical advantage $MA = L/E$, velocity ratio $VR=V_E/V_L=d_E/d_L$, input (W_I), output (W_0), efficiency (η), relation between η and MA, VR (derivation included): for all practical machines $\eta<1$; $MA<VR$.
Lever: principle. First, second and third class of levers; examples: MA and VR in each case. Examples of each of these classes of levers as also found in the human body.
Pulley system: single fixed, single movable, block and tackle; MA, VR and η in each case.
Gears; *types of gear, applications, analysis of gear pair, gear ratio, uses*

(vi) Principal of Conservation of energy.

Statement of the principle of conservation of energy; theoretical verification that $U + K$ = constant for a freely falling body. Application of this law to simple pendulum (qualitative only; [simple numerical problems].

I Force

Turning forces concept; moment of a force; forces in equilibrium; centre of gravity; [discussions using simple examples and simple numerical problems]

Elementary introduction of translational and rotational motion; moment (turning effect) of a force, also called torque and its cgs and SI units; common examples – door, steering wheel, bicycle pedal etc.; clockwise and anti-clockwise moments; conditions for a body to be in equilibrium (translational and rotational); principle of moment and its verification using a metre rule suspended by two spring balances with slotted weights hanging from it; simple numerical problems; centre of gravity (qualitative only) with examples of some regular bodies and irregular lamina.

GENERAL

Force[1]

In simple terms, a **force** is a push or pull acting upon an object as a result of its interaction with another object. In most cases, the action results in a 'change' of physical position of the object(s)

Alternatively,

the concept of force can be best comprehended by reference to Newton's first law of motion stated as
"an object will remain at rest or in a constant state of motion unless acted upon by net external forces".

If interpreted differently, the object may be imagined to be stationary, that is with no or 'zero' motion being at rest, and would continue to be in that state unless made to change its

[1]From Latin *fortis* meaning "strong". In English, the word "force" can assume numerous meaning and connotation; for example, coercion, strength, organised body of soldiers in army or police, influence, drive etc. to name few. Also, colloquially, "a force to reckon", "force the issue", force a person's hand" (make a person act prematurely or unwillingly) etc.

position. The agency or physical entity that can bring about the change of state of an object from rest or *uniform* motion can be termed or defined as FORCE. In terms of action, this may entail a push or pull of the object and hence force can be defined as a **push** or **pull** in a layman's language.

Some examples of push and/or pull force are illustrated below.

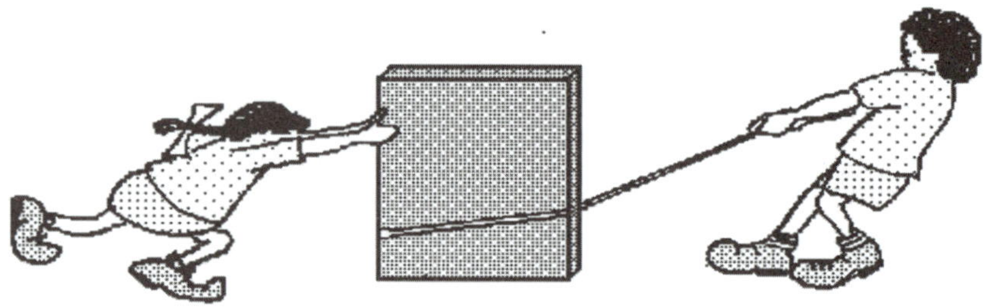

(a) push and pull forces

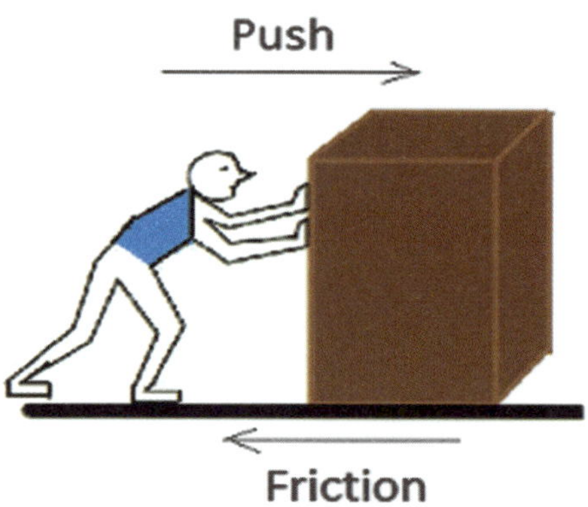

(b) push force
[observe the action of friction force]

(c) pull force

Fig.1.1 : Examples of pull and push forces

Note that the tendency of an object to be at rest or in uniform motion can also be recognised as the concept of inertia due to Galileo[2] , often termed simply the "Law of Inertia".

Contact and Non-contact Force

Contact force

In physics, a **contact force** is a force that acts at the point of contact between two objects, without coming physically in contact with them.

---

[2]Galileo Galilei Italian polymath: 1564-1642.

The Various Types of Contact Force

- Applied Force
- Frictional Force
- Air Resistance Force
- Spring Force
- Tension Force
- Normal Force

Applied Force:
Force which is applied to an object by another object. A person pushing his car is an example of applied force. When the person pushes the car, there is a force acting upon the car. The applied force is the force exerted on the car by the person.

Frictional Force:
Friction force is the result of two surfaces being pressed together closely.

This causes intermolecular attractive forces between molecules of different surfaces resulting in friction. The frictional forces depend upon the nature of the surfaces. The more rough the surface the more the frictional force.

Air Resistance Force:
The Air resistance force acts upon objects as they travel through the air. This force opposes the motion of the object in the air; for example, a skydiver diving in the sky or a train running at hundreds of km/hr speed.

Spring Force:
The Spring force is the force exerted by a compressed or stretched spring upon any object which is attached to it. Here the magnitude of the spring force is directly proportional to the amount of stretch or compression of the spring.

Tension Force:
The tension force is transmitted through a string, rope or cable. The tension force is directed along the length of the wire. This force pulls the objects equally on the opposite ends of the wire.

Normal Force:
This is also called the support force. If an object is resting upon a table then the table is exerting an upward force upon the object in order to support the weight of the object. If a person leans against a wall, the wall pushes horizontally on the person.

Some Images of Contact Force

A few images of contact force are illustrated in Fig.1. 2.

(a) a boy and girl pushing a child in a cart

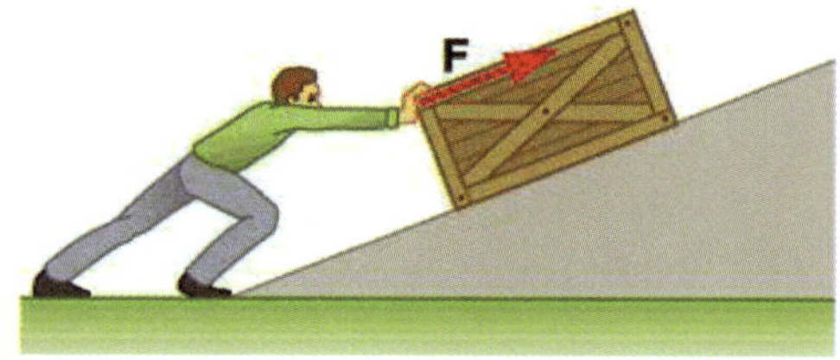

(b) a man pushing a box up an incline support

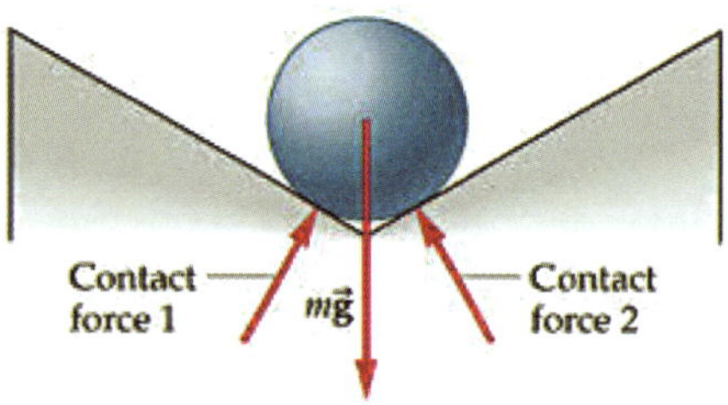

(c) a ball resting in a cut section

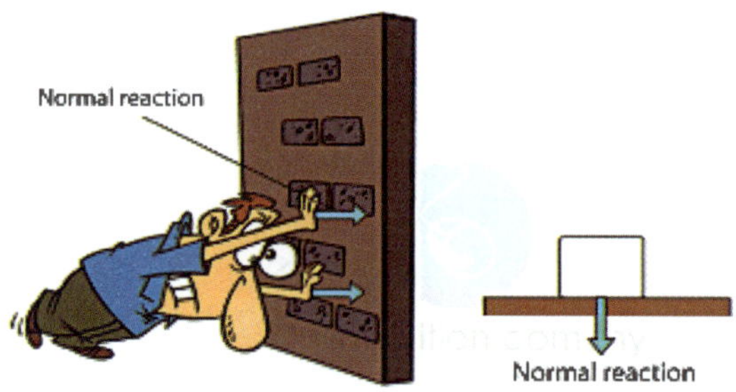

(d) a man pushing against a vertical wall

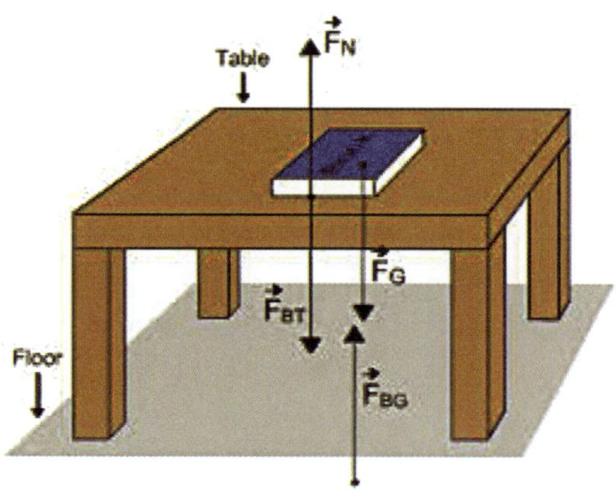

(e) a book resting on a table

(f) a boy kicking a ball

Fig.1.2 : Images of contact force

Non-contact force

The force which acts on an object without coming physically in contact with it is called **non-contact force**.

Some of the commonly known non-contact forces

- Gravity:

A well-known example is the force due to gravity or gravitational force; the force that is exerted by way of a pull on a body such as the one falling from a height. The force is given by

force = mass of the body x acceleration due to gravity

that is obtained in newton if the mass is in kg and acceleration, g, is m/s/s. The other example is that of the force of attraction that exists among all bodies that have mass. The force exerted on each body by the other through weight is proportional to the mass of the first body times the mass of the second body divided by the square of the distance between them.

- Magnetic Force:

A permanent magnetic has a field of its own and therefore can exert a force of push or pull, that is attraction or repulsion, on a piece of iron or steel, or even another magnet, that would depend on the distance of the piece from the magnet.

- Force due to Electromagnetism:

An electromagnet when excited, that is, when a current flows through its winding, exerts a pull or push, or a force of attraction or repulsion on a ferromagnetic body, similar to that of a permanent magnet. The force comes into play when the body is not in contact with the electromagnet, the strength of the force being inversely proportional to the distance, in general.

- Electrostatic Force:

A body possessing electrostatic charge can exert an attractive or repulsive force on a body at a distance by virtue of electrostatic induction.

- Nuclear Force:

Unlike gravity and electromagnetism, the strong nuclear force is a 'short- distance' force that takes place between basic particles within a nucleus. It is charge independent and acts equally between a proton and a proton, a neutron and a neutron, and a proton and a neutron. The strong nuclear force mediates both nuclear fission and fusion reactions.

Some images of non-contact force are illustrated in Fig.1.3.

(a) force due to gravitation

(b) inter-planetary gravitational force

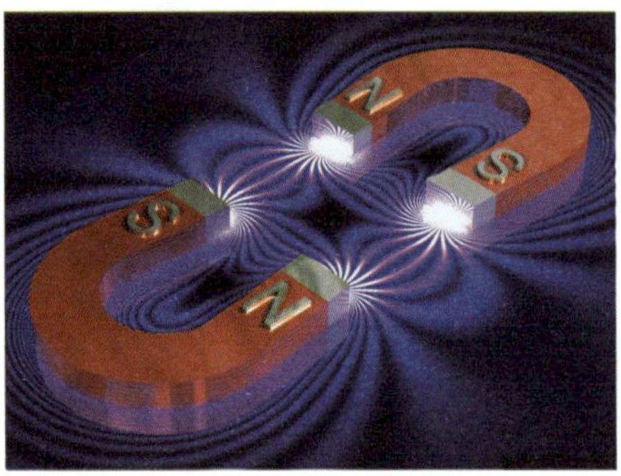

(c) force due to permanent magnet(s)

(d) magnetic pull of an electromagnet

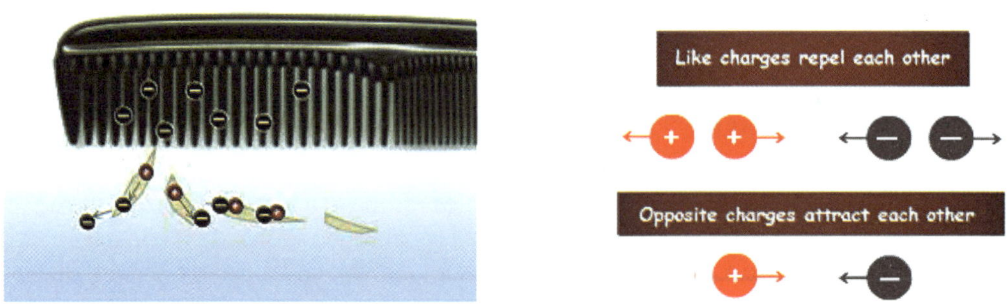

(e) electrostatic force

Fig.1.3 : Images of non-contact force

Friction

An important aspect often associated with force is **friction.** The tendency of friction is to oppose the applied force, in general; that is, to act in a direction *opposite* to the direction of force. Whilst friction is essential in relation to various activities in day-to-day life, including making it possible for one to walk, its detrimental effect, particularly in moving or rotating machines such as a motor, is to act in retarding the motion and thus lower the performance and efficiency of the machine; the frictional power loss being subtracted from the useful power produced in the machine. In such applications, it is much desirable to reduce friction by, for example, lubricating the moving parts like bearings; in the same vein apply oil to rusted hinges of a door to make it open or close smoothly.

Quantitative Interpretation of Force

Quantitatively, the first factor to specify the quantum of force required to change the state of an object or body would be the *mass* of the body, m: greater the mass, greater would be the force. The second aspect is the phenomenon that follows the application of the force. If the body is at rest in a given position, the force applied to it at a point or surface will result in its movement to acquire a new position, the body moving with a velocity. However, the velocity (v) is not reached instantaneously, but takes some time (t). The same arguments apply to a body already in motion at an initial velocity v_o that is changed to v_1 on application of the force. This change of *velocity* from initial (= 0 for body at rest) to final brings in the physical entity called acceleration, defined by "rate of change of velocity with time", or dv/dt, and denoted by f or a. Thus, the term force required to change the state of the body will be a product of the mass of the body and the acceleration caused by the force, or simply

$$F = m \times f$$

in any system of units.

Newton's Second Law of Motion

The concept of force is also brought out by Newton's second law of motion[3] stated as "The relationship between an object's mass m, its acceleration a, and the applied force F is F = ma both, acceleration and force being vector quantities, the latter directed *in the direction of acceleration*"[4].

---

[3]Sir Isaac Newton, FRS, English mathematician: 1643-1727.

[4]This is the most powerful of Newton's three Laws, because it allows quantitative calculations of force in the dynamics: how do velocities change when forces are applied. In contrast, according to Aristotle [Greek philosopher: 384(BC) – 322(BC)], the force is defined by F = m x v where v is the *velocity* acquired by the body under the action of the force. Thus, according to Aristotle there is only one velocity if there is a force, but according to Newton an object with a certain velocity maintains that velocity *unless* a force acts on it to cause an acceleration (that is, a change in the velocity).

Or

"The acceleration of an object is directly related to the net force acting on it and inversely proportional to its mass".

Unit of Force

The SI unit of force is newton, named after the great British scientist; the symbol of the unit being N.

By definition,

One unit of force or one newton (in SI units) = one unit of mass in kg x one unit of acceleration in m/s/s

and similarly in other systems of unit[5].

Example

Two bodies have masses in the ratio of 3 : 4. When a force is applied to the first body, it moves with an acceleration of 6 m/s/s. What will be the acceleration produced in the second body by the same force?

Let the mass of body A be 3 kg and that of body B be 4 kg.
The force acting on body to produce an acceleration of 6 m/s/s will be

$$F = 3 \times 6 \text{ or } 18 \text{ N}$$

The same force when applied to body B will result in an acceleration of

$$\alpha = 18/4 \text{ or } 4.5 \text{ m/s/s}$$

Translational and Rotational Motions

Translational motion

Translational motion is the motion by which a body shifts from one point in space to another. One example of translational motion is the motion of a bullet fired from a gun. The other common examples of translational motion are:

- a car moving on a road in a given direction

---

[5]For example, in CGS system the unit is called dyne defined as a mass of one gramme x acceleration of one cm/s^2.

- a train travelling on a 'straight' railway track

Rectilinear Motion

Rectilinear motion is another name for straight-line motion of a body or object. A body is said to experience rectilinear motion if *any two* particles of the body travel the same distance along two parallel straight lines as depicted in Fig.1.4.

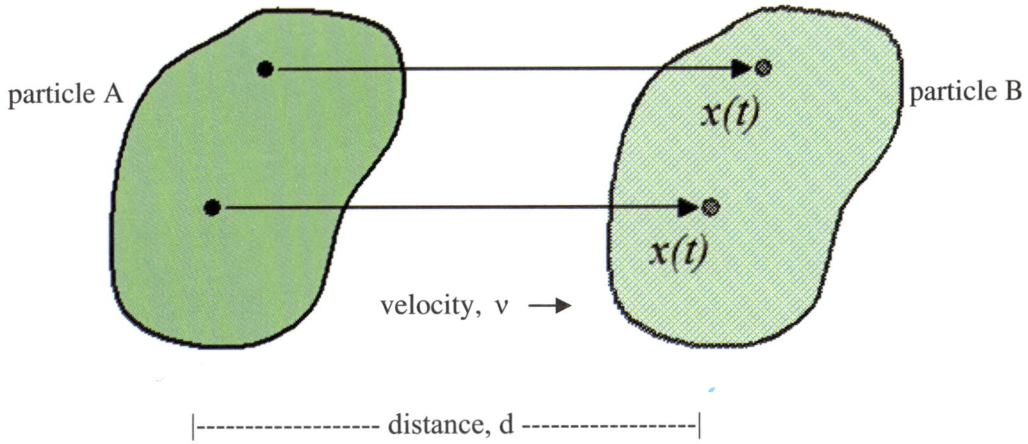

Fig.1.4 : Rectilinear motion

As shown in the figure, particles A and B of a body *move together* in direction x and reach a position defined by x(t) at time t, having moved a distance d units. For a body in rectilinear motion as a whole, the relationship of the distance in terms of a *constant* velocity, v, and time t is expressed by a linear equation for a rectilinear motion, given by

$$d = v \times t$$

in a given system of units.

It is important to note that a rectilinear motion is characterized, in general, by

- a constant velocity
- a varying velocity due to a constant acceleration
- a varying velocity accompanied by a varying acceleration
  with time, having their own implications in practice to relate the quantities with displacement, velocity and acceleration.

Equations governing rectilinear motion

In case of a *constant* acceleration, the FOUR physical quantities, viz. acceleration, velocity, time and displacement, can be related by using the equations

$$V_{(final)} = V_{(initial)} + at$$

$$D = V_{(initial)} + \frac{1}{2}at^2$$

$$V_{(final)^2} = V_{(initial)^2} + 2aD$$

$$D = \frac{1}{2}\,[V_{(final)} + V_{(initial)}]\,t$$

where

$V_{(initial)}$ is the initial velocity
$V_{(final)}$ is the final velocity
a is the acceleration
D is the displacement, and
t is the time

in appropriate system of units, pertaining to the body in motion.

These relationships can be demonstrated graphically as depicted in Fig.1.5. The gradient of a line on a displacement time graph represents the velocity. The gradient of the velocity time graph gives the acceleration while the area under the velocity time graph gives the displacement. Thus, in Fig.1.5

OA: initial velocity u at t = 0
OC: velocity v at time t
$= u + \alpha t$
α : (constant) acceleration
= slope of line AB
$[\tan\theta]$

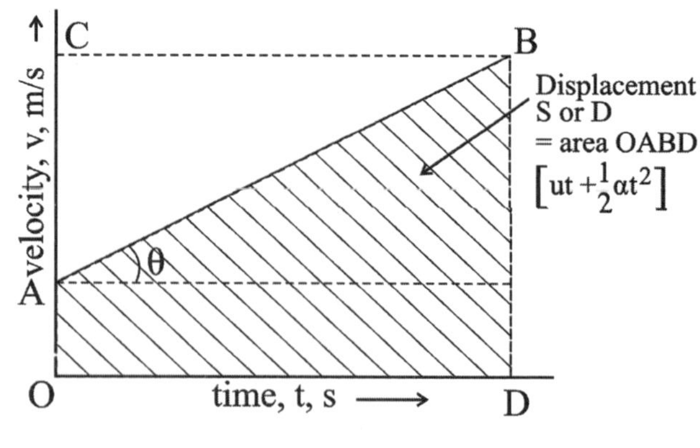

Fig.1.5 : Graphical representation of velocity time relationship and equations of motion in rectilinear mode

Thus, referring to the figure, line AB represents the motion with an initial velocity u at t=0 and increasing with a constant acceleration a. At the end of time t, the velocity is v. The slope of the line being (v-u)/t is simply the acceleration, giving $v = u + at$. The area of the trapezium OABD represents the distance traversed in time t, given by (u+v)/2 x t resulting in $s = ½ (u + v)/2$ as the distance. Other expressions involving u, v, a and s can be obtained by simple manipulation of the above equations.

Example

1. A car of mass 200 kg moving at 36 km/hr is brought to rest after covering a distance of 10 m. Calculate the retarding force acting on the car.

Mass of the car = 200 kg
Initial velocity, u = 36 km/hr; final velocity, v = 0.0
Distance covered, s = 10 m
Using the expression $v^2 - u^2 = 2 a s$
$0 - 10^2 = 2 \times a \times 10$, giving a = - 5 m/s/s [negative sign denoting retardation]
From F = m a,
retarding force = 1000 N

2. A small (toy) car of mass 500 g travels with a uniform velocity of 25 m/s for 5 s. The brakes are then applied and the car is uniformly retarded and comes to rest in further 10 s. Calculate (i) the retardation, (ii) the distance that the car would travel after the brakes are applied, (iii) force exerted by the brakes.

The mass of the car is 500 g or 0.50 kg
Initial velocity, u = 25 m/s; final velocity, v = 0.0
The time of travel being 5 s, the distance covered, s = 25 x 5 or 125 m
The time for car to come to rest, t = 10s
(i) from equation $\alpha = (v - u)/t$ by substitution
$\alpha = (0.0 - 25)/10$ or the retardation, $\alpha = 2.5$ m/s/s
(ii) using the equation $2 \alpha s = v^2 - u^2$ and substituting
$s = (25 \times 25)/2 \times 2.5$ or the distance to brake = 125 m
[Note that the velocity 25 m/s corresponds to 90 km/hour and sudden application of brakes to the car would still drag the car for a distance of *125 m before it come to rest*]
(iii) the force applied would simply be, $F = m \times \alpha$ or $0.5 \times 2.5 = 1.25$ N

3. A cricket ball of mass 100 g strikes the hand of a player with a velocity of 20 m/s and is brought to rest in 0.01 s. Calculate (i) the force applied by the hand of the player, (ii) the acceleration of the ball during the action.

Here the "initial" velocity of the ball, u, is 20 m/s. As the ball comes to rest in the player's hand its final velocity, v, is zero, the time, t, taken being 0.01 s.
Using the expression $v = u + \alpha t$ and substituting

Acceleration of the ball $\alpha = \frac{0 - 20}{0.01} = -2000$ m/s/s

[the negative sign indicates *retardation* of the ball]
Then, the force acting on the ball is

$$F = m \times \alpha$$
$$= 0.1 \times 2000 \text{ or } 200 \text{ N}$$

4. A lead bullet of mass 20 g, moving with a velocity of 350 m/s, comes to rest after penetrating 40 cm in a still target. Calculate (i) the resistive force offered by the target, (ii) the retardation produced.

Here, initial velocity of the bullet, u = 350 m/s
final velocity, v = 0
'distance' traversed, s = 40 cm or 0.4 m [penetration in the target]

∴ using the expression $v^2 - u^2 = 2 \alpha s$

$$\text{the retardation, } \alpha = \frac{0-(350)^2}{2 \times 0.4} \text{ or } 1.53 \times 10^5 \text{ m/s/s (magnitude alone)}$$

and the resistive force by the target
$F = 0.02 \times 1.53\times10^5$ or 3062.50 N

Speed-time graph of a typical electric train

An example of much importance involving the above quantities and equations is the depiction of "speed-time graph" of a typical electric traction or train movement, esp. the suburban services, as given in Fig.1.6.

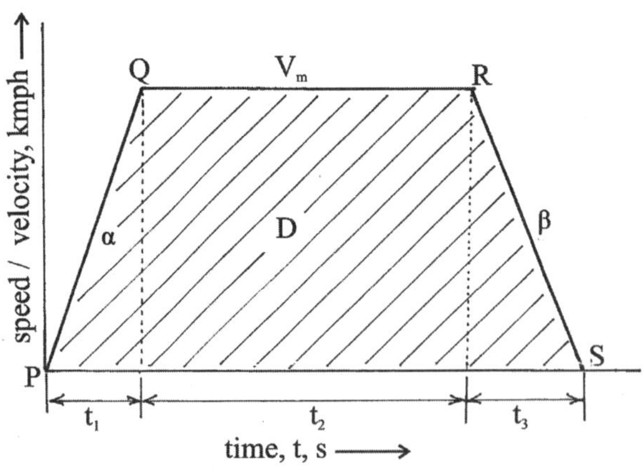

α : acceleration, kmphps
V_m: maximum, steady speed, kmph
β: deceleration, kmphps
D : distance travelled, km (= area under PQRS)

Fig.1.6 : Speed-time graph of electric traction

In the diagram shown in Fig.1.6, the time, t, in *seconds* is plotted against abscissa and train speed, v, in *km/s* is plotted against the ordinate. The train starts from rest at a given station, P, picks up speed, reaches a 'maximum' speed and travels at that speed for a distance. At that point brakes are applied so that the train retards and comes to halt at the following station, S. The entire movement is depicted in the graph by four points, PQRS, and three sections:

PQ – related to acceleration of the train from rest at a *constant* acceleration, α, in *km/hr/s*;
QR – denoting the train travel at a constant (or 'maximum') speed, v, in *km/hr*, reached at point B or instant of time;
RS – the interval expressing deceleration (or retardation) of the train, β, in *km/hr/s*.

The distance PS on the time axis denotes the total time of travel of the train *in seconds* whilst the area of the graph, bound by sections P Q R S and time line PS, or area of the quadrilateral PQRS, will work out to be the distance between the stations P and S *in km*, depicted by the hatched or 'cross' lines.
Such speed-time diagrams are of immense importance in planning of train movements pertaining to a number of stations along a railway route. The parameters involved in the train movement, viz., acceleration, maximum speed, retardation are worked out with great care, including the halt of the train at intermediate stations (in seconds) to arrive at the most efficient and cost-effective overall movement of a given train; and there may be several trains moving on a given track with optimum time intervals in between the consecutive trains.

Example

An electric train has a maximum speed of 52 kmph and an average speed of 42 kmph on a level track between two stops 1,400 m apart. It is accelerated at 1.7 kmphps and braked at 3.3 kmphps. Draw the speed-time curve for the run.

Acceleration $\alpha = 1.7$ kmphps

∴ time of acceleration

$= V_m/\alpha = 52/1.7$ or 30.6 s
Retardation, $\beta = 3.3$ kmphps

∴ time of retardation or braking

$= V_m/\beta = 52/3,3$ or 15.8 s
Distance of run = 1,400 m = 1.4 km

∴ time of run = distance/average speed 1.4/42 hr = (1.4/42) x 3600 s or 120 s

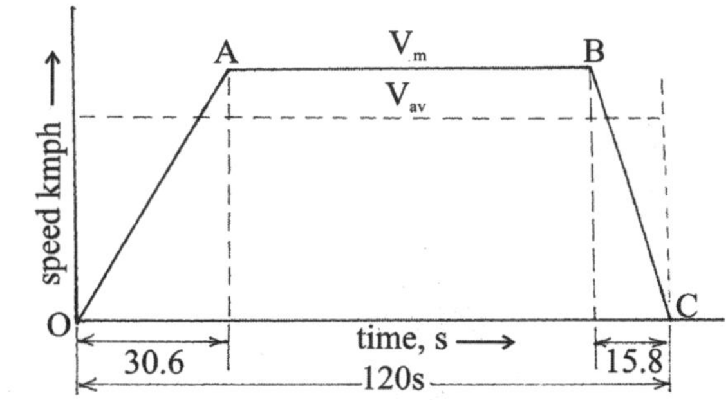

The speed-time curve is shown in the adjoining figure

Rotational Motion

As the name implies, rotational motion relates to motion of an object or body around an axis or point whilst maintaining constant distance from the axis of rotation. It deals only with 'rigid bodies'. In mechanics, a rigid body is an object that retains its overall shape, meaning that the particles that make up the body remain in the same position relative to one another. A rotating wheel of a motor is a common example of a rigid body that exhibits a rotational motion; on a much larger scale an example of rotational motion is the earth rotating about its axis.

Translational Motion versus Rotational Motion

There is a strong analogy between rotational motion and standard translational motion. Indeed, each physical concept used to analyse rotational motion has its translational concomitant. This is illustrated in the following table.

Table 1.1: Comparison of translational and rotational motion

Translational Motion		Rotational Motion	
Displacement	D	Angular displacement	ϕ
Velocity	$v = \frac{dD}{dt}$	Angular velocity	$\omega = \frac{d\phi}{dt}$
Acceleration	$a = \frac{dv}{dt}$	Angular acceleration	$\alpha = \frac{d\omega}{dt}$
	$= \frac{d^2D}{dt^2}$		$= \frac{d^2\phi}{dt^2}$
Power	$P = Fv$	Power	$P = T\omega$

TURNING FORCE CONCEPT

This concept is applicable when a force is made to act on a body that is free to rotate, usually about a pivot or axis of rotation as depicted in Fig.1.7 where a force F (N) is acting at a distance D (m) at the end of a 'bar' from its pivoted end or "turning point".

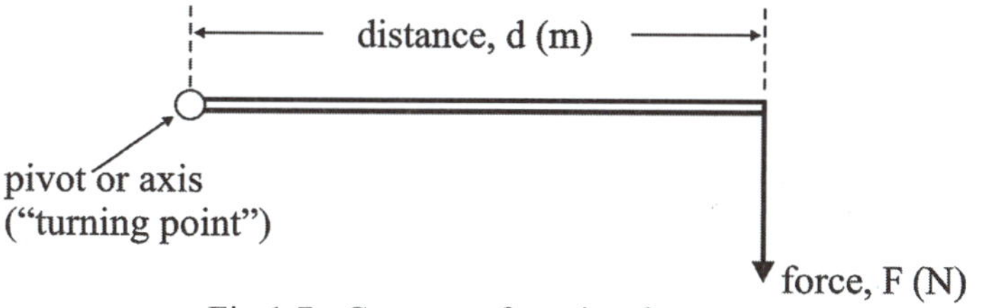

Force F is known as turning force.

Fig.1.7 : Concept of turning force

MOMENT OF A FORCE

More scientifically, the concept of turning force relates to a term called "moment of a force" that can be understood as *turning effect* of the force, defined as product of the **force** multiplied by the perpendicular distance from the line of action of the **force** to the pivot, point or the axis about which the object will turn.
Note that a more technical term often used, esp. in engineering, to express moment of a force is **torque,** defined once again as

Torque = force x its perpendicular distance from the point of (free) rotation of the body to which the force is applied

A simple example of production of torque is in the use of tightening (or opening) of a nut by means of a wrench as illustrated in Fig.1.8

Here F represents the force applied by the hand in a clockwise direction to imply tightening of a right-handed bolt fitted with the nut, N; an anticlockwise application of the force would result in opening of the nut. Point P signifies the axis of rotation of the nut, whilst d denotes *the perpendicular* distance of the force from the axis of rotation.
The, moment of the force or torque = F x d

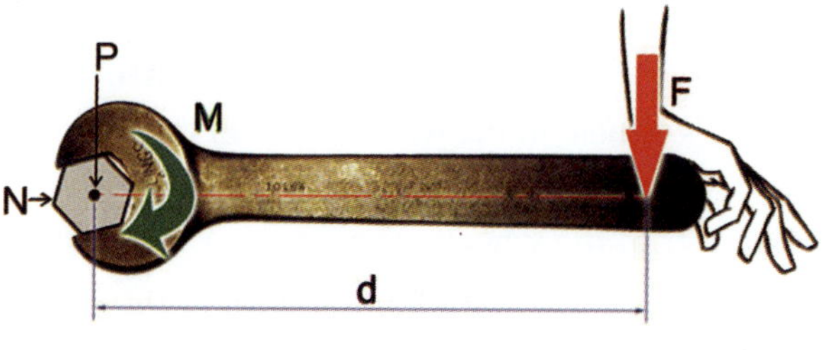

Fig.1.8 : A wrench tightening a nut

Unit of torque

In the above example, and in general, if the force F is expressed in newton (N) and distance d in metre (m), the torque is obtained in newton-metre or simply Nm. Thus, the SI unit of torque or moment of a force is Nm.

Likewise, in CGS units when the force is given in dyne and distance in cm, the unit of moment of the force would be dune-cm.

Other common examples of moment of force

A door being opened (or closed) about its hinges

As shown in fig.1.9, H, H represent two hinges fitted on the door at appropriate distance each from the top and bottom edge. F is the force applied on the external surface of the door (implying closing action) at an angle θ from the vertical as shown such that its component perpendicular to the door surface would be F sin θ and *its* perpendicular distance from the axis of rotation, that is the hinges, is r as indicated. The moment of force in this case would be given by

$$F \sin\theta \times r \text{ units}^{6}.$$

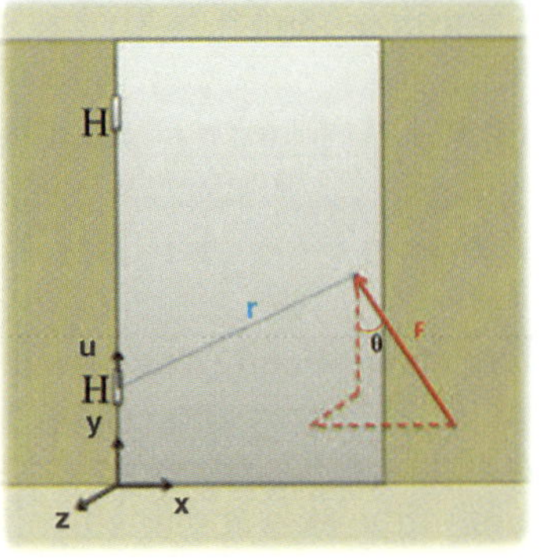

Fig.1.9 : Closing of a door

Steering wheel of a car

A photographic illustration of a steering wheel common to all motor vehicles is given in Fig.1.10.

Note the position of hands on the steering wheel which should correctly be the "8-O'clock" position for efficient manoeuvre of the vehicle. Here, the forces, F, on the wheel are applied by the driver's hands on the wheel's rim in the opposite directions at points P P, separated by a distance d; usually nearly equal to the diameter of the wheel. The force on the right hand is applied in anti-clockwise direction (assuming a "right-hand" driven vehicle) to turn the car to the left and *vice versa.*

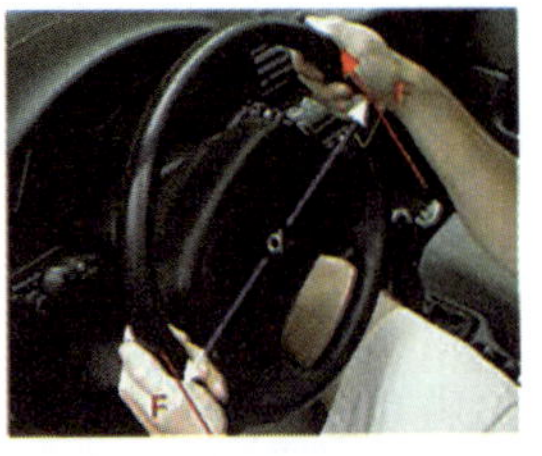

Fig.1.10 : Torque(s) applied to steering wheel of a car

This is a special case of production of torque by the force F, at a perpendicular distance of d/2 from the axis of rotation of the wheel; in one direction on the 'right' side and in the opposite direction on the left with both the forces being applied simultaneously. If both the forces are assumed to be of same magnitude – a situation not

[6]What is the importance of the location of hinges with respect to door edges? If the door is very heavy, should the number of hinges be increased to 3, or even 4?

necessarily obtainable in practice – the phenomenon of production of *two* torques in the above fashion is termed as resulting in a "couple"[7] as explained in mechanics.
This is dealt with later.

Bicycle pedal(s)

A schematic of pedal arrangement in a typical bicycle is shown in Fig.1.11.

The picture in this case shows two 'key' wheels or sprockets that are responsible for producing forward movement of the bicycle as the rider pushes on the two pedals, marked P P. Considered individually, each pedal , the right one in the given instant, when pressed by the rider with a force F produces a moment or torque T given by F x D units, acting about the axis of rotation at the axle A.

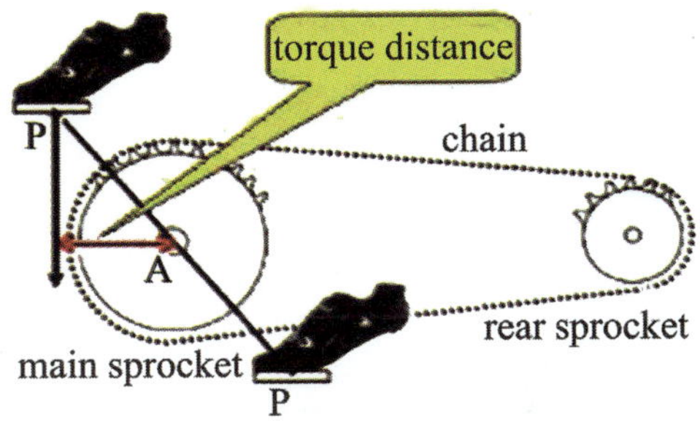

Fig.1.11 : Torques produced by pedals in a bicycle

The other pedal, or the left one, is very nearly idle being in the position shown in the figure till it reaches the upward position, similar to the right pedal, by virtue of forward motion of the bicycle and pressed by the rider downward, resulting in a torque as in the previous stage.

The pedals thus produce a forward torque alternately leading to forward motion of the bicycle. The torque(s) are then 'transferred' by means of the chain to rear sprocket[8] that is actually connected to the rear wheel of the bicycle through the rear axle. Observe that although there are two pedals fitted at the same distance from the axis of rotation in opposite direction, they do not form a couple in the usual sense as, for example, in the case of the steering wheel of a car.

Example

1. A uniform metre scale is balanced on a fulcrum at 40 cm mark when 'forces' of 25 g wt. and 10 g wt. are suspended at 5 cm and 75 cm marks of the scale, respectively. Calculate the mass of the metre scale.

The configuration of the problem is given in the figure below.

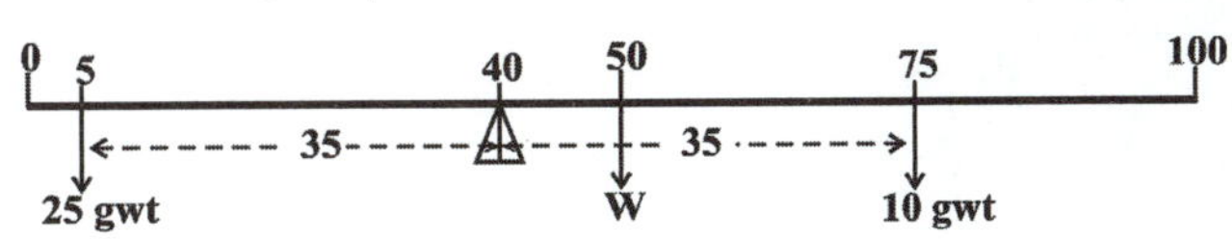

Here, the forces expressed as g wt. (acting downward) are in effect masses in gramme, multiplied by

---

[7]To be understood from the simple dictionary meaning "two" or a "pair".
[8]The rear sprocket is designed to "free-wheel" such that the produced torque(s) result in motion of the bicycle *in forward direction only* and NO movement in the reverse direction.

acceleration due to gravity, 9.81 m/s/s in SI units. If the masses were expressed in kg the forces would be obtained in newton. In the configuration shown, the forces acting at respective distances produce moments in clockwise and anticlockwise directions.
Let the mass of the scale be M g, the force being M g wt., assumed to be concentrated at mid point and acting downward, at a distance of 10 cm from the fulcrum and to its right as shown.

Therefore, clockwise moment = M g wt. x 10 + 10 g wt. x 35
= (350 + 10 M) g wt.-cm

and anticlockwise moment = (25 x 35) g wt.-cm

Equating the two at balance

(350 + 10 M) g wt.-cm = 875 g wt.-cm

whence

M = 52.5 g

2. A uniform metre scale weighing 100 g is pivoted at its mid-point, O. Two forces of 150 g wt. and 250 g wt. are suspended from the scale as shown in the figure below. Calculate (i) total clockwise moment, (ii) total anticlockwise moment about O. Also, calculate the distance from O where a force of 100 g wt. may be suspended to balance the scale horizontally.

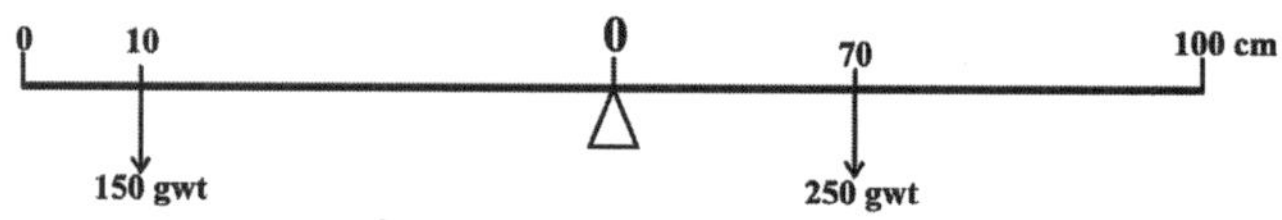

(i) Total clockwise moment = 250 x 20 or 5000 g wt.-cm
(ii) Total anticlockwise moment = 150 x 40 or 6000 g wt.-cm

The anticlockwise moment is greater than the clockwise moment by 1000 g wt.-cm. Hence to balance the scale the 100 g wt. force should be suspended at a distance of 1000/100 or 10 cm *to the right of* O.

3. A uniform metre rule of mass 100 g is balanced horizontally on a fulcrum at 40 cm mark. An unknown mass 'm' is suspended from the rule at the 20 cm mark. Find 'm'. To which side the rule will tilt if mass 'm' is moved towards 10 cm mark. What is the resultant moment in this case? How can the rule be balanced by making use of another mass of 50 g?

The configuration related to the problem is as shown

The rule causes a clockwise moment of 100x10 or 1000 g wt.-cm
Correspondingly, the mass 'm' should provide a counter-clockwise moment of m x 20 gwt.-cm to balance the load

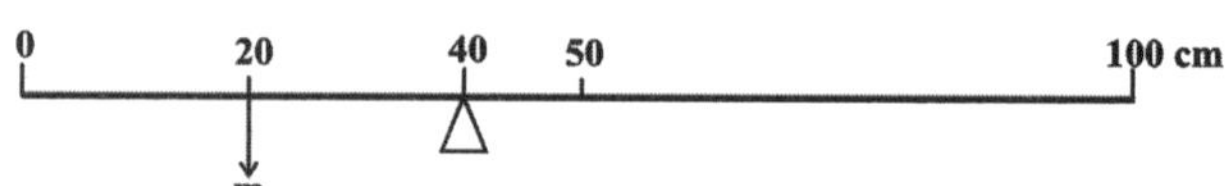

$\therefore$ 20 m = 100 x 10

giving m = 50 g

If the mass 'm' is moved to 10 cm mark, the anticlockwise moment will be 50x30 or

1500 gwt.-cm [see the figure]. Since this is greater than the clockwise moment, the rule would tilt left or in anticlockwise direction.

Let the distance at which the mass of 50 g be placed to the right of the fulcrum to balance the rule in the last condition. Then the new balance equation will be

$$50 \text{ x } 1000 = 1500$$

whence $x = 10$ cm

that is, the mass (of 50 g) should be suspended at "50 cm" mark of the scale.

4. A wheel of diameter 2 m is shown in the figure with axle at O. A force F = 2 N is applied at B in the direction shown. Calculate the moment of the force about (i) centre O; (ii) point A.

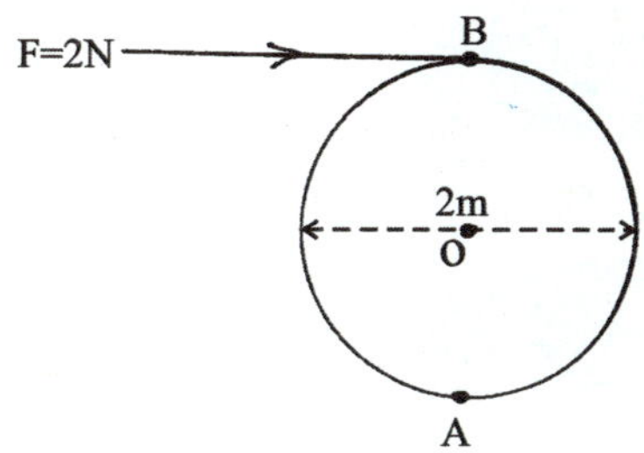

(i) the moment at O is given by

moment = force x perpendicular distance
= 2 N x 1 (m)
= 2 Nm

(ii) moment at point A = 2 N x 2 (m)
= 4 Nm

Bending Moment

A particular variation of moment of force is the "bending moment" (BM), an entity of significant importance in the design of weight-carrying protruding beams encountered in building construction and other engineering structures. The aspect is illustrated in Fig.1.12 where the beam B is rigidly fitted to a wall, carrying a weight W at the end. The moment of this weight, or force F, is given by

BM = F x D

where D is the perpendicular distance of F from the wall as shown. Observe that the moment is proportional to D, varying from zero at the wall surface to a maximum at the point of location of F as depicted by the lower diagram in Fig.1.12. The BM will lead to bending of the beam if the latter is not suitably designed for the purpose.

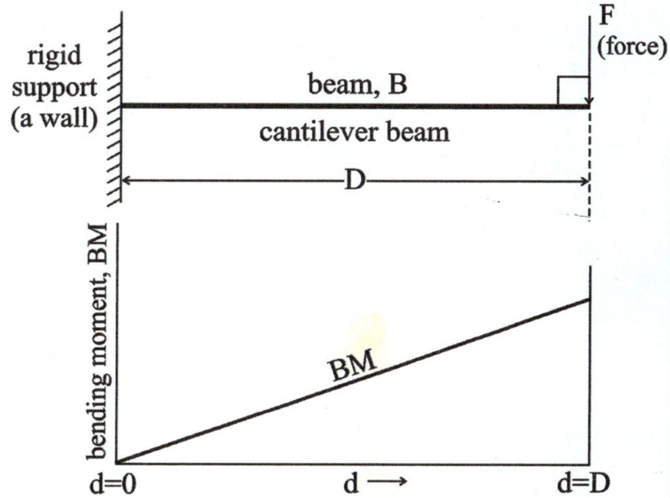

Fig.1.12 : Bending moment of a cantilever beam

Couple

A couple is a pair of forces, *equal in magnitude*, oppositely directed, and displaced by a perpendicular distance between them as depicted in Fig.1.13. The forces may be coplanar or positioned differently.

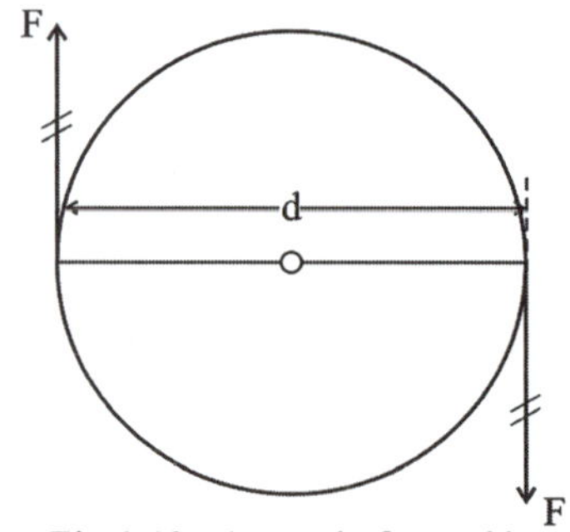

Fig.1.13 : A couple formed by two (coplanar) forces

This is also called a "simple couple" in which each force has a turning effect or moment, or a torque, about an axis which is *normal* (perpendicular) to the plane of the forces. The SI unit for the couple is newton-metre (Nm) as for torque.
If the two forces are F and $-F$, the negative sign implying the force being oppositely directed, then the magnitude of the torque is given by

$$\tau = Fd$$

where

τ is the torque
F is the magnitude of each of the forces
d is the perpendicular distance between the forces, also called the *arm* of the couple.

An example of useful application of couple of forces is illustrated in Fig.1.14 for changing stepnee or spare wheel of a car using (box) spanner or wrench.

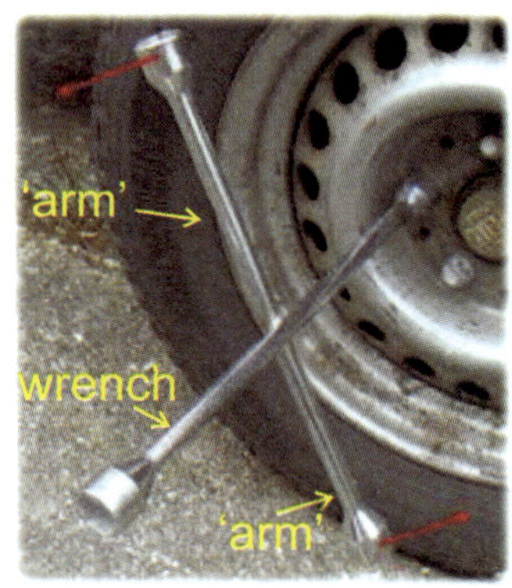

Fig.1.14 : Changing stepnee or spare wheel of a car

Note the position of wrench fitting the nut and the two 'arms' at tight angles to the axis of rotation.

Example

Two forces, each of 5 N act vertically upwards and downwards respectively at the two ends of a uniform metre scale which is suspended at its mid-point. Determine the magnitude of the resultant moment of these forces about the mid-point.

[ICSE: 2014]

The metre scale on its fulcrum is shown in the figure.

Resultant moment of the two forces
= moment of couple

= force x perpendicular distances between them
= 5 x (2 x 0.5)
= 5 Nm [anti-clockwise]

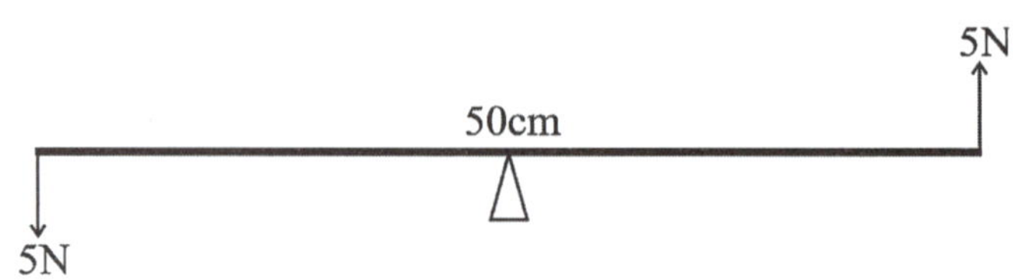

Clockwise and Anticlockwise Moments

- If the moment of a force turns or rotates the body in clockwise direction, then it is called as clockwise moment.
- If the moment of a force turns or rotates the body in anti-clockwise direction, then it is called anti-clockwise moment.

Refer to Fig.1.8 which shows the use of a wrench for opening or tightening a nut fitted on a bolt. The movement of hand in the figure *downward* would result in rotation or turning of the wrench clockwise; and this means a clockwise moment or torque. This also indicates tightening of the nut on a "right-handed" bolt. If the wrench is turned towards left or anticlockwise; this would result in an anticlockwise moment or torque, leading to opening of the nut.

Verification of Principle of Moment

The principle of moment can be explained in terms of balancing of clockwise and anticlockwise moments in a system of forces such that the net moment would be zero. Specifically, assume a 'rigid' rod or beam suspended at its mid-point from a rigid support, or supported on a fulcrum at mid-length. Then if two weights, W_1 and W_2 are suspended from the beam at distances D_1 (on the right) and D_2 (on the left) from the point of support, the weight (or force) W_1 would result in a clockwise moment of $W_1 \times D_1$ Nm whilst the anticlockwise moment resulting from the weight on the left would be $W_2 \times D_2$ Nm, assuming the forces to be in newton and distances in metre. Now if the two moments are equal such that there is no net moment and the beam remains horizontal, that is in the original condition, it would corroborate the principle of moment.

The principle can be verified by means of a simple experiment using a metre scale and a few weights that might be suspended from the scale variously, similar to the example discussed

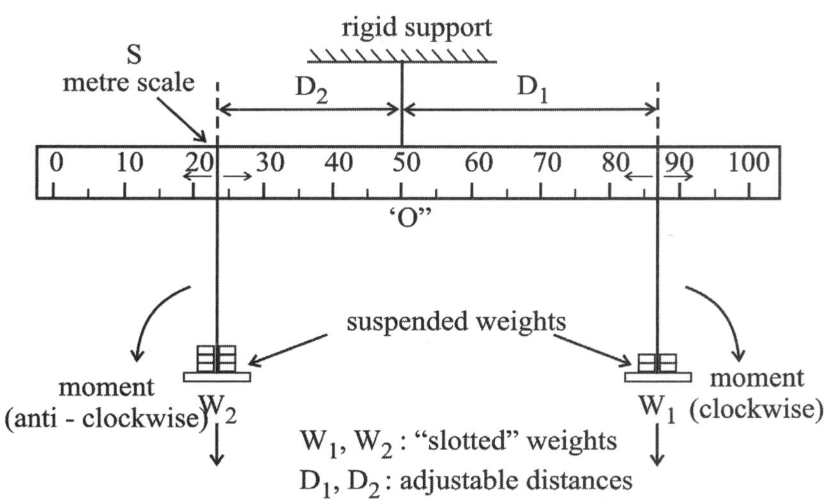

Fig.1.15 : Experiment for verification of principle of moments

above. . The schematic of the set-up is shown in Fig.1.15 in which S is the metre scale suspended from a rigid support at its mid-length. The scale carries two "pans" having provision to add "slotted" weights of varying masses on either side of the "zero" position of the scale, each being adjustable from the mid-point and the distance measured from the mid-point.

For given weights W_1 and W_2, their relative distances from either side being D_1 and D_2 are adjusted such that the metre scale is horizontal as it would be without any weights. Under this conditions

the clockwise moment due to $W_1 = W_1 \times D_1$

and anticlockwise moment due to $W_2 = W_2 \times D_2$

and $W_1 \times D_1 = W_2 \times D_2$

This is verification of the principle of moments.

MOMENTUM

An important quantity in kinematic that is associated with mass AND velocity in the motion of a body is identified as **momentum**[9], defined as

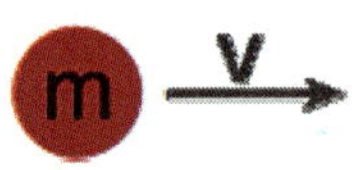

$$\text{momentum} = \text{mass} \times \text{velocity}$$
$$p = m \times v$$

A commonest example of momentum of a body is a moving train, having a mass of hundreds of killogramme (or a few tonnes), consisting of mass of the coaches and the engine of the train as well as all the passengers, the velocity of the train reaching over a hundred km/hr. The train thus acquires tremendous momentum whilst in motion resulting in a situation that makes it extremely difficult to stop the train in an emergency within a desirable distance.

The most common symbol for momentum in physics is p. The SI unit for momentum is kg m/s when mass is expressed in kg and velocity in m/s. Clearly, if velocity is in km/hr, the momentum will be given in kg km/hr.

Momentum is a *vector quantity* since it involves the quantity velocity that itself is a vector quantity.

Observe that momentum is a function of both mass and velocity and can be 'controlled' by either; see, for example, Fig.1.16.

---

[9]It is interesting that the word "momentum" also finds frequent use in idiomatic English to refer to impetus to continuity of efforts in an activity or 'project'.

Two bodies moving with (same) velocity, V

small mass, smaller momentum **large mass, larger momentum**

Fig.1.16 : Momentum of bodies of different mass and same velocity

Example

A force of 10 kg wt. is applied on a body of mass 100 g initially at rest for 0.1 s. Calculate (i) momentum acquired by the body, (ii) distance traversed by the body in 0.1 s. Assume acceleration due to gravity, g = 10 m/s/s.

The mass of the body is 100 g or 0.1 kg
The force applied is 10 kg wt. This amounts to 10 x g or 10 x 10 = 100 N

$\therefore$ acceleration acquired by the body, α = force/mass or 100/0.1 = 1000 m/s/s
From equation $v = u + \alpha t$ and with initial velocity being 0.0 (at rest), the final velocity
$v = 0.0 + 1000 \times 0.1$ or 100 m/s
Therefore (i) momentum = m x v or 0.1 x 100 = 10 kg-m/s
(ii) from the equation $2 \alpha s = v^2 - u^2$,
the distance traversed $s = (v^2 - u^2)/2\alpha$ and substituting
$= (100 \times 100)/2 \times 1000$ or 5 m

Change of momentum

This is defined as the difference of final and initial momentum of a body. Thus if mass of a body is M_1 at time T_1 and its velocity is V_1, its momentum is $p_1 = M_1 \times V_1$. If the mass is M_2 and velocity V_2 at time T_2, the momentum would be $p_2 = M_2 \times V_2$. Then the difference $\Delta p = M_2V_2 - M_1V_1$, assuming V2 > V1, is the change of momentum in the time interval $\Delta T = T_2 - T_1$.

In most cases, the mass of the body may remain unchanged whilst the velocity may change – increasing or decreasing – on various counts; for example in the case of a moving train referred to earlier, the speed of the train may be *reduced* as a station approaches before applying brakes and the final momentum may be less than that before reduction of speed, the change of momentum may then be termed 'negative'.

Rate of change of momentum

A term variously in use in kinematics is the rate of change of momentum and is defined as the ratio of "change of momentum and time interval" or simply

$$\Delta p/\Delta T$$

with the appropriate unit associated with it.

FORCES IN EQUILIBRIUM

Static Equilibrium

A concept of much importance in mechanics or physics when dealing with forces is the idea of **equilibrium** or **balance of the forces**. In general, an object or body can be acted on by several forces at the same time. A force being a vector quantity has both a magnitude and a direction associated with it. If the magnitude and direction of the forces, or their 'appropriate' components acting on the body are exactly balanced, then there is no **net force** acting on the body and it is said to be in **equilibrium**. In the absence of a net force acting on an object in equilibrium, an object at rest will stay at rest, and an object in motion will continue to be in motion according to Newton's first law of motion.

Two examples in practice from common experience are illustrated in Fig.1.17.

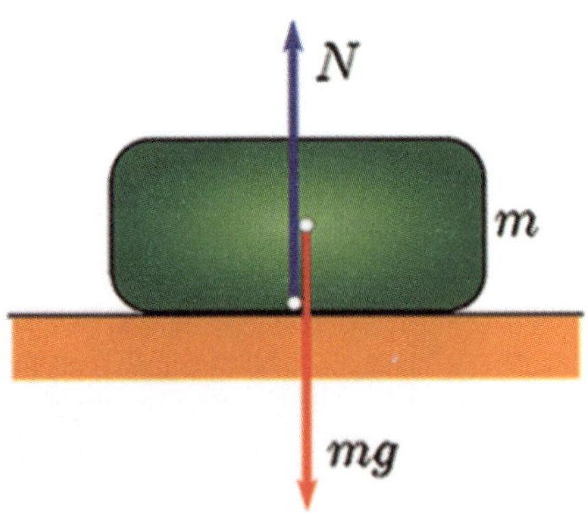

(a)

(b)

Fig.1.17 : Images of static equilibrium

In Fig.1.17(a). a body of mass m is at rest on a table top. If acted upon vertically upward by a force, N, it may have a tendency to lift up. However, if the magnitude of N is exactly equal to the body's weight, mg, and the direction of the two forces match, the body will continue to remain "at rest" on the table. This represents a simple example of "static equilibrium"

Fig.1.17(b) shows a nature's creation in stones, precariously remaining balanced over ages, exhibiting another example of static equilibrium.

In general, the forces are considered to be balanced if the rightward forces are balanced by the leftward forces and the upward forces are balanced by the downward forces. This however does not necessarily mean that all the forces are equal to each other. Consider the two objects pictured in the 'force diagram' shown in Fig.1.18. Note that the two objects are at equilibrium because the forces that act upon them are balanced; however, the individual forces are not equal to each other. The 50 N force is not equal to the 30 N force.

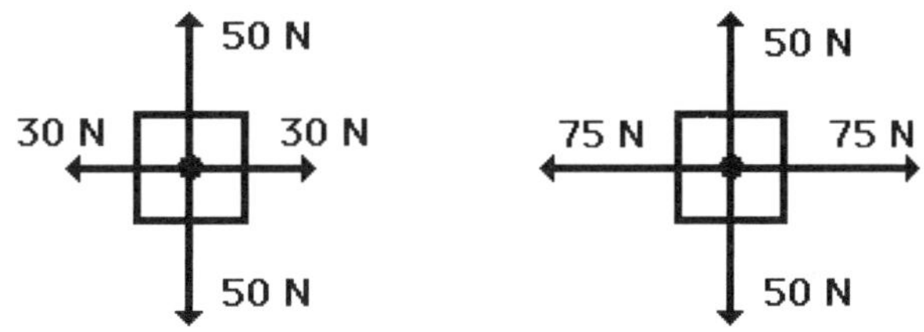

These two objects are at equilibrium since the forces are balanced.

Fig.1.18 : Horizontal and vertical forces on two bodies and equilibrium

Clearly, if there is an imbalance between the forces on opposite sides such that one is greater than the other, the body would begin to move in an appropriate direction and would no longer be in the state of equilibrium.

Triangle of Forces

A very common occurrence in practice, requiring static equilibrium, for example in trusses in engineering design of building construction, is the balance of *three* forces that may be simply coplanar but in different directions. The balancing of forces in such cases follows the concept of "triangle of forces". As a simple illustration of this concept, consider the diagram of Fig.1.19.

Shown in the figure is a body of mass M, suspended by two strings S, S tied to two parallel, vertical walls at points P. Both the strings are equal in length and tied to the walls at the same height from the ground. Then the forces F_1 and F_2 balance the weight of the body, expressed as a vertically downward force F_3, such that the body hangs in the middle, equidistant from the walls, the three forces being *coplanar*.

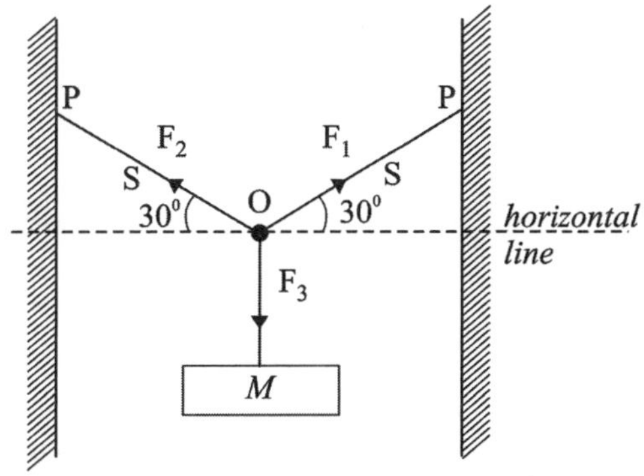

Fig.1.19 : Concept of triangle of forces

Note that in this special case of three forces, the vertical components of forces F_1 and F_2 *added together* exactly balance the body's weight, that is force F_3, whilst the horizontal components along the horizontal line, balance each other; cancelling being oppositely directed[10].

What would happen if the strings were unequal in lengths or were tied at different heights on the walls?

Dynamic Equilibrium

If instead of being in static equilibrium under the action of balancing forces (or their components), a body is in motion at steady speed, for example a train, continues to remain in that state, following Newton's first law of motion, the body is said to be in "dynamic equilibrium". The various forces acting on the body again balance to maintain dynamic equilibrium; for example, in the case of the train a 'driving' force in the direction of its motion balancing against the force of friction on the track and that due to wind in opposite direction[11].

A classic example of dynamic equilibrium is the stable position of heavenly bodies in the universe, relative to each other, under the action of several forces.

CENTRE OF GRAVITY

The **centre of gravity** of a body is that point through which the resultant of the system of parallel forces formed by the *weights* of all the particles, or sub-parts, constituting the body passes for all positions of the body. It is denoted as "**CG**" or, simply "**G**".
Alternatively

The centre of gravity is a point on any object where all the weight of the object is concentrated. The centre of gravity is the position of the object where we assume that the whole weight of the body is concentrated, in other words we say that centre of gravity is the assumed location of the whole weight of the body. It is always inside the body.

The concept of centre of gravity is explained qualitatively by reference to Fig.1.20

---

[10]In general, if the two string lengths S_1 and S_2 were unequal, suspended from different heights, the suspended weight take up a resultant position such that the three co-planer forces represented by F_1, F_2 and F_3 would balance as sides of a triangle.

[11]If the train is moving up an incline, there would also be opposing force due to component of its weight along the incline that would have to be overcome by the driving force.

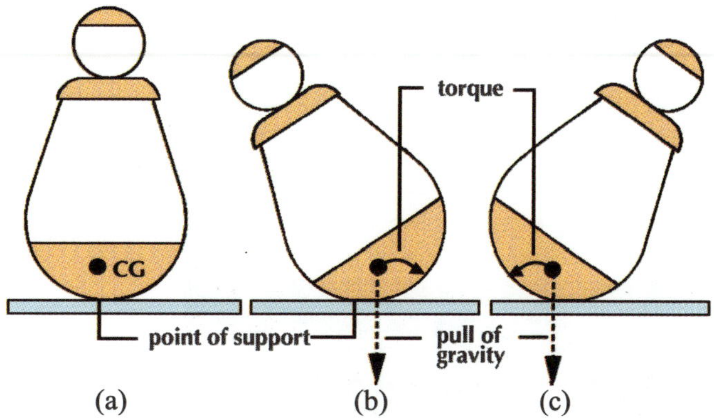

Fig.1.20 : The concept of centre of gravity

In the figure, a jar, partly filled with some solid or liquid, is shown supported on a table top. Fig.1.20(a) shows the jar standing upright, its bottom being resting at point P on the table and its CG passing through P, the point lying on a vertical line dividing the jar symmetrically as indicated[12].

In Fig.1.20(b), the jar is shown tilted slightly to the left, the tilt having been effected by a manual push, such that the CG is shifted to the right with respect to the point of support, not passing through it. The jar would exhibit a 'restoring' torque in clockwise direction, about the CG, in opposition to the push applied initially; if the push would be removed the jar would revert to the position shown in Fig.1.20(a) and CG aligning with the vertical as before. The condition shown in Fig.1.20(c) is the reverse of that depicted in Fig.1.20(b).

Note that in all the three cases the "pull of gravity" is through CG.

The table below lists the position of CG for a few bodies of common shape(s).

Table 1.2: Shape of body vs. centre of gravity

Shape of body	Position of CG
Thin uniform bar	Middle point of the bar
Circular ring	Centre of the ring
Circular disc	Centre of the disc
Solid or hollow sphere	Centre of the sphere
Cubical or rectangular block	Point of intersection of the diagonals
Triangular lamina	Point of intersection of the medians
Square, parallelogram and rectangular lamina	Point of intersection of the diagonals
Cylinder	Middle point of the axis
Cone or pyramid	On the line joining the apex of the centre of the base at a distance equal to ¼ of the length of this line from the base

[12]This also shows condition of stable equilibrium of the jar resting on the table.

An example of calculation of CG

Based on "moment of forces"

Consider a seesaw of length 4 m and of weight 100 kg on which are sitting two kids of 40 kg and 60 kg weight, respectively, at 0.5 m distance from both the ends as shown in Fig.1.21.

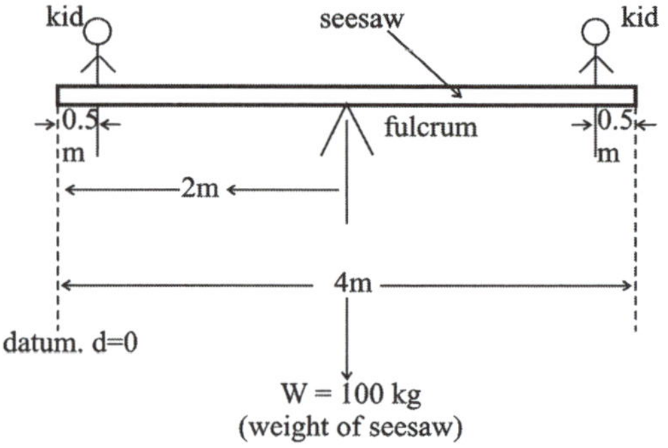

(a) seesaw with kids (b) schematic

Fig.1.21 : Calculation of centre of gravity

First, choose any arbitrary point called datum. Let it be at one end of the seesaw. So that the distances of both the kids from the datum are 0.5 m and 3.5 m, respectively. The distance of seesaw CG alone from the datum line is 2 m. Now the moment of the seesaw from the datum is $100 \times 2 = 200$ Nm, moment of the first kid (at left) is $40 \times 0.5 = 20$ Nm and the moment of the second kid (sitting to the right) is 60×3.5 m= 210 Nm, the total moments are $200 + 20 + 210 = 430$ Nm. The total weights are $40 + 60 + 100 = 200$ kg. The distance of the centre of gravity from the datum is then obtained as $430 / 200 = 2.15$ m.

Note that if both kids weighed equally, the CG of the seesaw would be at the mid-point – the location of the fulcrum - or 2 m from the left as expected.

Theoretical Approach for Calculation of Centre of Gravity

To illustrate the method, consider a solid object of arbitrary shape and size of which the CG is to be determined. Assume that the object can be divided into small 'particles' or sub-sections, n in numbers each having weights $w_1, w_2, w_3, \ldots, w_i, \ldots, w_n$, located in three-dimensional space with respective coordinates given by (x_1, y_1, z_1), (x_2, y_2, z_2), (x_3, y_3, z_3), $\ldots$, (x_i, y_i, z_i), $\ldots$, (x_n, y_n, z_n) etc. with respect to a (arbitrary) datum 0, 0.

Then, the coordinates of the CG of the body with respect to the chosen datum are given by

$$x_{CG} = \frac{\sum_{i=1}^{n} w_i x_i}{w_{total}} \qquad y_{CG} = \frac{\sum_{i=1}^{n} w_i y_i}{w_{total}} \qquad z_{CG} = \frac{\sum_{i=1}^{n} w_i z_i}{w_{total}}$$

where w_{total} represents total weight of the particles given by
$w_{total} = w_1 + w_2 + w_3 + \ . \ . \ . \ + w_i, \ . \ . \ + w_n$

To derive just x_{CG} for an arbitrary body is depicted in Fig.1.22.

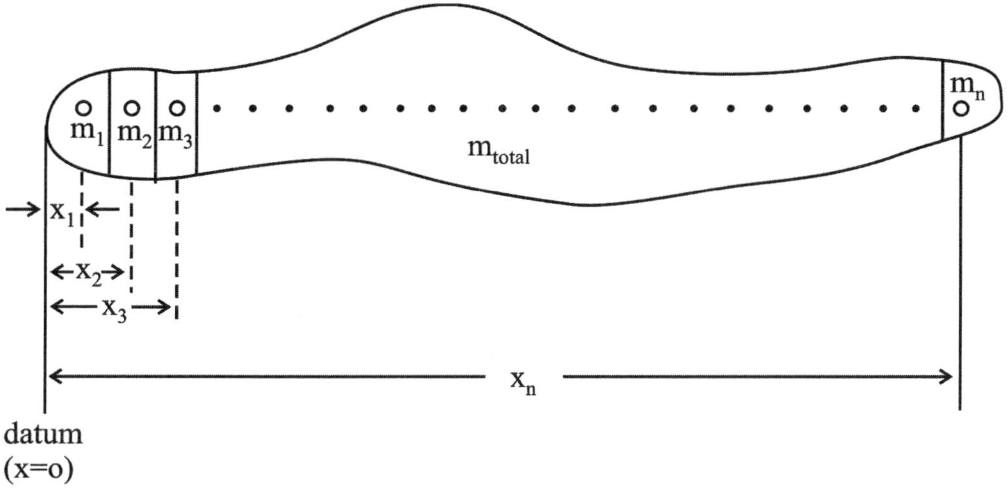

Fig.1.22 : To derive x_{CG} of a body

$$X_{CG} = \frac{\sum_{i=1}^{i=n} m_i x_i}{m_{total}}$$

Centre of Mass

In physics, the **centre of mass** of a distribution of masses in space is the unique point where the weighted relative position of the distributed masses sums to zero or the point where if a force is applied, causes it to move in the direction of force without rotation. The distribution of mass is balanced around the centre of mass and the average of the weighted position coordinates of the distributed mass defines its coordinates.

Alternatively

The centre of mass is the point where all of the mass of the object is concentrated. When an object is supported at its centre of mass, there is no net torque acting on the body and it will remain in static equilibrium

EXERCISES

1. A force of 50 dynes acts on a rigid body such that the perpendicular distance between the fulcrum and the point of application of the force is 75 cm. Calculate the moment due to the force.

[3750 dyne-cm

2. A force of 50 N produces a moment of 10 Nm in a rigid body. Calculate the perpendicular distance between the point of application of the force and the turning point.

[0.20 m

3. A couple of 15 N force acts on a rigid body such that the arm of the couple is 85 cm. Calculate the moment of the couple in SI units.

[12.75 Nm

4. A uniform metre scale is balanced at 20 cm mark from its "zero" mark (at left end) when a force of 100 g wt. is suspended from the "zero" end. Draw the configuration of the problem and calculate the mass of the scale, assumed concentrated at its mid-point.

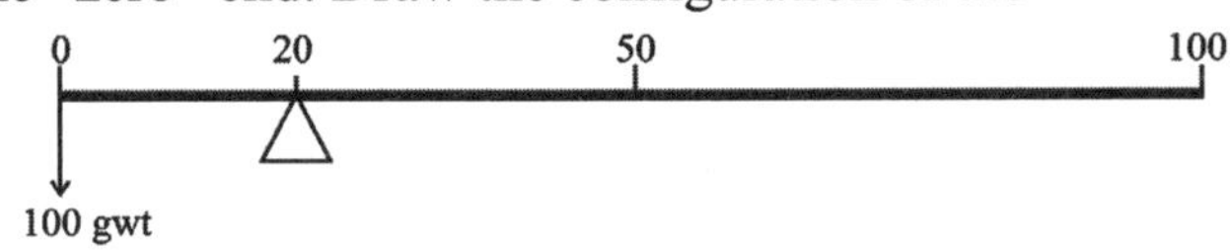

[66.67 g

5. A regular metre scale of mass 50 g is placed on a fulcrum at 40 cm mark from the zero at left end as shown. A force of 100 g wt. is suspended at 5 cm mark of the scale to the left of fulcrum. Calculate the distance and location of a force of 80 g wt. to balance the scale.

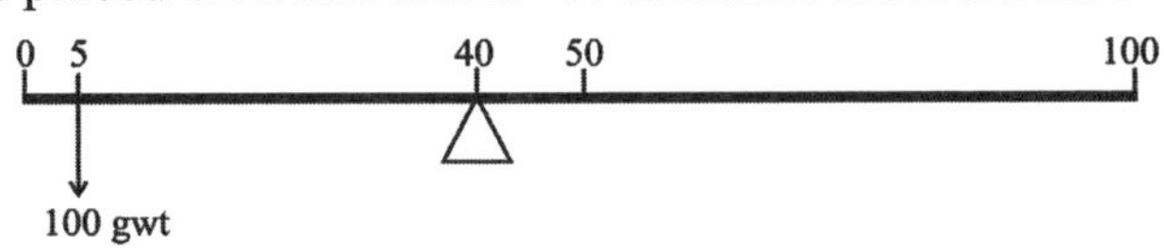

[77.5 cm to the right of the fulcrum

6. A uniform wooden beam AB, 80 cm long is supported on a triangular wedge as shown. Calculate the maximum force (g wt.) that can be placed on end A to balance the beam. Assume the mass of the beam to be concentrated at its mid point.

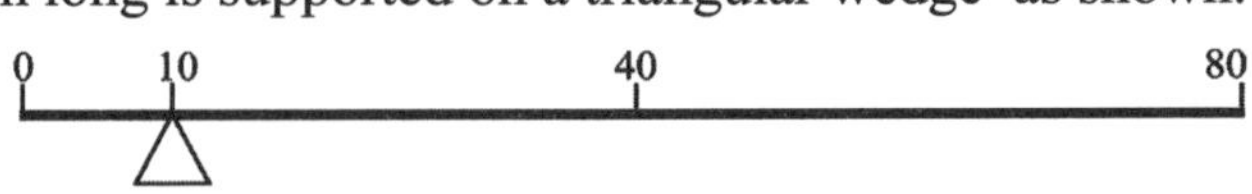

[750 gwt.

7. A body of mass 1.5 kg is dropped from a height of 12 m. What is the force acting on it during the fall? Assume g = 9.8 m/s/s.

[14.7 N

8. Calculate the acceleration produced in a body of mass 50 g when acted upon by a force of 20 N.

[400 m/s/s

9. A body of mass 50 g is moving with a velocity of 10 m/s. It is brought to rest by a resistive force of 10 N. Find (i) the retardation, (ii) the distance that the body will travel after resistive force is applied.

[200 m/s/s; 25 cm

10. A uniform metre rule balances horizontally on a knife edge placed at the 58 cm mark when a weight of 20 gwt. Is suspended from one end. Draw the diagram and find weight of the rule.

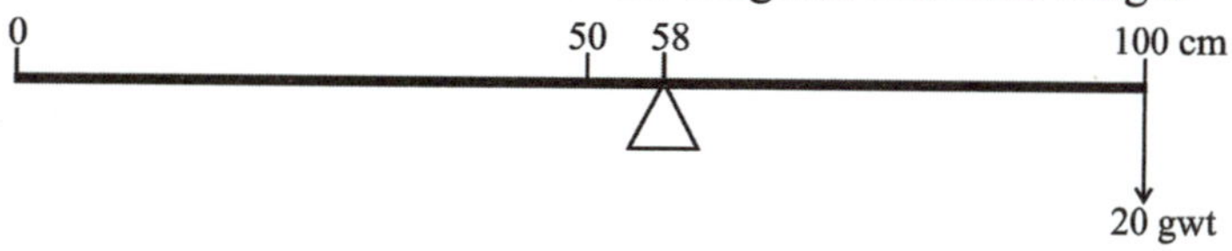

[106 gwt.

11. When a boy of 20 kgwt. sits at one end of a 4 m long seesaw as shown. Where should the man of 40 kgwt. sit to balance the seesaw?

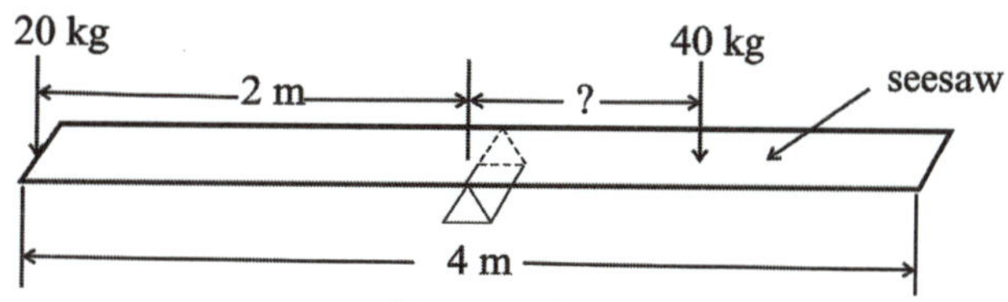

[1 m to the right of the fulcrum

12. The figure shows a uniform metre rule placed on a fulcrum at its mid point and having a weight of 40 gwt. at the 10 cm mark and a weight of 20 gwt. at the 90 cm mark. (i) Is the rule in equilibrium? If not, how will it turn? (ii) How can it be balanced by using additional weight of 40gwt.?

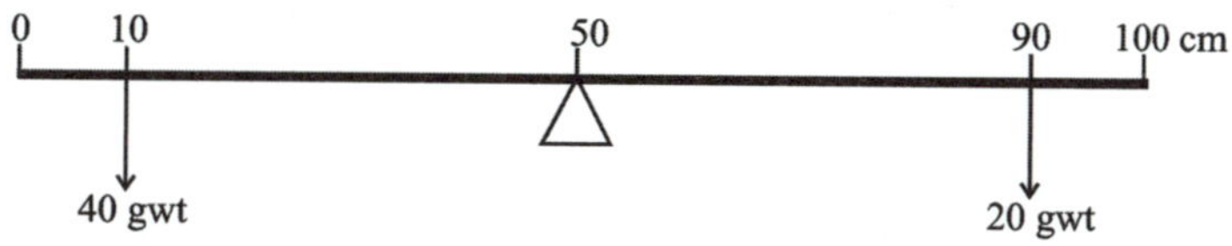

[(i) it is not in equilibrium, will turn anticlockwise, (ii) at 70 cm mark

13. The figure shows the velocity-time graph for a scooter, having a total mass of 150 kg. From the graph calculate

a) the acceleration
b) the distance covered
c) the force acting in first 4 seconds.

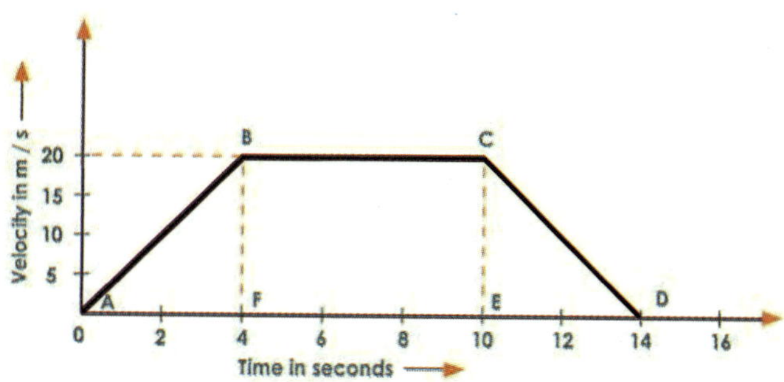

[5 m/s/s; 40m; 750 N

14. A force of 50 dynes acts on a rigid body such that the perpendicular distance between the fulcrum and the point of application of the force is 75 cm. Calculate the moment due to the force.

[3750 dyne-cm

15. A force of 50 N produces a moment of 10 Nm in a rigid body. Calculate the perpendicular distance between the point of application of the force and the turning point.

[0.20 m

16. A couple of 15 N force acts on a rigid body such that the arm of the couple is 85 cm. Calculate the moment of the couple in SI units.

[12.75 Nm

17. A uniform metre scale is balanced at 20 cm mark from its "zero" mark (at left end) when a force of 100 gwt. is suspended from the "zero" end. From the configuration of the problem and calculate the mass of the scale, assumed concentrated at its mid-point.

[66.67 g

18. A regular metre scale of mass 50 g is placed on a fulcrum at 40 cm mark from the zero at left end. A force of 100 g wt. is suspended at 5 cm mark of the scale to the left of fulcrum. Calculate the distance and location of a force of 80 g wt. to balance the scale.

[77.5 cm to the right of the fulcrum

19. A uniform wooden beam AB, 80 cm long is supported on a triangular wedge as shown. Calculate the maximum force (gwt.) that can be placed on end A to balance the beam. Assume the mass of the beam to be concentrated at its mid point.

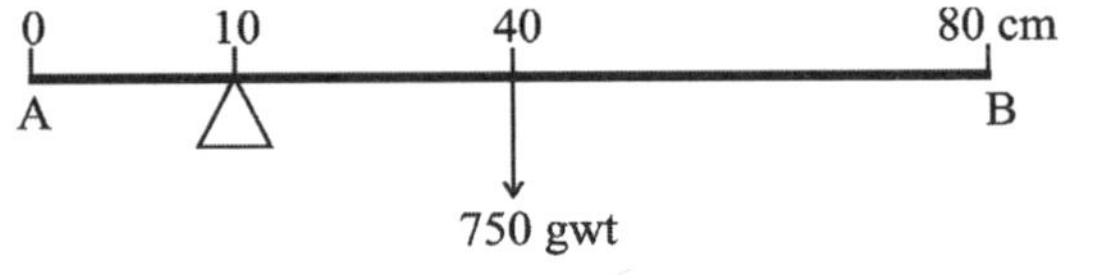

[750 gwt.

20. A body of mass 1.5 kg is dropped from a height of 12 m. What is the force acting on it during the fall? Assume g = 9.8 m/s/s.

[14.7 N

21. Calculate the acceleration produced in a body of mass 50 g when acted upon by a force of 20 N.

[400 m/s/s

22. A body of mass 50 g is moving with a velocity of 10 m/s. It is brought to rest by a resistive force of 10 N. Find (i) the retardation, (ii) the distance that the body will travel after resistive force is applied.

[200 m/s/s; 25 cm

II Uniform Circular Motion

Example of constant speed, though acceleration (force) is present. Differences between centrifugal and centripetal force

Circular motion is a common type of rotational motion.

Newton's first law states

that an object in motion remains in motion at constant velocity unless acted upon by an outside force. If the force is applied perpendicular to the direction of motion, only the direction of velocity will change. If a force constantly acts perpendicular to a moving object, the object will move in a circular path at constant speed. This is called **uniform circular motion**.

The circular motion of a *rigid* body occurs when every point in the body moves in a circular path around a line called the **axis of rotation.** A schematic of such motion is shown in Fig.2.1

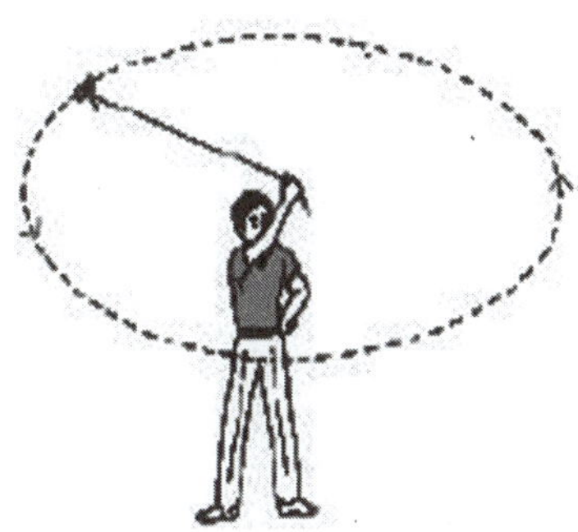

Fig.2.1 : Schematic of circular motion

This is further illustrated in Fig.2.2

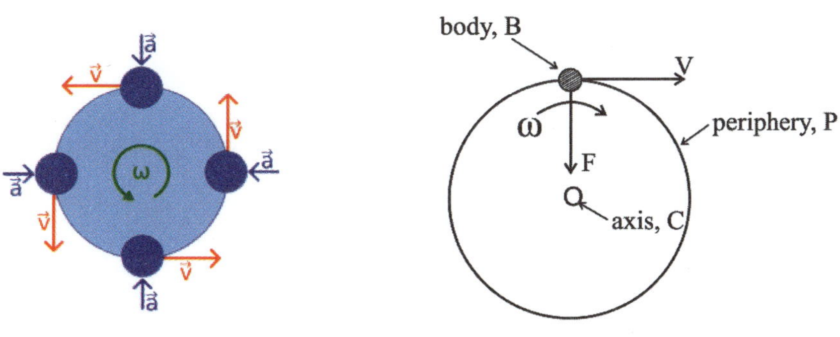

(a) (b)

Fig.2.2 : Force and velocity in circular motion

Referring to Fig.2.2(a) and (b), body B is seen rotating in a circular motion traversing the periphery P about an axis C at the centre of the circular path. Observe that *at every point* of its motion, the body is acted upon by a force resulting in a *constant* tangential velocity v (say, in m/s), the *angular* velocity being ω rad/s as indicated. Since the object's velocity is constantly changing direction as the body goes round, the moving object is undergoing acceleration a as marked in Fig.2.2(a).

Examples of circular motion, shown in Fig.2.3, include an artificial satellite orbiting the Earth *at constant height* thus maintaining a circular orbit; a 'giant' wheel in action; a stone which is tied to a rope and is being swung in a circle by a child; a car moving round a circle in a race tracker; an electron moving perpendicular to a uniform magnetic field.

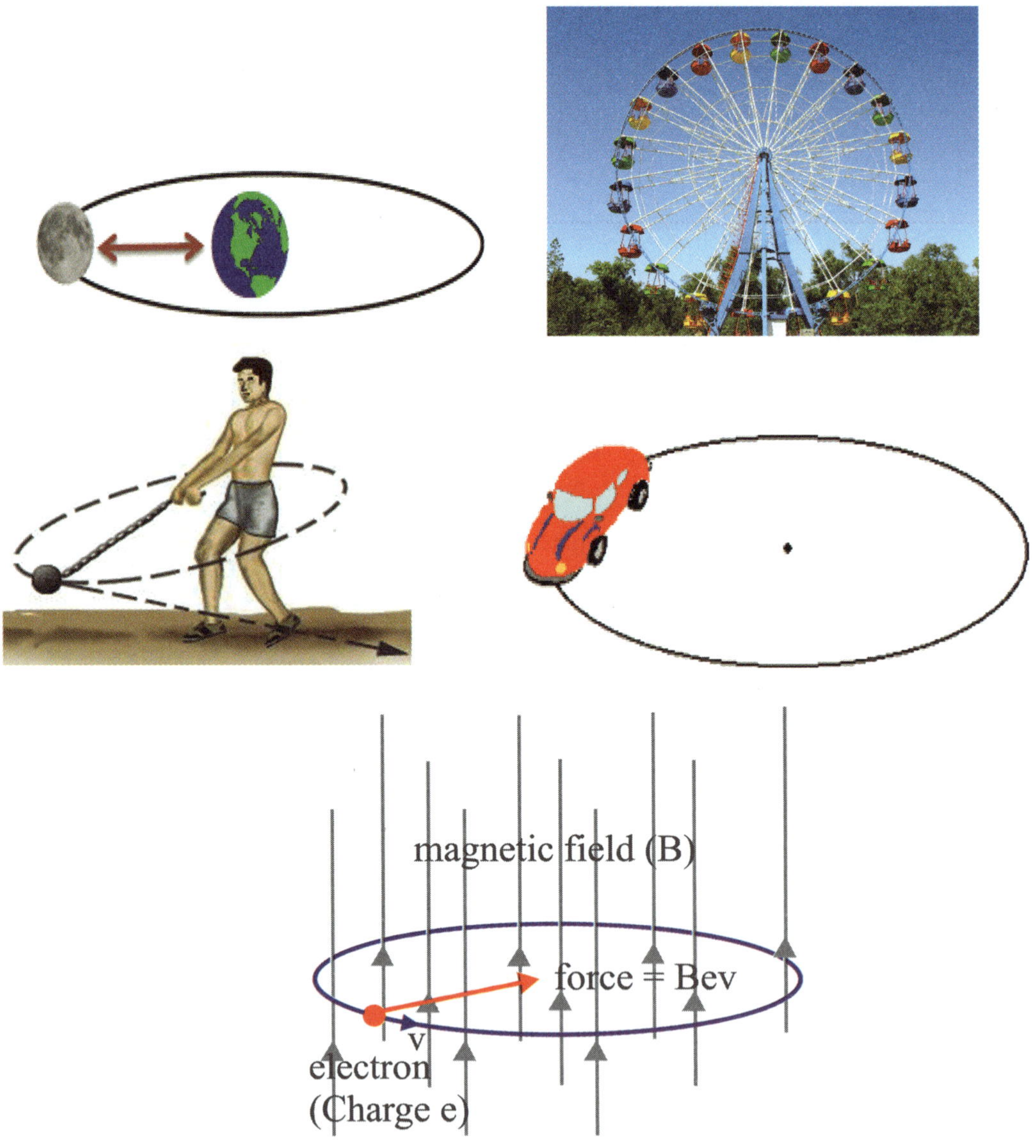

Fig.2.3 : Examples of circular motion

Mathematics of Uniform Circular Motion

[See also Table 1.1, Chapter I]

Consider an object moving in a circle with a uniform speed round a fixed point O at centre as shown in Fig.2.4.

If the object moves from A to B so that the radius OA moves through an angle θ, its *angular* speed ω about O is defined as the change of the angle per second. Thus, if the time taken by the object to move from A to B is t s, the angular speed would be[1]

$$\omega = \frac{\theta}{t} \text{ rad/s}$$

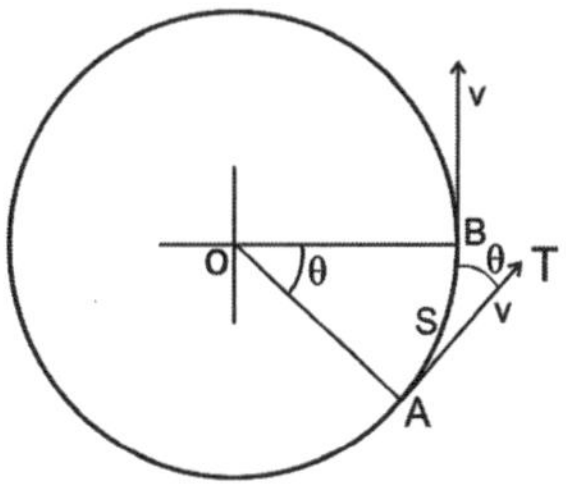

Fig.2.4 : Motion of an object in a circle

If S is the length of the *arc* AB, then by definition of an angle in radians as S/r

$$S = r\,\theta$$

where r denotes radius of the path of the circular motion.
Dividing by t, the time taken to move from A to B

$$\frac{S}{t} = r\,\frac{\theta}{t}$$

But S/t = speed v of the rotating object, and θ/t is the angular velocity.

Therefore

$$v = r \times \omega$$

Example

A *model* car moves round a circular track of radius 0.3 m at 2 revolutions per second. Calculate

(a) its angular speed

(b) the speed v of the car

Find also the angular speed if the car moves with a uniform speed of 2 m/s in a circle of radius 0.4 m.

---

[1]It is to be noted that angular distance, esp. in circular motion, is measured/stated in **radians** such that a rotation by 180° in a plane or space is taken equivalent to π radians or 3.1415927 (approx.). Accordingly,

1 degree = 3.1415927/180 or 0.01745 rad,

both being present only during motion.

(a) The car is moving with 2 rps. Therefore, angle covered in one second = 2π radians. Hence, its angular speed

$$\omega = 2 \times 2\pi \text{ or } 4\pi \text{ rad/s.}$$

(b) The speed $v = r \times \omega$

$$= 0.3 \times 4\pi = 1.2\pi \text{ or } 3.8 \text{ m/s.}$$

When the car is moving in a circle of radius 0.4 m, the angular speed will be

$$\omega = \frac{v}{r} = \frac{2(m/s)}{0.4(m)} = 5\,rad/s$$

Centripetal and Centrifugal Force

A body in a uniform circular motion with a peripheral velocity at a radius R from the axis of rotation is acted upon by TWO forces called

- centripetal force, and
- centrifugal force

both being present only during the motion.

Centripetal force

According to Newton's first law of motion, there must be forces acting on an object moving in a circular path since it does not move off in a straight line. The second law of motion (F = ma) also indicates forces since an unbalanced force is required to change the motion of an object. An object moving in a circular path is continuously being accelerated since it is continuously changing direction. This means that there is a continuous unbalanced force acting on the object that pulls it out of a straight-line path. The force that pulls an object out of a straight-line path and into a circular path is called a **centripetal** force.

Centrifugal force

Centrifugal force (Latin for "centre fleeing") describes the tendency of an object following a uniform circular motion to "fly outwards", away from the center of the curve. It is only a fictitious force that results from inertia — the tendency of an object to resist any change in its state of rest or motion.
In other words, it is a force that causes an object moving in a circular path to move out and away from the centre of its path.

The concept of centrifugal force, usually difficult to grasp, can be best explained by reference to the following example.

A stone on a string

Consider a stone being whirled round on a string as depicted in Fig.2.5.

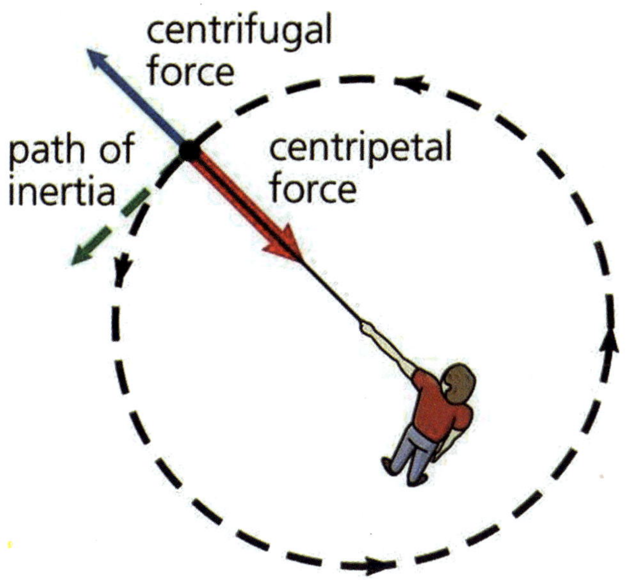

Fig.2.5 : Centrifugal and centripetal force

The only real force acting on the stone is the tension in the string. There are no other forces acting on the stone so there is a net force on the stone.

Were it not for this net force acting on the stone, the stone would travel in a straight line, according to Newton's first law of motion. In order to keep the stone moving in a circular path, this force must be continuously applied to the stone. As soon as it is removed (for example if the string breaks) the stone would move in a straight line.

As discussed, this (inward) force is the "centripetal" force under given system of uniform circular motion.

Now, to use Newton's laws of motion, it is necessary to 'invent' a new force that acts on the stone and is equal and opposite to the tension, or the inward centripetal force, in the string. This force would act in the outward direction and is called the centrifugal force. With this new (fictitious force), the net force on the stone is zero and the stone remains stationary in the rotating frame of reference and to continue its uniform circular motion.

It follows that whilst centripetal force is a *real* force that counteracts the centrifugal force and prevents the object from "flying out," keeping it moving instead with a uniform speed along a circular path, the centrifugal force represents only a fictitious force that comes into being as long as the body is in uniform circular motion.

Comparison between centripetal and centrifugal forces

This is listed in the following table.

	Centrifugal Force	Centripetal Force
Meaning	Tendency of an object following a curved path to fly away from the centre of curvature. Might be described as "lack of centripetal force."	The force that keeps an object moving with a uniform speed along a circular path.
Direction	Along the radius of the circle, from the centre towards the object.	Along the radius of the circle, from the object towards the centre.
Example	Mud flying off a tyre; children pushed out on a roundabout.	Satellite orbiting a planet
Is it a real force?	No; centrifugal force is the inertia of motion.	Yes; centripetal force keeps the object from "flying out"

III Work, Energy and Power

Work, energy, power and their relation with force.
Definition of work $W = F S \cos \theta$; special cases of $\theta = 0^o$, 90°; $W = mgh$; definition of energy; energy as work done; various units of work and energy and their relation with SI units (erg, calorie, kWh and eV);· definition of power, $P = W/t$; SI and cgs units; other units. kilowatt (kW), megawatt (MW) and gigawatt (GW) and horse power (1 hp = 746 W) [simple numerical problems on work, power and energy]

WORK

The word "work", *per se*, has many connotations[1]. However, in general work can be understood to refer to physical or mental work. Here, it would refer to physical work or the work related to stream of science that is physics; or more specifically mechanics.

Definition of Work

Work is done when a force is applied to a body to move it through a distance *and the force is in the same direction as the distance moved.*

In classical physics terms, **work** is done on an object when a force is exerted on the object causing it to move some distance. The amount of work actually done may have little relationship to the amount of effort applied.
For example, if a car stuck in a snow drift is pushed, a lot of force (and effort) may be exerted, but if the car does not budge, no work has been done! In order for work to be done on an object, the object *must move* some distance as a result of the force that is applied.

[1]For example, a version of the Oxford Dictionary lists the word work as follows:
n. **1** application of mental or physical effort to a purpose; use of energy. **2** task to be undertaken.**3** thing done or made by work; result of an action.**4** employment or occupation etc., esp. as a means of earning income. **5**. Literary or musical composition. **6**. action or expression of a specified kind and so on, amongst many more forms of usage and expressions.

Alternatively "work results when a force acts upon an object to cause a displacement (or a motion) or, in some instances, even hinder a motion".

Three variables are of importance in this definition - force, displacement, and the extent to which the force causes or hinders the displacement. Each of these three variables find their way into the equation for work. That equation is

Work = Force • Displacement • Cosine(phi)
W = F • d • cos (phi)

Or, as shown in Fig.3.1

$W = F \times d \cos \phi$

where

W is the work;

F the force acting on the body

d the displacement of the body due to F *in the direction of force*, and

ϕ is the angle between the direction of applied force and displacement of the body.

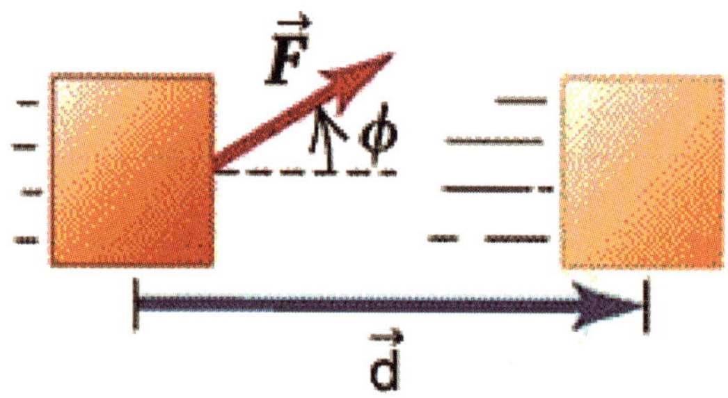

Fig.3.1 : Relationship between force, displacement and work

This is shown conceptually in Fig.3.2.

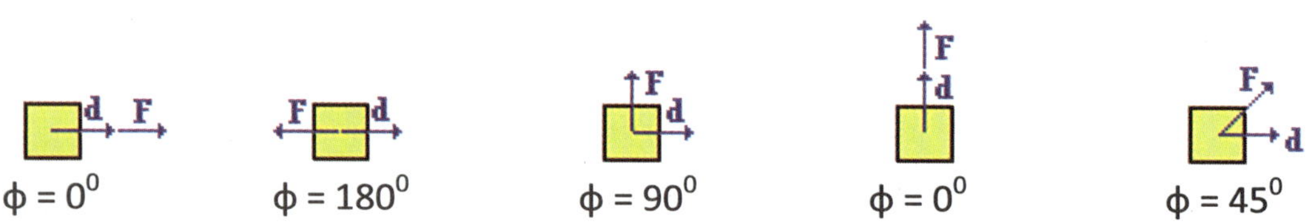

< ---------Direction of displacement of interest ---------->

(a) maximum work (b) no 'useful' work (c) no work (d) work at 'right' angle (e) work= max work x cos 45^0

Fig.3.2 : Force, displacement and work

If the force is in the same direction as the displacement, then the angle ϕ is 0 degree. If the force is in the opposite direction as the displacement, then the angle ϕ is 180 degrees. If the force is up and the displacement is to the right, then the angle is 90 degrees.

Interestingly, note that whilst both force and displacement are *vector* quantities, that is each having given magnitude and direction, **work is a *scalar* quantity.**

Units of Work

Since the standard metric or SI unit of force is the newton and the SI unit of displacement is the metre, then the standard metric or SI unit of work is a newton-metre, defined as a **joule**[2] and abbreviated with a J.

Thus, one joule of work is done when a force of one newton is applied through a distance of one metre *in the same direction.*

In CGS system, one unit of work results when a force of one dyne when applied to a body moves it by one cm and the unit is called **erg**.

One joule = 10^7 ergs[3]

Example

1. A force is applied on a body of mass 20 kg moving with a velocity of 40 m/s. The body attains a velocity of 50 m/s in 2 s. What is the work done on the body?

From the equation $v = u + \alpha t$, the acceleration of the body during its motion is given by

$$\alpha = (v - u)/t \text{ or } (50 - 40)/2 = 5 \text{ m/s}$$

Then force applied on the body, $F = m \times \alpha$ or $20 \times 5 = 100$ N

Also, distance travelled from $s = (v^2 - u^2)/2\alpha$

or $= (50^2 - 40^2)/(2\text{x}5)$ or 90 m

Then work done = force x distance

$$= 100 \times 90 \text{ or } 9{,}000 \text{ J}$$

2. A body moves through a displacement of 4 m while a force F of 12 Newton acts on it as shown in the adjoining figure

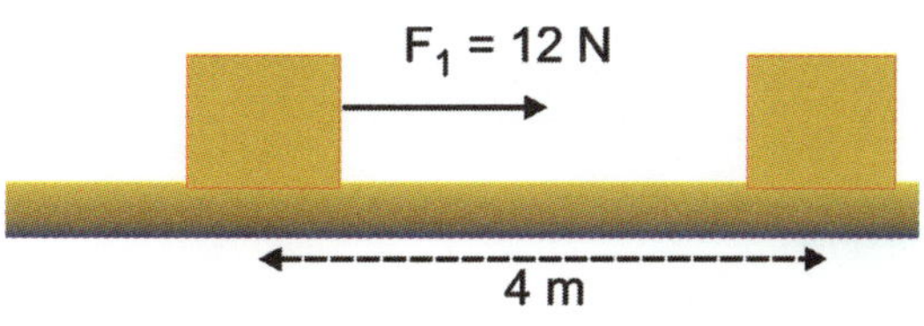

What is the work done by the force on the body?

$$\text{work} = \text{force} \times \text{displacement}$$

or $$W = F \times S$$

---

[2]Named after English physicist James Prescott Joule: 1818-1889.

[3]The conversion can be worked out by changing the force in newton (N) into dyne and distance in metre into cm.

W = 12 x 4
W = 48 joule

3. A block is pulled by a constant force of 40 newton. If the work done by the force on the block is 680 joule what is the displacement of the block?

The displacement of the block

W = F x S
680 = 40 x S
S = 680 / 40
S = 17 metre

4. Two forces that are F_1 = 10 N and F_2 = 5 N act on a body in a frictionless floor as shown. The displacement of the body is 5 m, what is the work done by the forces on the body!

$W = (F_1 + F_2) \times S$
W = (10 + 5) x 5
W = 15 x 5
W = 75 joule

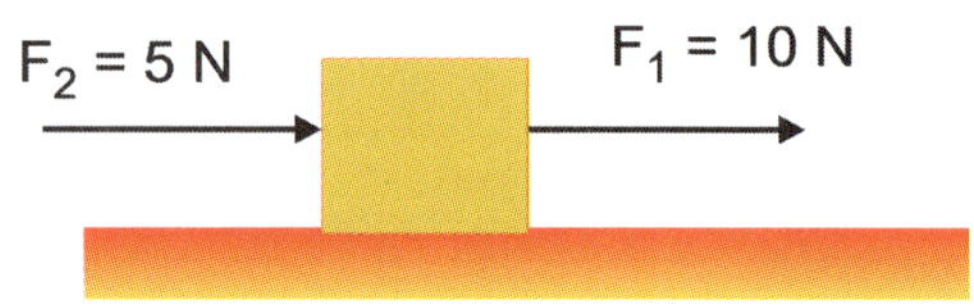

Work under Influence of Gravity

This comprises a special case of work done on or by a body when acted upon *vertically* by gravitational force; for example, a water fall as shown in Fig.3.3(a).

Consider a body of mass M kg falling vertically down from a height at point A as shown in Fig.3.3(b). The body is being acted on by acceleration due to gravity at g m/s/s and thus the body's mass and g together act as a force equal to Mxg N. If the body having been released from its position A and *falls freely* under pull of gravity by a distance D m, the 'work' done by the body will be (Mg) x D Nm or joule.
The numerical value of g in SI units is very nearly 9.8 m/s/s, usually rounded to 10 m/s/s.

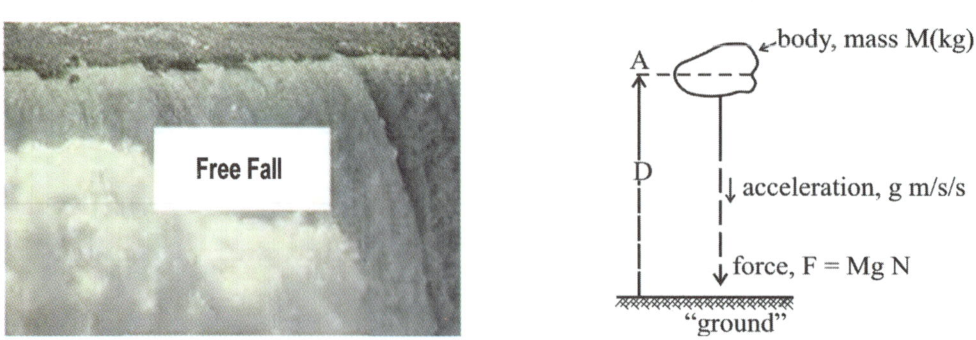

(a) (b)

Fig.3.3 : Free fall of a body

More images of work under the influence of gravity are given in Fig.3.4.

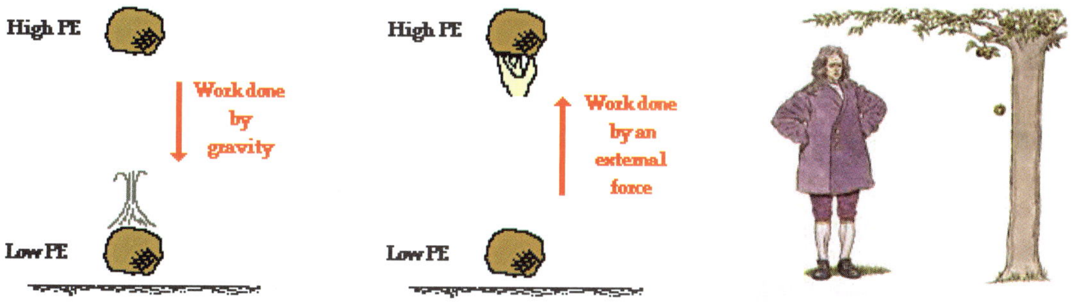

Fig.3.4 : Images of work under the influence of gravity

A graphical comparison of work during free fall and work against gravity is illustrated in Fig.3.5. As before, during free fall a body of mass m falling from a height of h under effect of gravity corresponds to a work done = m x g x h as shown in Fig.3.5(a).

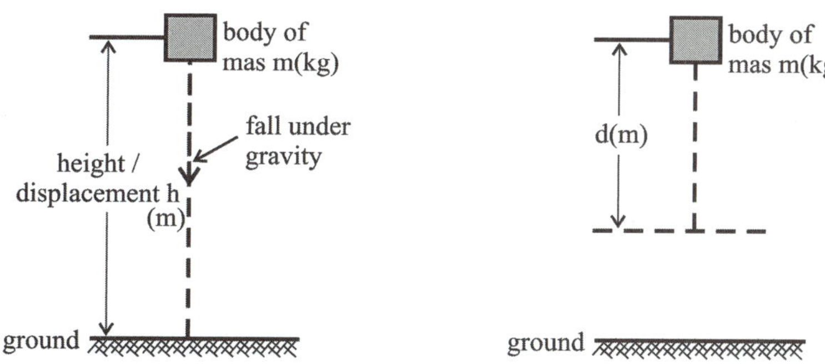

(a) work during free fall
(gravitational) force on the body
F = m g N
distance or height h, work
done due to "free fall"
W = F x h or mgh J

(b) work against gravity
A body of mass m being
brought down (say, manually) by a
work done, W = mgd J

Fig.3.5 : Work under the influence of gravity

Note that in Fig. 3.5(b), if the body were lifted upwards (say manually, again) by the same distance or height, the work done would be the same, that is mgd J, but *against* gravity and expressed by W = - mgd J.

With the gravitational force or acceleration due to gravity being in opposition, the work to be done on the body will be (mg) x d J to lift the body by a height of d m. A case in practice would be a "lift crane" lifting a box of given mass in a workshop up to a shelf of given height.

Example

1. A ball is thrown vertically upwards with an initial velocity of 30 m/s. Find (i) the time taken to reach its highest point, (ii) the distance then travelled. Assume g = 10 m/s/s.
(i) the initial velocity, u = 30 m/s

At highest point as the energy is exhausted the ball momentarily comes to rest and so v = 0. Also, g = -g or -10 m/s/s
From $v = u + g t$
$0 = 30 + (-10) \times t$, giving $t = 3$ s
(ii) the distance traversed is given by $D = u t + \frac{1}{2} g t^2$

∴ substituting $D = 30 \times 3 + \frac{1}{2} \times (-10) \times 3^2$
$= 45$ m

2. The work done by the human heart is 1 J per beat. Calculate the power of the heart if it beats 72 times per minute.

The heart beats 72 times per minute. Therefore, total work done = 1 x 72 or 72 J
The time taken is 60 s
The power of heart = 72/60 or 1.2 W

3. A lift is moving down with an acceleration of 3 m/s/s. A ball is released from 1.7 m above the lift floor. Assuming g to be 9.8 m/s/s, how long will the ball take to hit the lift floor?

Since the ball is falling within the lift, its downward acceleration relative to that of the lift would be (9.8 – 3) or 6.8 m/s/s
The ball is starting from rest; therefore u = 0
Using the expression $s = u t + \frac{1}{2} a t^2$
$1.7 = 0 + \frac{1}{2} \times 6.8 \times t^2$
whence $t = 0.7$ s

4. A machine raises a 'load' of 800 N through a height of 15 m in 5 s. Calculate the power exerted by the machine.

The work done by the machine is force x distance or 800 x 15 = 12,000 J
The time taken for doing the work is 5 s

∴ power = work done/time or 12,000/5.0 = 2400 W

5. For how long must an electric motor-pump rated at 2 HP operate so as to pump 5 m^3 of water from a depth of 15 m. Assume 1 m^3 of water being equal to 1,000 kg and g = 10 m/s/s.

The work done by the pump against gravity, $W = m \times g \times h$ J

Substituting,

$$W = 5{,}000 \times 10 \times 15 \text{ or } 7{,}50{,}000 \text{ J}$$

Since work done or energy used up is power x time
the time taken by the pump, t = 7,50,000/(2x750), assuming 1 HP = 750 W = 500 s

6. An electric motor-pump having an efficiency of 60% is rated at 2 HP. Calculate the maximum quantity of water it can lift to a height of 5 m in 40 s. Assume 1 HP = 750 W and g = 10 m/s/s.

The motor is rated for 2 HP or 2x750 = 1,500 W
Since its efficiency is 60%, the available power from the motor/pump

$$P = 1{,}500 \times 0.60 \text{ or } 900 \text{ W}$$

$\therefore$ the energy supplied by the pump in 40 s = 900 x 40 or 36,000 Ws or J
When used to pump water, this represents gravitational potential energy
or

$$m\,g\,h = 36{,}000$$

$\therefore$ quantity or mass of water that can be lifted to a height of 5 m will be

$$m = 36{,}000/(10 \times 5) \text{ or } 720 \text{ kg}$$

7. A boy lifts up a book from the floor to the table as shown. The mass of the book is 300 gram and the height of the table is 80 cm.

Find the work done by the boy when the acceleration due to gravity is 10 m/s^2

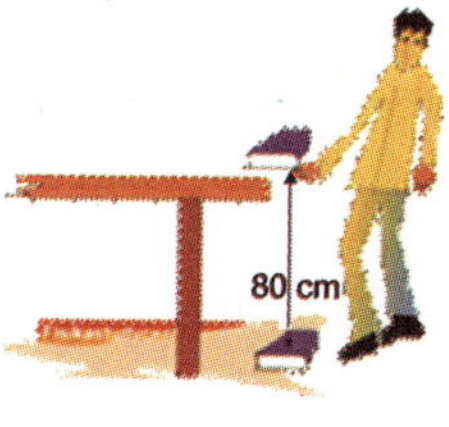

Finding the work done by calculating the difference of gravitational potential energy of the book, where the gravitational potential energy is $E_p = m \times g \times h$

$$W = \Delta E_p$$
$$W = m \times g \times h$$
$$W = 0.300 \times 10 \times 0.80$$
$$W = 2.4 \text{ joule}$$

Work up an incline

A situation pertaining to "doing work" usually encountered in practice is work up an incline; for example, a man walking upwards on a sloppy ground or a crane pulling a car on a road with a gradient.

The situation is as depicted in Fig.3.6.

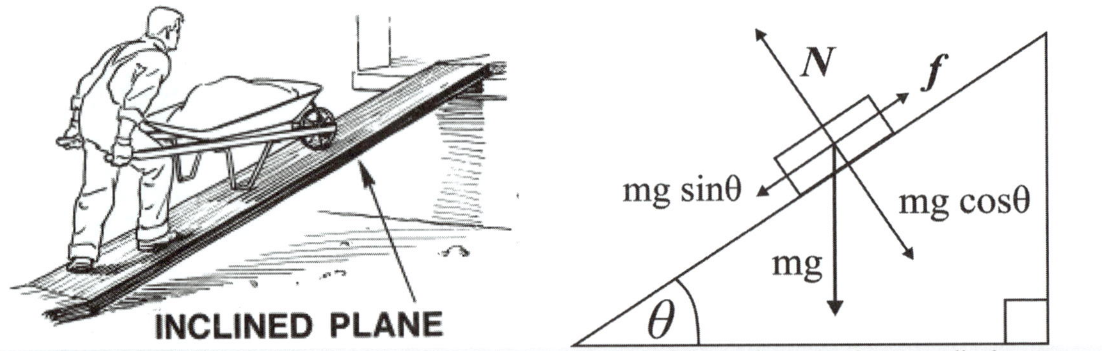

Key: N = normal force that is perpendicular to the plane, W = mg, where m = mass, g = gravity, and θ (theta) = angle of inclination of the plane

Fig.3.6 : Work done up an incline and related mathematics

Mathematics of Work Up an Incline

The first peculiarity of inclined plane problems is that the normal force, that is the force perpendicular to surface on which a 'body' is resting, is NOT directed in the direction that one is accustomed to. Up to this point in the course, only normal forces acting in an upward direction, opposite the direction of the force of gravity have been considered. But this is only because the objects were always on horizontal surfaces and never upon inclined planes. The truth about normal forces is not that they are always upwards, but rather that they are always directed perpendicular to the surface that the object is on. This is illustrated graphically in Fig.3.7.

Fig.3.7 : Inclined plane and normal force

Note that *normal* force is always directed perpendicular to the surface

Example

A coolie is pushing a box weighing 1500 N up an inclined plane 7.5 m long on to a platform, 2.5 m above the ground.
(i) Calculate the mechanical advantage of the inclined plane.
(ii) Calculate the effort applied by the coolie.

[ICSE, 2013]

The configuration of the activity is as shown in the figure.

(i) The mechanical advantage is given by
MA = length up the incline/height

= 7.5/2.5 or 3

(ii) The mechanical advantage is also related to load and effort as

MA = load/effort

∴ effort = load/MA

= 1,500/3 or 500 N

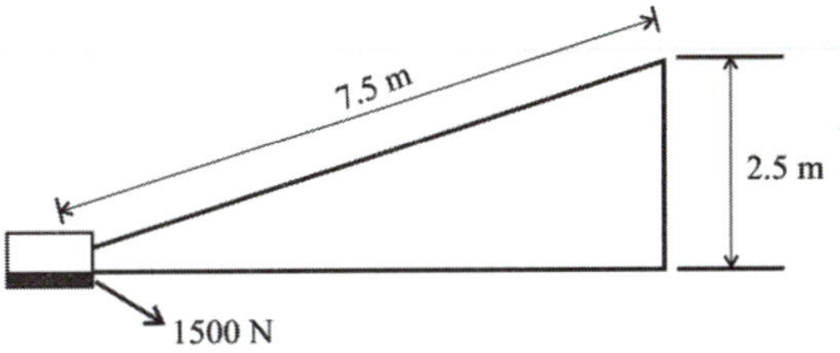

The Components of the Gravity Force

The task of determining the *net* force acting upon an object on an inclined plane involves resolving one (or more) of the forces acting on the body into perpendicular components, usually in horizontal vertical components, so that they can be easily added to the other forces acting upon the object. Here, the process of analysing the forces acting upon objects on inclined planes will involve resolving the weight vector, or weight of the body, (F_{grav}) into two perpendicular components - one directed parallel to the inclined surface and the other directed perpendicular to the inclined surface as shown in Fig.3.8.

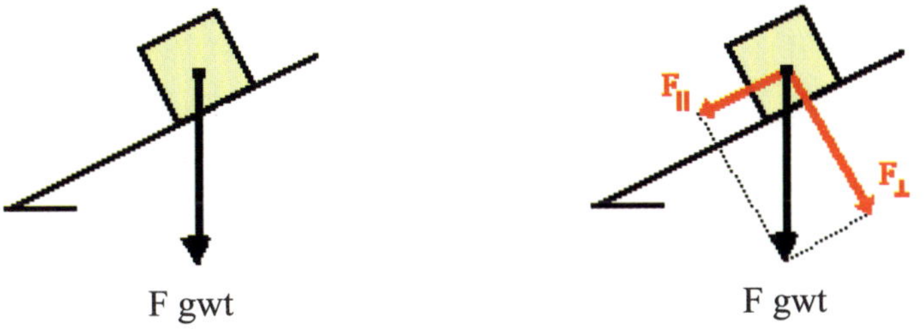

Fig.3.8 : Weight of a body on inclined plane resolved in two perpendicular components

The perpendicular component of the force of gravity is directed opposite the normal force and as such balances the normal force. The parallel component of the force of gravity is not balanced by any other force. This object will subsequently accelerate down the inclined plane due to the presence of an unbalanced force. It is the parallel component of the force of gravity that causes this acceleration. The parallel component of the force of gravity is the net force as shown in Fig.3.9.

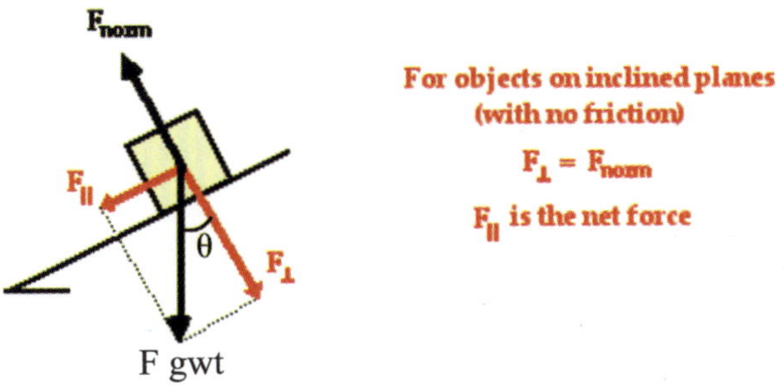

Fig.3.9 : The two components of force of gravity

The magnitude of the two components of the force of gravity are given by the equations

$$F_I = m \times g \times \cos\theta$$
$$F_{II} = m \times g \times \sin\theta$$

where m = mass of the 'body'

g = acceleration due to gravity

and θ = angle between F_I and F gwt

In the absence of friction and other forces (tension applied etc.), the *acceleration* of an object on an incline is the value of the parallel component (m x g x sine of angle), divided by the mass (m). This yields the equation

$$\mathbf{a = g \cdot \sin\theta}$$

(in the absence of friction and other forces)

See also Fig.3.4.

Example

1. An elastic band is pulled out steadily through a small distance of 10 cm. The force exerted can be assumed to be increasing from 0 to 10 N. What is the work done on the band?

Here, the force is proportional to the extension of the band,
the work done = average force x distance moved

$$= \tfrac{1}{2}(0 + 10)\text{ N} \times 0.1\text{ m}$$
$$= 5 \times 0.1 \text{ or } 0.5\text{ J}$$

2. A force of 20.8 N is applied parallel to the incline to lift a 3.00-kg loaded cart to a height of 0.450 m along an incline which is 0.636 m long. Determine the work done upon the cart.

The method of solving this problem involves using the equation

$$W = F \times d \times \cos\theta$$

where F=20.8 N, d=0.636 m, and $\theta = 0$ degree.

[The angle θ represents the angle between the force and the displacement vector; since the force is applied parallel to the incline, the angle is zero.]

Substituting and solving yields

$$W = 20.8 \times 0.636 \times \cos 0 = 13.2\text{ J}$$

ENERGY

Definition

What is energy?

Energy is the amount of work a physical 'system' is capable of performing so energy can be defined as that which changes the position, physical composition or temperature of an object. In general, energy can be defined as

Energy = work (done) x time duration

In 'mechanical' form or in kinematics, there are essentially two categories of energy:

- **kinetic energy**, and
- **potential energy**

The difference between them is whether the energy is being transferred (kinetic) or stored (potential). In general, they are inter-convertible.

This is discussed later.

Units of Energy

The SI unit of energy is joule symbolised by letter J.
However, since energy is used in different forms such as mechanical, electrical or thermal to perform numerous types of work, there are many different units used to express and measure the quantities of energy used. For example, whilst the joule is a measure of energy in general – electrical or mechanical - the calorie or British thermal unit (BTU) is typically used for measuring thermal energy.

Recall that the SI unit of work is also joule (J).

The various aspects of "unit of energy" are discussed later.

Some common units and their relationships

joule(J): The amount of work or energy exerted when a force of one newton results in a displacement of one metre.

calorie(cal): The energy needed to raise the temperature of one gram of water by 1 °C[4].

British Thermal Unit (BTU or Btu):The energy needed to raise the temperature of one pound of water by 1°F[5].

Conversion Factors

1 J = 0.2388 cal
1 cal = 4.1868 J
1 BTU = 1.055 kJ = 0.252 kcal

Electrical unit of energy

The power rating in electrical usage, for example for lamps and appliances etc. in domestic use and electric motors in industry, is specified in watt (as discussed later). The unit of energy is then specified as

watt-second (or joule), Ws	:	power of 1 watt 'used' for a second
watt-hour, Wh	:	1 watt power consumed in one hour
kilowatt-hour, kWh	:	1 kilowatt power consumed in one hour

The last one represents the 'standard' unit of electricity consumption in domestic and most commercial premises.

Unit of Energy as Electron-volt (eV)

In some cases, the unit of energy is expressed in "electron volt", abbreviated as eV. The kinetic energy gained by an electron that has been accelerated through a potential difference of 1 volt is called an *electron-volt* (or eV). Since the energy gained in moving a charge Q through a PD V is equal to QV, it follows that

$$1 \text{ eV} = \text{charge of an electron} \times 1$$

$$= (1.6 \times 10^{-19}) \times 1 \text{ joule} \quad \text{or} \quad 1.6 \times 10^{-19} \text{ J}$$

Being a small unit, electron-volt is a useful unit of energy in atomic physics. For example, an electron in an X-ray tube moving through, or accelerated by, a PD of 50,000 V would gain energy equal to 50,000 eV.

---

[4]The "nutritional" calorie is often expressed as kilocalorie (kcal),
[5]The unit is used extensively in engineering applications; mainly heating and cooling devices such as air conditioners.

POWER[6]

What is Power?

Energy is a measure of how much work can be done *without any consideration of how long it takes to accomplish the work.* Power is the measure of how quickly the work is done; so it is a time-dependent function.

Thus, consider two marathon runners. They would do the same amount of *work* when they run the marathon of 10 km. However, the runner that finishes the marathon in 2 hours 30 minutes runs twice as 'powerfully' (or faster) as the runner that finishes in 5 hours. In other words the former runner puts in twice as much 'effort' (in common language) than the latter: this trait is understood as **power**.

Power is the physical quantity that relates to energy in terms of time. It can be defined as **the amount of energy (used) per unit of time.**

It follows the that difference between energy and power is that whilst power may be measured at any point in time, energy has to be measured during a certain time period such as a second, a minute or an hour.

Or

Power = Work(energy)/Time
= (Force x Distance)/time or Force x Velocity

Units of power

The 'common' unit of power is joules per second or J/s when work is measured in joules and time in seconds. The basic unit of power, 1 J/s is called a watt (W) which is also the SI unit of power[7]

Hence, by definition, a watt is the consumption of one joule of energy per second.

Or

one watt (1 W) = one joule (1 J)/one second (1 s)

or

1 W = 1 J/s

---

[6]Like work, the word power is in common use variously in English; for example being associated with authority vested in a CEO or influence of a great leader or artist.

[7]Named after the British 'scientist' James Watt who made significant contributions in the improvement of steam engines.

Example

A boy of 450 N of weight climbs a wood steps with a length of 3 m. The boy takes 6 seconds to get to the end of the steps. What is the power used by the boy in this activity?

The power (P), the work (W) and the time (t) relationship:

$$P = W/t$$

where W = (weight of the boy) x (the boy's displacement) = 450 x 3 = 1350 joule

Then

$$P = W/t$$
$$P = 1350/6$$
$$P = 225 \text{ watt}$$

Horsepower : the unit of power for 'machines'

Historically, most machine such as "engines" and motors, mostly involving some kind of motion, have been associated with unit of power specified in *horse power*, abbreviated as HP, or simply hp; for example a diesel engine or an electric motor. This is related to the SI unit of power as specified later.

Genesis of "Horse Power" as Unit of Power

In the times of "industrial revolution" – 17th and 18th century – it was common to 'employ' horses to do much work; for example pulling carts, even to ferry passengers for long distances (up to several hundred miles when the horses would be replaced at regular intervals). A very common requirement, esp. in England, was to pull up mined products such as coal from the bottom of mines and this was undertaken using horses as illustrated in Fig.3.10.

Fig.3.10 : Horse pulling up a load

It was James Watt who whilst working on the design of his steam engines came upon the idea of relating the work done by a 'normal' horse in a given time, pulling a given load or weight from a given depth of the mine and the corresponding work performed by an engine, naming the power of the either as a horse power. Much later, the unit of work being pound-feet/minute in those days was equated to watt as

1 HP = 746 W

(or 745.70 correctly)

Metric Horse Power

The *metric* unit of horsepower differs slightly from the earlier derived value, defined as one metric horsepower being the power needed to lift a mass 75 kilograms by 1 meter in 1 second, such that

1 metric HP = 735.5 W

At present, it is common to use only watt as the unit of power in ALL usage; however, some machines, for example electric motors, may be specified both in watt as also HP.

Example

6.4 kJ of energy causes a displacement of 64 m in a body in the direction of force in 2.5 seconds. Calculate (i) the force applied (ii) power in horse power (hp). (Take 1hp = 746 W).

[ICSE, 2009

The energy in the direction of force is given by

energy = force x distance

∴ 6.4 x 1,000 = force x 64

whence force = 6,400/64 or 100 N

Also, energy = power x time

∴ 6,400 = power x 2.5

whence power = 6400/2.5 or 2560 W

and power in horse power = 2560/746 or 3.43 hp

Multiples of Unit of Power

In numerous applications the unit of power as watt is rather small to specify capacity or rating of an appliance or equipment; for example, a microwave oven in domestic use or a generator or motor in industrial usage. In such cases, following multiples of watt are commonly employed

- 1000 watt = 1 kilowatt, denoted as 1 kW
- 1 million or 10^6 watt = 1 megawatt, denoted as 1 MW
- 1 trillion or 10^9 watt = 1 gigawatt, denoted as 1 GW

Clearly,

1 MW = 1000 kW

and 1 GW = 1000 MW

Sub-multiples

Contrary to the above, there are many appliances in common use where the power requirement may be less than a watt; for example, a cell-phone charger or an adapter for a small radio set. In such cases, the power may be expressed in milliwatt, where

1 milliwatt = 1 watt/1000 or 10^{-3} W, denoted as mW

Yet again, in electronic circuitry, it is usually required to deal with power that may be less than 1 mW. Here, the power may be expressed in microwatt, where

1 microwatt = 1 watt/1,000,000
= 10^{-6} W, denoted as μW

Note that all the multiples and sub-multiples of the unit of power are '*standard*' expressions in line with SI units.

Comparison of Work, Energy and Power

Work	Energy	Power
refers to an *activity* involving a *force* and movement in the *direction* of the force. A force of 20 newton pushing an object 5 meters in the direction of the force does 100 joules of work.	is the *capacity* for doing work. Energy is needed to accomplish work - it is like the "currency" for performing work. To do 100 joules of work, it is necessary to expend 100 joules of energy.	is the *rate* of doing work or the rate of using energy, which are numerically the same. If 100 joules of work is done in one second (using 100 joules of energy), the power is 100 watt

Example

An elevator must lift 1000 kg a distance of 100 m at a velocity of 4 m/s. What is the average power the elevator exerts during this trip?

The work done by the elevator over the 100 meters is given by

$W = m\,g\,h = (1000)(9.8)(100) = 9.8 \times 10^5$ Joules.

The total time of the trip can be calculated from the velocity of the elevator:

$$\text{time, } t = \frac{\text{distance, h}}{\text{vrelocity}} = \frac{100(\text{m})}{4(\text{m/s})} = 25 \text{ s}$$

Thus, the average power is given by, $P = \dfrac{W}{t} = \dfrac{9.8 \times 10^5 \text{ (J)}}{25(\text{s})}$

$$= 3.9 \times 10^4 \text{ W or } 39 \text{ kW}$$

EXERCISES

1. If the power of a motor is 40 kW, at what speed can it raise a load of 20,000 N?

[2 m/s

2. An energy of 6.4 kJ causes a displacement of 64 m in a body in the direction of force in 2.5 s. Calculate (i) the force applied, (ii) power in HP, assuming 1 HP = 746 W.

[100 N; 3.43 HP

3. A person exerts a force of 150 N in pulling a cart at a constant speed of 10 m/s. Calculate the power exerted.

[ICSE, 2015]

[1,500 W

4. A block displacement cause of a force in a horizontal surface is 13 meter and the work done by the force on the block is 15.6 joule. Find the magnitude of the force!

[1.2 N

5. A work done on a block through two forces in opposite directions is 120 joule. The force in forward direction, F1, is 36 N. If the displacement of the block is 5 m, determine the magnitude of force in reverse direction, F2.

[12 N

6. In two minutes the energy dissipated by a lamp is 3000 joule. Find the power of the lamp.

[25 W

IV Types of Energy

Different types of energy (e.g. chemical energy, mechanical energy, heat energy, electrical energy, nuclear energy, sound energy, light energy).
Mechanical energy: potential energy $U = mgh$ *(Derivation included); gravitational PE, examples; kinetic energy* $K = \frac{1}{2} mv^2$ *(derivation included); forms of kinetic energy: translational, rotational and vibrational – only simple examples [numerical problems on* K *and* U *in case of translational motion]; qualitative discussions of electrical, chemical, heat, nuclear, light and sound energy; conversion from one form to another; common examples.*

Some Common Forms of Energy

- Mechanical Energy
  - Kinetic
  - Potential

- Electrical Energy
  - Alternating
  - Direct
- Chemical Energy
- Thermal Energy
- Nuclear Energy
  - Fission
  - Fusion
- Gravitational Potential Energy

- Solar Energy
- Wind Energy
- Tidal Energy
- Geothermal Energy
- Biomass Energy

The last five are known to constitute "renewable sources (or forms) of energy", to be dealt with in detail later.

TYPES OF ENERGY

Mechanical Energy

Potential energy, PE

Potential energy is essentially considered as **"stored" energy** or energy of *position* that has the potential to do work[1].

In general, potential energy is associated with bodies or objects held at a height *against the force due to gravity* and thereby having acquired potential energy by virtue of its position in space. Owing to the gravitational force being implicit to the energy, this form of energy is known as **gravitational potential energy.**

It follows that energy can be stored in an object by lifting it up. The amount of potential energy would be

$$\mathbf{PE = m\,g\,h}$$

where m is the mass of the object, h is the height of the object and g is the force due to gravity acting on the object.

Gravitational potential energy, GPE

Gravitational potential energy is energy an object possesses because of its position in a gravitational field; thus, it is the energy associated with gravity or gravitational force. Since the force required to lift a body is equal to its *weight*, it follows that the gravitational potential energy is equal to its weight times the height to which it is lifted.

$$PE_{gravitational} = \text{weight x height} = mgh$$

Clearly, the maximum value of GPE would relate to conditions on the earth's surface where the acceleration due to gravity, g – a component of weight – is maximum, being about 9.8 m/s/s.

Observe that the zero datum of gravitational potential energy can be chosen at any point (like the choice of the zero of a coordinate system), and hence the potential energy at a height h

---

[1]In general usage of English language, the word potential is meant to capability of coming into being or any entity possessing capacity for use or development.

above that point is equal to the work which would be required to lift the object to that height *with no net change in kinetic energy*.

Recall that the expression for work (done) under the influence of gravity was also derived as mgh in which mg denoted the gravitational force "working on a body" to raise it to a height of h. Thus

$$GPE = W$$

and this being a direct consequence of the work-energy principle, discussed later.

This brings in the concept of kinetic energy vs. (gravitational) potential energy.

Consider

a. when a man running goes uphill and slows down to come to a stop doing work and having acquired some kinetic energy; where does this energy go?
b. a box is lifted from the ground using energy in one's arm muscles to reach a height h and hold the raised hand there; what happens to the energy that has been used?

The answer is that in both the cases the energy by virtue of motion, that is the kinetic energy, turns into gravitational potential energy

Or $$KE \rightarrow GPE$$

This important concept is illustrated in the following examples.

Example

1. A ball of mass 0.5 kg is thrown vertically upward with a velocity of 8 m/s. Calculate the maximum potential energy (PE) gained by the ball.

As the ball moves upward, it possess a kinetic energy, KE of

$$\frac{1}{2}\, m\, v^2$$

That is $$\frac{1}{2} \times 0.5 \times 8^2 \text{ or } 16 \text{ J}$$

At a certain height when the ball comes to rest (due to gravitational retardation), it would have acquired maximum *potential* energy. This would be equal to the maximum KE with which it started as per the "law of conservation of energy".

Hence, maximum potential energy gained by the ball will be 16 J.

2. A ball of mass 200 g falls from a height of 5 m. What will be its KE when it reaches the ground? Assume g = 9.8 m/s/s.

Using the equation $v^2 - u^2 = 2\, g\, s$ and substituting

$u = 0.0$ (as the ball starts from rest from the given height) and $s = 5.0$ m

the velocity of ball as it reaches the ground will be

$$v^2 = 2 \times 9.8 \times 5.0 \text{ or } 98.0\ (m/s)^2$$

Then $$KE = \frac{1}{2} m v^2 \text{ or } \frac{1}{2} \times 0.2 \times 98.0 = 9.8 \text{ J}$$

Kinetic energy, KE

Kinetic energy is the energy of motion of *any* object or physical entity (for example the motion of a car on the road or a train on railway track or even waves, electrons, atoms, molecules etc.)

Suppose that an object of mass M is moving with a velocity u and is gradually brought to rest in a distance s by a *constant* force F acting against it. The kinetic energy originally possessed by the body would, by definition, be equal to the work done against force F and, hence

$$\text{kinetic energy (KE)} = F \times s$$

But $F = M a$, where a is the acceleration of the body whilst in motion.

Therefore $$F \times s = M a s$$

Now, from the expression $v^2 = u^2 + 2 a s$ (discussed in Chapter I), since $v = 0$ and a is negative in this case

$$0 = u^2 - 2 a s \quad \text{or} \quad a s = u^2/2$$

and hence

$$\text{kinetic energy} = M a s = \frac{1}{2} M u^2$$

Or, in general,

Kinetic energy = ½ x mass of moving body x velocity^2

When the mass, M, is in kg and velocity, u, is in m/s, the KE is obtained in joule (J). Thus, a car of mass 1000 kg when moving with a velocity of 36 km/hr or 10 m/s would possess a kinetic energy of

$$KE = \frac{1}{2} \times 1{,}000 \times 10^2 \text{ or } 50{,}000 \text{ J}$$

Example

Calculate the change in the KE of a moving body when its velocity is reduced to 1/3 of its initial velocity.

[ICSE, 2014]

Let the mass of the body be m and its initial velocity v m/s
Then its (initial) kinetic energy $(KE)_1 = \frac{1}{2} m v^2$ units
When the velocity is reduced to v/3, its kinetic energy will be

$$(KE)_2 = \frac{1}{2} m (v/3)^2 \text{ units } (1/9) \times (\frac{1}{2} m v^2) \text{ units}$$

This shows that the kinetic energy in the second case is *reduced* by 9 times
[this is in line with the fact that KE of any moving body is proportional to square of its velocity in motion]

Uniform circular motion and kinetic energy

The discussion of kinetic energy above pertains to the *translational* motion of a body on a plane surface at an angle (θ) in general with respect to a datum, usually the ground etc.[2] When the body describes a circular motion at a constant velocity; for example a stone tied to a string and whirled around above a performer's head in a *horizontal* plane[3], the basic equation of kinetic energy will still hold but will undergo slight change as follows.

Assume a body of mass m executing a circular motion with a peripheral velocity v m/s at a radius r m such that the body's angular velocity will be ω rad/s, related to v by $v = \omega r$ m/s.
The kinetic energy of the configuration will be given by

$$KE = \frac{1}{2} m v^2$$

Substituting for v ($= \omega r$)

$$KE = \frac{1}{2} m (\omega r)^2$$
$$= \frac{1}{2} I \omega^2$$

where I [$= m r^2$] is called **moment of inertia** of the body that comes into play owing to its rotational motion.

Relation between momentum and kinetic energy

At times, it is desirable to express the kinetic energy of a body in terms of its momentum when moving with a velocity, v. The relation between the two can be derived as follows. If the momentum of the body is p and its mass m, then

$$p = m \times v \quad \text{or} \quad v = p/m$$

[2]The discussion also disregards any frictional force between the body and the surface.
[3]If the plane deviates from the horizontal, the force(s) due to gravity would come into play.

Since kinetic energy possessed by the body is given by

$$KE = \frac{1}{2} m v^2$$

Then, substituting for v from above

$$KE = \frac{1}{2} m (p/m)^2$$

giving $$KE = (p^2/2m)$$

Note that if a massive body and a light body have the same momentum, the light one will have a lot more kinetic energy. If a light body and a heavy one have the same velocity, the heavy one would have more kinetic energy.

Example

A body of mass 50 kg has a momentum of 3,000 kg-m/s. Calculate (i) KE, (ii) velocity of the body.

Let the velocity of the body be v m/s

Since momentum is $p = m \times v$

$$v = p/m$$

$$= 3{,}000/50 \text{ or } 60 \text{ m/s}$$

Then $KE = \frac{1}{2} m v^2$

$$= \frac{1}{2} \times 50 \times 60^2 \text{ or } 90{,}000 \text{ J}$$

Work energy principle

This principle, often used in kinematics problems, can be stated as

"The change in the kinetic energy acquired by an object during motion is equal to the net work done on the object".

Derivation of the theorem of work energy principle

In the case of a *translational* motion, the displacement, s, of a body of mass m when acted upon by a force F moving with an *initial* velocity v_1 and *constant* acceleration a can be expressed by the equation

$$s = \frac{v_2^2 - v_1^2}{2a}, \; v_2 \text{ being final velocity of the body,}$$

from the equation of motion $v^2 = u^2 + 2as$.

Now, the work done, W, is given by

$$W = F \times s = m\,a\,s$$

Substituting for s from above

$$W = m\,a\,[(v_2^2 - v_1^2)/2\,a]$$

Or

$$W = \frac{mv_2^2}{2a} - \frac{mv_1^2}{2a}$$

= change in kinetic energy (ΔKE)

Sudden Impact

A special case of the above expression is when $v_2 = 0$; for example, a car running with a velocity v_1 suddenly comes to rest by, say, hitting a tree, in which case

$$W = -\frac{mv_1^2}{2a}$$

and the work done is 'lost' in causing an **impact** against the tree.

This is illustrated conceptually in Fig.4.1 below

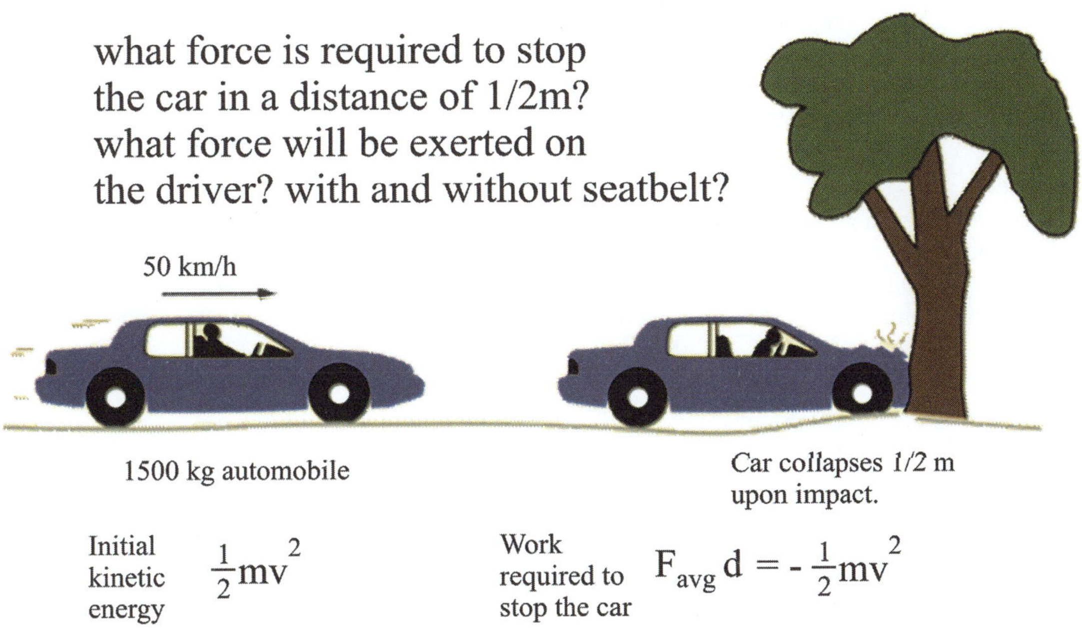

Fig.4.1 : Concept of impact and kinetic energy

For a straight-line collision, the net work done is equal to the average force of impact times the distance travelled during the impact.

Example

A force is applied on a body of mass 20 kg moving with a velocity of 40 m/s. The body attains a velocity of 50 m/s in 2 s. Calculate the work done by the body.

[ICSE, 2013]

The mass of the body is, $m = 20$ kg
Initial velocity of the body, $u = 40$ m/s
Final velocity, v $v = 50$ m/s
The time of motion, $t = 2$ s
The change in kinetic energy of the body in 2 s

$$\Delta KE = \tfrac{1}{2}(m v^2 - mu^2)$$
$$= \tfrac{1}{2} \times 20 \times (50^2 - 40^2)$$

According to "work-energy principle", the change in kinetic energy of the body is also the work done, WD, by the body

$$\therefore \quad WD = \tfrac{1}{2} \times 20 \times (2500 - 1600) \text{ or } 9{,}000 \text{ J}$$

Vibrational kinetic energy

This may be visualised as a form of kinetic energy associated with vibrations of a body or object where a translational motion is not encountered. In other words, energy acquired by an object or system by virtue of vibrations in a given mode. An example of this form of KE may be vibrations of a tuning fork, or a taught string held horizontally in which vibrations are induced by plucking the string in the middle and releasing to vibrate freely[4].

Kinetic energy vs. potential energy

Consider

a. a running person going uphill at some speed and coming to a stop and having acquired some kinetic energy;
b. a person lifting a box from the ground a height h and hold the raised hand there;
c. a car moving up an incline at a constant speed and moving down the slope after a while

In all these cases the energy by virtue of motion, that is the kinetic energy. turns into gravitational potential energy towards the end of the motion.

Or $$KE \rightarrow GPE \rightarrow KE$$

As brought out earlier (see page 63).

This important concept is illustrated by two more examples in Fig.4.2.

---

[4]It is not easy to work out an expression for the kinetic energy in this case.

Fig.4.2 : The concept of potential energy vs. kinetic energy

Similarly, a waterfall possesses both potential and kinetic energy. The water at the top of the fall possesses potential energy. As the water flows over the edge of the falls, its energy is changed into kinetic energy. See Fig.4.3.

Fig.4.3 : Waterfall and energy

In simple kinematics or physics, the above examples demonstrate a very important aspect of energy, known as the law of *conservation of energy*, dealt with later.

As stated before, energy is obtainable in various forms. Some of these are described as follows.

Electrical Energy

This is by far the most important form of energy today. Electricity or electrical energy can be truly considered as *the* life line of mankind today as almost everything in life depends on this form of energy in one way or the other.

Electrical energy is the energy that is caused by movement of electric charges. Since the charges are in motion, this is a form of kinetic energy. The more the charges (in number) and faster they move the more electrical energy they produce.

An example of tremendous source of electrical energy, albeit not exploitable as of now, is varied forms of lightning as illustrated is Fig.4.4

Fig.4. 4 : Lightning as an example of tremendous electrical energy[5]

Advantages of Electrical Energy

Electrical energy has numerous advantages over other forms. Some of these are

- Cleanest, with no environmental repercussions
- Most efficient and cost-effective in many applications
- Easy to control
- Efficiently 'transportable'

Means of generating electrical energy

Generation Using Conventional Means

The means used in large generating stations, having installed capacities of thousands of megawatt, are called conventional means for these consist of well-designed, time-tested equipment to generate electricity. The basic principle of generation, even in generators rated at many hundred megawatt to over a thousand megawatt per unit, derives from the laws of electromagnetic induction enunciated by the British scientist, Michel Faraday, in 1831. The laws that relate to a conductor (more likely a coil) when exposed to a *time-varying* magnetic field induces an EMF are slightly modified for the 'rotating' electric generators in that the induced EMF is based on the *flux-cutting rule* given by

[5]Whilst lightning could be a gigantic source of electrical energy if it were possible to harness it, there is another side of it: often an extremely high dis charge, *esp. in the vicinity of residential premises*, might cause irreparable damage to some appliances in the house such as a TV, a ceiling fan or electronic gadgets; for example, an adapter used to power the appliance at low (DC) voltage.

EMF = (non time-varying) magnetic field x conductor (or a set of coils) x velocity of conductor with respect to the field

or, simply, $E = B\,l\,v$

where B is the flux density, l the 'length' of conductor and v the velocity, *all three being mutually perpendicular to each other*. This is also expressed by the well-known **Fleming's right-hand rule,** depicted in Fig.4.5.

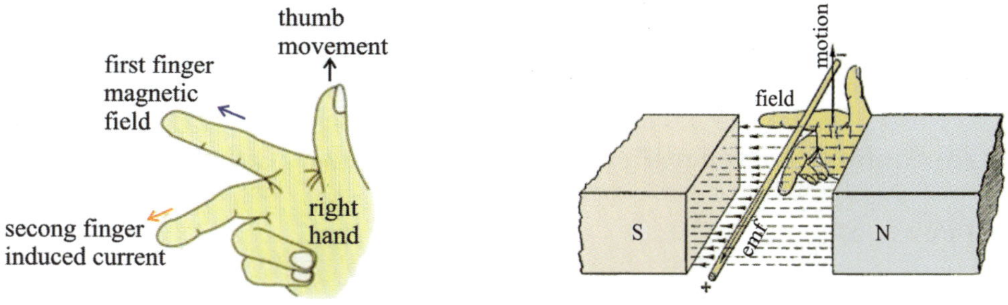

Fig.4.5 : Fleming's right-hand rule

Schematic of a conventional generator

Any form of conventional generating system used to generate electrical energy in a power house or generating station primarily comprises

- an appropriately conceived rotating device rigidly coupled to the generator and known as a **prime mover**, and
- the **generator** which when driven by the prime mover at a given, constant speed actually generates electrical energy.

A schematic of this basic arrangement is shown in Fig.4.6.

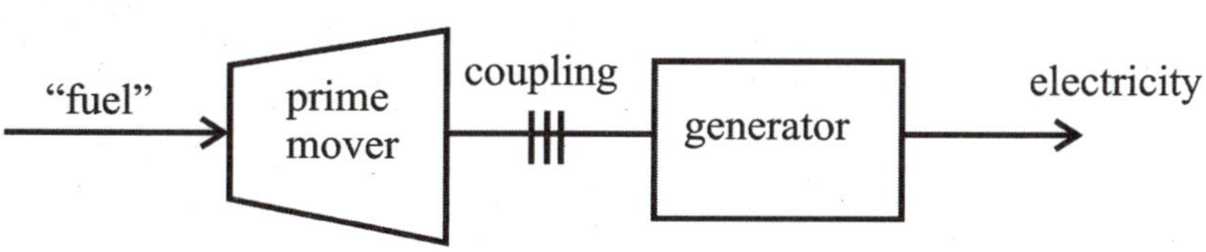

Fig.4.6 : Schematic of the basics of conventional power generation

The prime movers

The choice of a prime mover to connect to the generator would depend on the type of power, station. In **thermal** power stations, the prime mover is essentially a **steam turbine,** the rotation of which is caused by suitably input steam into the turbine. The steam may be produced, for example, by the heat from

- burning of coal in a boiler

- fission of nuclear material such as uranium in a reactor
- burning of natural gas, or biomass, in a chamber
- geothermal source

as depicted schematically in Fig.4.6.

The other main form of a prime mover is a **hydro turbine** that is used in a hydro station to drive the (hydro) generator.

An important difference between a steam and hydro turbine is rotational speed of the two: whilst a steam turbine may usually rotate at a few *thousand* rpm, the speed of rotation of a typical hydro turbine may be limited to a few tens or hundreds rpm.

Two main 'stations' of generation of conventional electrical energy

Thermal Power Station

A **thermal power station** houses a power plant in which heat energy is converted into electrical energy. In most of the existing plants, the heat is used to produce steam from water in a boiler, which is input to a mechanical device called a (steam) turbine that is connected to a generator (also called an alternator since it generates *alternating* current or AC power), The schematic of thermal generation is given in Fig.4.7(a) whilst a photo of a typical steam turbine is shown in Fig.4.7(b).

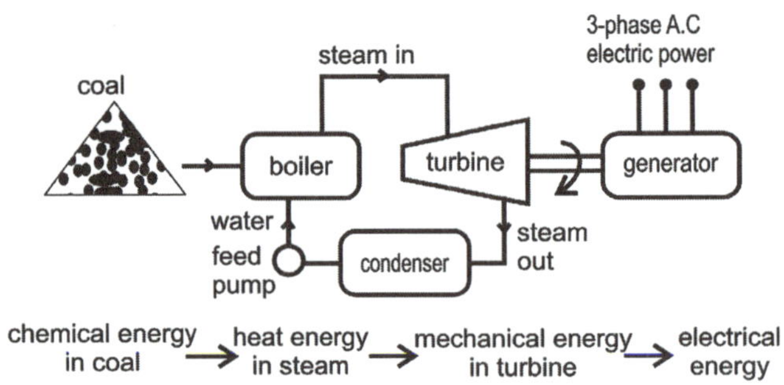

(a) schematic of thermal electric generation

(b) a typical steam turbine

Fig.4.7 : Details of thermal power generation

Hydro Power Generation

A **hydroelectric generating station** produces electricity by using gravitational force of falling water to propel the turbines, which, in turn, drive the generators.

The essentials of a typical hydro power plant are illustrated in Fig.4.8(a) and the details of a hydro turbine, connected to the generator are shown in Fig.4.8(b).

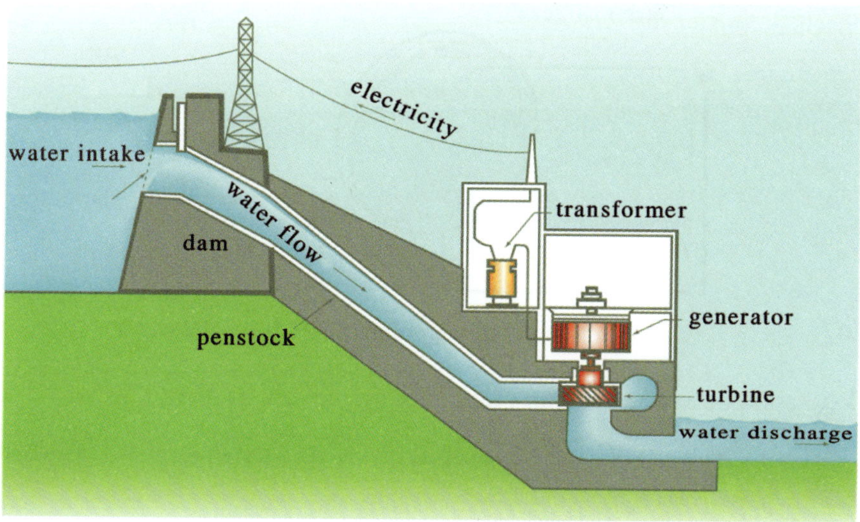

(a) essential features of a hydro station

Observe that a hydro plant exhibits an example of conversion of potential energy of water, stored at a great height in a reservoir, into kinetic energy as it flows down through huge pipes (called a "penstock").

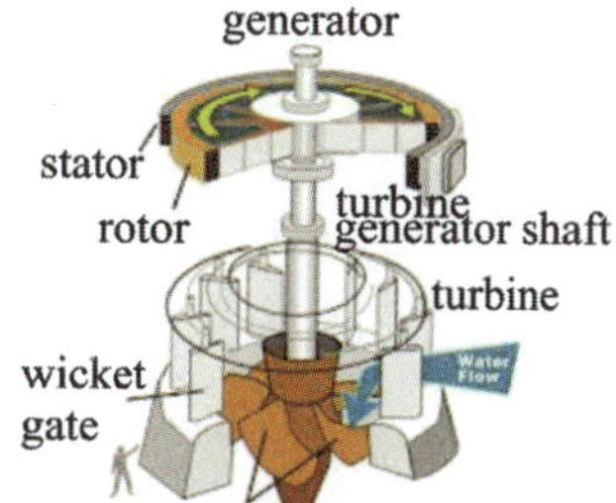

(b) cut-away section of a hydro turbine

Fig.4.8 : Generation of electrical energy in a hydro plant

Generation of electrical energy by renewal or nonconventional means

Using Solar Energy

The Sun offers 'direct' and inexhaustible source of energy, esp. in a tropical country like India with over 300 days of clear sky and sunshine, that can be utilised to generate electrical energy. This is commonly achieved by making use of a kind of semiconductor device called **photo-voltaic cell (PVC).** The principle of operation of a typical PVC is illustrated in Fig.4.9.

When sunlight, consisting of little particles of solar energy called **photons** 'good' number strikes the exposed surface of the cell, the electric field provides momentum to the electrons stimulated by the photons, resulting in a flow of current (externally) as depicted schematically in fig.4.9.

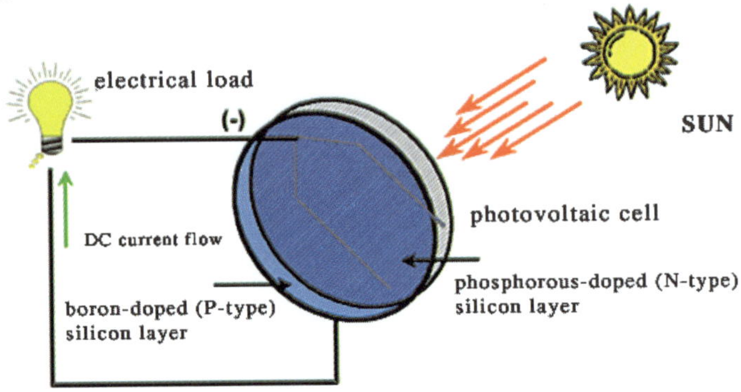

Fig.4.9 : Principle of operation of a photo-voltaic cell

The four key 'components' of a solar electricity system are as shown below.

The assembled system is illustrated in Fig.4.10.

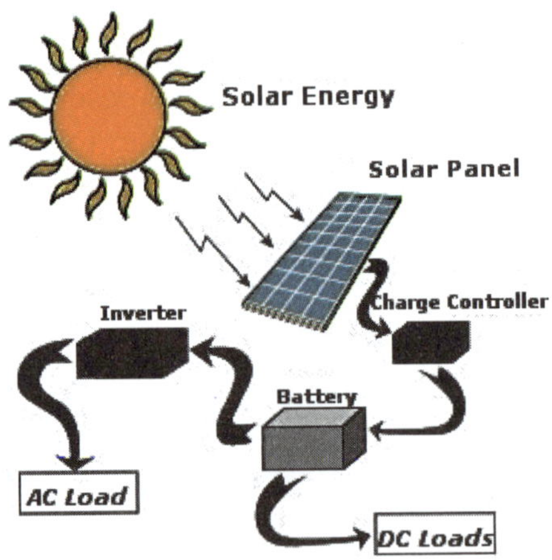

Fig. 4.10 : The essential components of a solar electricity system

Wind Energy

The kinetic energy contained in wind can be harnessed by employing "windmills" for various useful applications; for pumping water or grinding grains etc. A far more important use of windmills in modern time is to use them for generation of electricity.
This is achieved by using a large **wind turbine**, comprising huge blades or propellers (usually three in number) that are mechanically coupled or connected to a generator to

generate electricity. As the wind crosses through the blades, it rotates the shaft connected to the generator. A common type of windmill and an electric generator housed within are illustrated in Fig.4.11.

(a) a windmill of high capacity

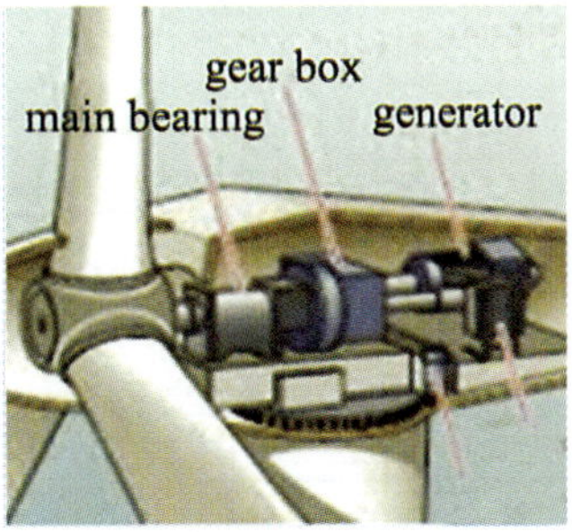

(b) location of electric generator inside a windmill

Fig.4.11 : Windmill and generation of electricity

Once commissioned at an appropriate location, considering 'average' available wind velocity and other aspects such as aesthetic and environmental ones, the electricity generated using a windmill is 'free' since no fuel such as coal etc. is needed to spin the turbine. However, under usually fluctuating wind condition, the generated electric power may not be steady with time and it may be imperative to employ suitable means of energy storage such as batteries.

Nuclear Energy

The key for nuclear energy or electric power generation is the energy generated, largely as heat energy, during *fission* of a radio-active substance such as uranium-235 or plutonium-239[6]. The radioactive substance is constitutes the fuel like coal in a thermal power plant. The heat produced during fission, a controlled chain reaction, is utilised to raise steam from water used in the set-up. The steam is arranged to rotate a steam turbine similar to that in a coal-fired thermal power station. The turbine is coupled to an AC generator or alternator, through appropriate gearing if necessary, that generates electricity, to be transmitted and utilised by consumers as usual.

Such a set-up is known as a nuclear power plant, usually designed to generate power in the range of hundreds to thousands of MW. The schematic of a typical nuclear power plant is given in Fig. 4.12 showing all essential stages of power generation.

---

[6]In some cases even thorium, available in abundance in some countries including India forms a relatively economic choice in place of uranium-235.

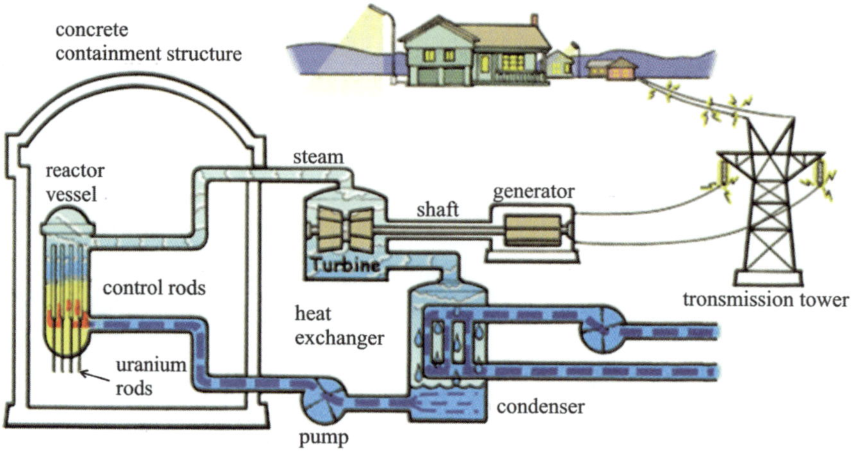

Fig.4.12 : Schematic of a nuclear electric power plant

The control rods, usually made of cadmium when uranium or plutonium is used as fuel, act to moderate or 'dampen' (or even to stop) the (chain) nuclear reaction so that it does not get out of control. The whole assembly of fuel and control rods etc. is submerged in (heavy) water to help keep the core cool.

Currently, approximately 17% of electricity is produced worldwide by nuclear power plants. In India, nuclear power constitutes the fourth largest source of electricity generation as of now after thermal, hydro and renewable sources (largely solar and wind). As of 2016, India has 22 nuclear reactors in operation in 8 nuclear power plants, having an installed capacity of 6780 MW and producing over 30,000 GWh of electricity. The latest plant is the Kudankulam nuclear power plant in Tamil Nadu having a final installed capacity of 2,000 MW.
See Fig.4.13.

Fig.4.13 : A view of Kundankulam nuclear power plant in Tamil Nadu

Advantages of Nuclear Electric Power

1. The nuclear power plant may be more cost effective compared with thermal in areas where coal field is far away.

2. There is no problem of long distance fuel transportation, storage and handling and ash handling as in thermal power plants.

3. Manpower required for the operation of nuclear power plant is less; the cost of operation may thus be less.

4. Nuclear plant occupies less space than thermal power or hydropower plant of same installed capacity which may reduce the cost of civil construction.

5. The capital cost in structural materials, piping and storage are less than thermal plants of the same capacity.

Disadvantages

1. Danger of nuclear radiation, esp. due to accidental leakage or blow-up, thus resulting in en-mass danger of diseases caused by increased environmental radio activity and pollution.

2. Problem associated with the safe disposal of the radioactive waste materials.

3. A nuclear plant has to be operated at full load throughout for a good efficiency; so part load operation becomes inefficient.

4. Capital cost of small size plants is very high, running into tens of crores of rupees per MW, esp. as compared to, say, a solar electricity plant.

5. It may take *years* before a nuclear plant may become operational (go "critical") and synchronised to the grid; the long wait may be a deterrent and not acceptable to many industries and bulk consumers supposedly to be served by the plant.

6. In recent times, the plans of commissioning a nuclear plant have been attracting lot of controversy and public unrest and disapproval in light of possible danger(s) resulting from mal-operation and accidents taking place in or about the plant as witnessed in the past in some countries[7].

See also Unit 6, Chapter II.

---

[7]Recall, for example, three major disasters: "Three Mile Island" in 1979; "Chernobyl" in 1986 and, in recent past, "Fukushima" in 2011.

Chemical Energy

Chemical energy is the energy stored in the bonds of chemical compounds (atoms and molecules). It is released in a *chemical* reaction, often producing heat as a by-product (exothermic reaction). A classic case of chemical energy application is the processing of food in a human being. Other examples of *stored* chemical energy are batteries, biomass, petroleum, natural gas and coal. See Fig.4.14 for some examples.

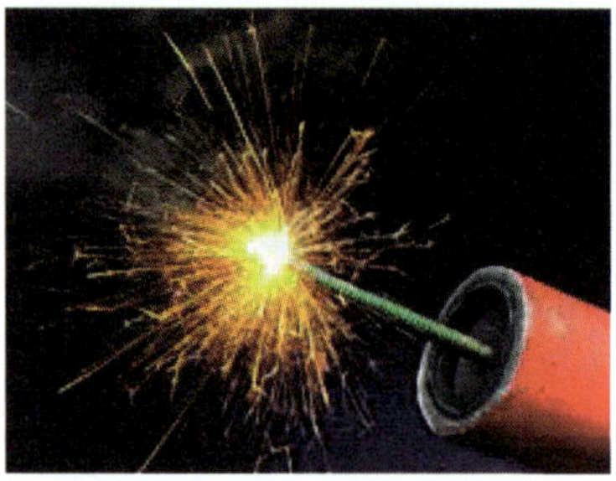

chemical energy released in a fire cracker

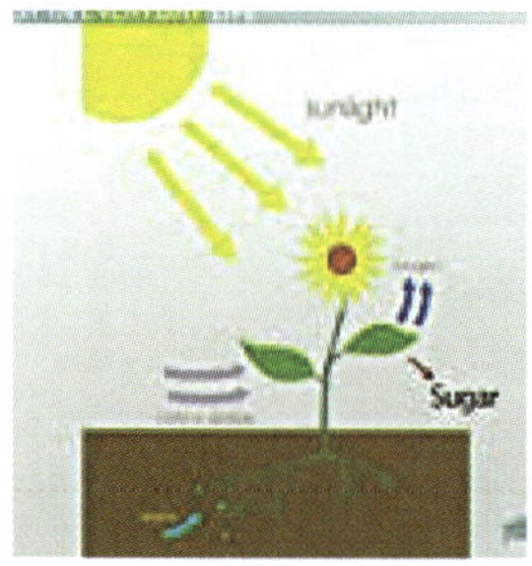

chemical energy in photosynthesis

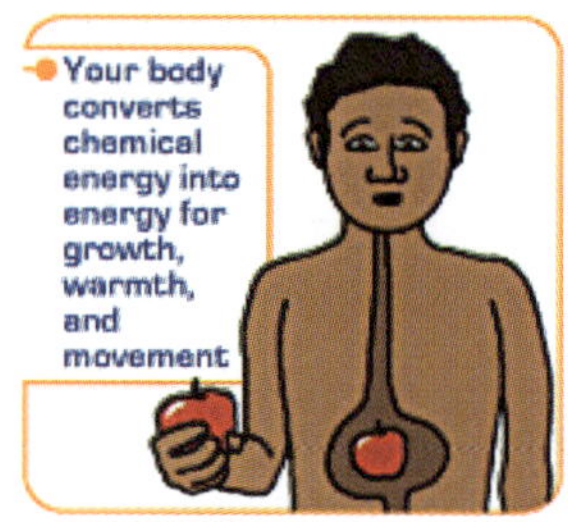

chemical energy in "food processing"

Fig.4.14 : Examples of chemical energy

Heat Energy

In physics, heat is the energy that may *spontaneously* (and 'invisibly') pass between a system and its surroundings in some way other than through work or the transfer of matter. When a suitable *physical* pathway exists, heat flows from a hotter to a colder body; the transfer taking place by contact between the source and the destination body, known as conduction.

Also, heat is a form of energy that can be transferred from one object to another or even created at the expense of the loss of other forms of energy; for example chemical energy converting into heat energy.

Heat Sources

The tremendous source of heat energy is the Sun itself where the heat is produced by nuclear fusion at massive scale. The heat energy from the Sun then reaches the Earth's surface, mainly by radiation.
The other examples of heat energy would be

- simple process of burning combustible material such as wood
- friction between two objects moving relative to each other
- heat produced by flame in a gas burner

See Fig.4.15

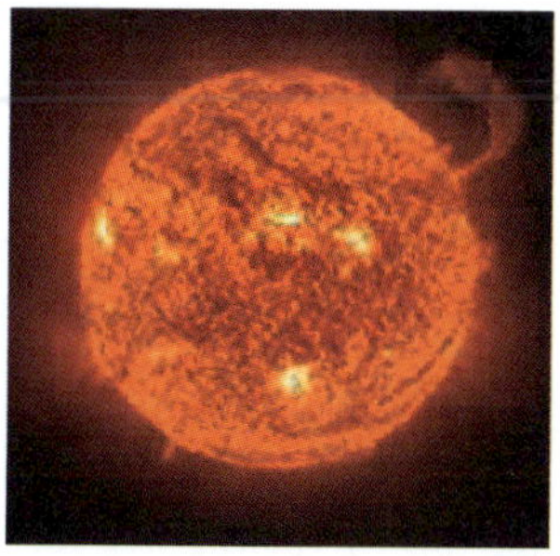

the Sun as source of tremendous heat energy

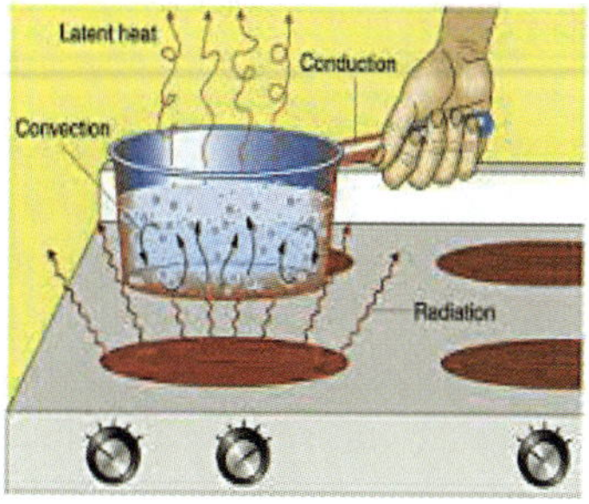

heat energy in cooking gas

Fig.4.15 : Examples of sources of heat energy

Light Energy

Light is a form of electromagnetic radiation, having a specific, rather narrow, spectrum in terms of waveform and frequency defining upper and lower limits. It refers to "visible light" that is visible to human eye, responsible for the sense of sight.

Hence, the most fantastic property of light energy is that it *enables people to see*!
In physics, it can also be identified as a type of kinetic energy that can travel with a *constant* velocity in space, or vacuum, being 3×10^8 m/s (very approximately).

The main source of light on Earth is the Sun. Sunlight provides the energy that green plants use to create sugars, mostly in the form of starches, which release energy into the living things that digest them. This process of *photosynthesis* provides virtually all the energy used by living things. Historically, another important source of light for humans has been fire, from ancient campfires to modern kerosene lamps. With the advent of electricity or electric power, light for most illumination is most effectively provided by a variety of electric lamps and devices. Some species of animals generate their own light, a process called bio-luminescence; for example, fireflies. See Fig.4.16.

an electric bulb for illumination

light energy from the Sun used by plants in photosynthesis

Fig.4.16 : A common source of light and photosynthesis

Sound Energy

Sound energy is a form of energy associated with the vibrations, or oscillations in general, of a matter or device; for example a tuning fork or a bicycle bell. Sound energy is primarily associated with the sense of hearing in all living beings.

The sound (energy) can emanate from any object or device that may be vibrating, naturally or by other means, and reach the 'destination' by any suitable medium; for example air. A given sound can be identified by the frequency (or a combination of various frequencies, called harmonics) and amplitude or 'loudness'.

The level of sound energy is specified in SI unit called Bel, or decibel abbreviated as db. Two examples of sources of sound energy are shown in Fig.4.17.

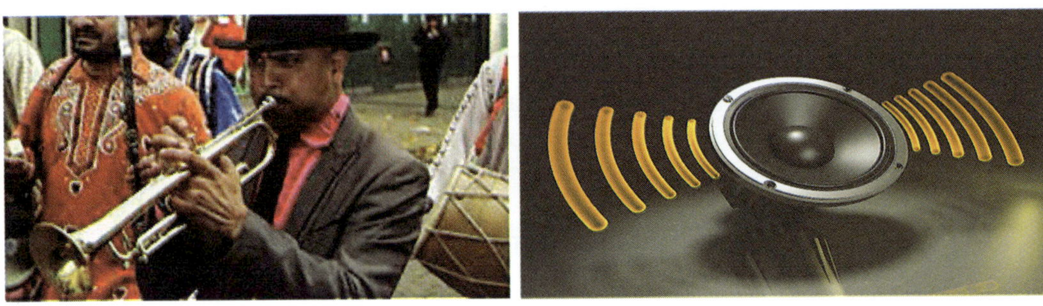

sound produced by a musical instrument　　　　sound from a loudspeaker

Fig.4.17 : Sources of sound energy

Conversion of Energy from One Form to the Other

In accordance with the law of conservation of energy, it is possible to convert one/given form of energy into the other. Some common examples are

■ Mechanical ▸ electrical

• Example(s) : using steam, hydro or gas turbines coupled to AC generators (commonly referred to as alternators), or diesel/petrol engines driving 'small' AC generators

■ Electrical ▸ mechanical

• Example(s): using electrical supply to drive a variety of electric motors for various domestic and industrial applications

■ Electrical ▸ thermal or heat

• Example(s): heaters and cooking range in domestic use or resistance ovens and furnaces in assorted industries

■ Electrical ▸ chemical
- Example: electrolysis

■ Electrical ▸ light
- Example(s): an incandescent bulb, Fig.4.18, a cine projector in cinema hall where light is produced by arcing across two electrodes and more recently a light emitting diode (LED)

Fig.4.18 : An incandescent bulb

■ Electrical ▸ sound
- Example(s) : an electromagnetic type microphone or a loudspeaker

■ Chemical ▸ electrical
- Example(s) : a dry cell or an accumulator

■ Nuclear ▸ thermal
- Example : nuclear fission in a reactor or fusion

■ Solar ▸ light
- Example : general illumination during day time

■ Solar ▸ electrical
- Example : using a photovoltaic cell

■ Solar ▸ thermal
- Example : heating water using solar water heater

■ Solar ▸ chemical
- Example : photosynthesis in plants

■ Wind ▸ mechanical
- Example : wind turbine to generate electricity

■ Tidal ▸ mechanical
- Example : device(s) designed to convert tidal movement into electricity

EXERCISES

1. A body of mass 5 kg is moving with a velocity of 10 m/s. What will be the ratio of its initial and final KE if the body's mass is doubled and velocity is halved?

[KE_1 vs. KE_2= 2 : 1

2. A girl of mass 35 kg climbs up from the first floor of a building at a height of 4 m above the ground to the third floor at a height of 12 m above the ground.
What will be the increase in her gravitational potential energy? (g = 10 m/s/s)

[ICSE, 2013]
[2,800 J

3. A body of mass 0.2 kg falls from a height of 10 m to a height of 6 m above the ground. Find the loss in potential energy taking place in the body. [g = 10 m/s/s].

[ICSE, 2012]
[8 J

4. A moving body weighing 400 N possesses 500 J of kinetic energy. Calculate the velocity with which the body is moving. (g = 10 m/s/s).

[ICSE, 2012]
[5 m/s

5. A body of mass 50 g has a momentum of 3,000 kg m/s. Calculate
(i) the kinetic energy of the body
(ii) the velocity of the body

[ICSE, 2010]
[90,000 J; 60 m/s

6. A body of mass 5 kg is moving with a velocity of 10 m/s. What will be the ratio of its initial kinetic energy and final kinetic energy, if the mass of the body is doubled and its velocity is halved?

[ICSE, 2009]
[2 : 1

7. A body of mass 1.50 kg is dropped from the 2nd floor of a building which is at a height of 12 m. What is the force acting on it during its fall? (g = 9.80 m/s/s).

[ICSE, 2008]
[14.7 N

V Machines as Force Multipliers

Machines as force multipliers; load, effort, mechanical advantage, velocity ratio and efficiency; simple treatment of levers pulley systems showing the utility of each type of machine.
Functions and uses of simple machines: terms – effort E, load L, mechanical advantage MA = L/E, velocity ratio VR = $V_E/V_L = d_E/d_L$, *input* (W_I), *output* (W_O), *efficiency* (η), *relation between* η *and MA, VR (derivation included); for all practical machines* $\eta < 1$; $MA < VR$.
Lever: principle; first, second and third class of levers; examples: MA and VR in each case; examples of each of these classes of levers as also found in the human body.
Pulley systems: single fixed, single movable, block and tackle; MA, VR and η *in each case.*

What is a Machine?

A **machine** is a device that makes work easier by transmitting or transforming energy. They have been used since ancient times to help people move or lift heavy objects by applying small force, bring substances like water from one place to another or perform any task to make day-to-day life easier, or to 'enlarge' the magnitude of force in a work situation. Today, machines are used for numerous tasks, including transportation, communication and entertainment.

Machines operate work easier by decreasing the force necessary to move an object or increasing the speed of an object. They can also make work easier by changing the direction in which force must be applied to do work. Different types of machines accomplish this in different ways. Examples of machines are levers, pulley(s), inclined planes, wedges, screws, wheels and axles and gears.

Force Multipliers

Force multipliers, also known as simple machines, are devices that are able to provide greater output force than input force. Typically, these include levers, pulleys and various types of gear.

COMMON TERMS RELATED TO MACHINES

Load

Load is a term frequently used in physical science and engineering to mean the force exerted on a surface or body. In general, it may be expressed as

- the weight lifted by a machine or any 'agency'
- the quantity that can be carried at one time by a specified means
- a mass or weight supported by 'something'

Thus, the bags full of grains carried by a bullock cart or the quantum of washed cloths on the back of a mule, constitute load.

Fig.5.1 illustrates two examples of load.

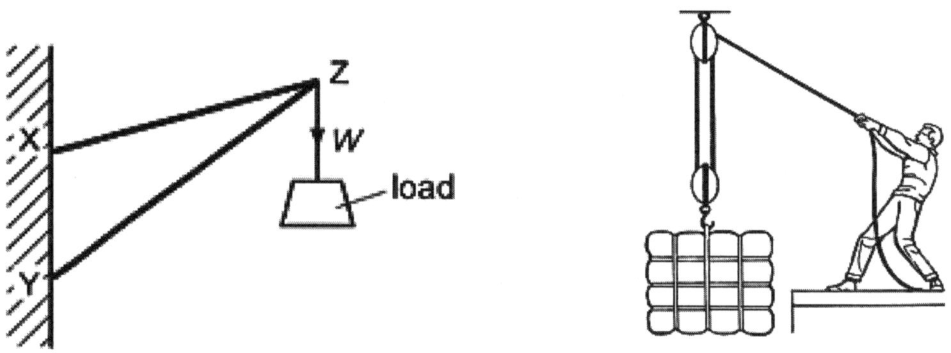

(a) a load hanging from a support

(b) a man lifting a load

Fig.5.1 : Examples of load

Effort

The power directly applied to a machine to lift a load is called **effort** or power, usually denoted by the letter P.

Also called 'effort force' sometimes is the force that moves an object over a distance by overcoming a resistance force.

An example of effort force is lifting the handles of a wheelbarrow filled with rubbish etc. The hands act as the effort force (lifting force), while gravity acts as the resistance force that must be overcome to lift the wheelbarrow. To get the wheelbarrow moving, the hands are still the effort force. However, the hands now act as both a lifting and pushing force whilst gravity and friction at wheels base act as the opposing forces.

Mechanical Advantage, MA

The ratio of load or 'weight' lifted by a machine to the effort or (effort force) applied on a machine is called **mechanical advantage** of the machine.

Greater the value of mechanical advantage of a machine, more easier is the work done and more 'efficient' the machine.

Mathematically,

$$\text{Mechanical Advantage (MA)} = \text{load/effort}$$

Being a ratio of two similar quantities, MA has no unit.

Velocity Ratio, VR

As the term implies, **velocity ratio** of a simple machine is the ratio of the velocity of effort, V_E (in a motion or 'action') to the velocity of load, V_L. That is,

$$\text{Velocity Ratio, VR} = \frac{\text{velocity, } V_E}{\text{velocity, } V_L}$$

If during the 'travel, the distance travelled by the effort is D_E and the distance travelled by the load, *in the same time, t, and not involving friction* is D_L,. then since

$$D_E = V_E \text{ x } t \qquad \text{and} \qquad D_L = V_L \text{ x } t$$

It follows that velocity ratio is also given by

$$VR = \frac{D_E}{D_L}$$

Or

$$\text{Velocity Ratio (VR)} = \frac{\text{distance travelled by effort}}{\text{distance travelled by load}}$$

Again, being the ratio of two velocities or two distances (in identical units), VR has no unit.

Efficiency

The ratio of output of a machine to the input of machine is called its efficiency.

If a machine overcomes a load ‘L’ and the distance travelled by the load is D_L, the work done by the load is $L \times D_L$. It is also called output work or useful work. Therefore,

$$\text{output work} = L \times D_L$$

Likewise if the effort applied to overcome the load is E and the distance covered by effort is D_E, the work done by effort is $E \times D_E$. It is also called input work. Therefore,

$$\text{input work} = E \times D_E$$

The efficiency of the machine would then be defined as the ratio of useful work done by machine (output work) to the total work put into the machine (input work).

Or

$$\text{efficiency}\,(\eta) = \frac{\text{output}}{\text{input}} = \frac{L \times D_L}{E \times D_E} \times 100\%$$

$$= \frac{L}{E} \times \frac{D_L}{D_E}$$

$$= \frac{L}{E} \times \frac{1}{D_E / D_L}$$

But $\qquad L/E = MA \quad$ and $\quad D_E/D_L = VR$

Hence, it follows that

$$\eta = \frac{MA}{VR} \times 100\%$$

Clearly, for an ideal or ‘perfect’ machine, work output will be equal to the work input and the efficiency of the machine would be 100 %. In practice, however, the work output of a machine is always less than work input due to the effect of friction or losses etc. and the efficiency is always less than 100 %.

SIMPLE MACHINES

LEVER

A lever is a simple machine which is used to lift heavy bodies or heavy **load** in an easy way. It may simply consist of a rigid bar capable to rotate about a fixed axis in perpendicular direction called **fulcrum** as illustrated in Fig.5.2.

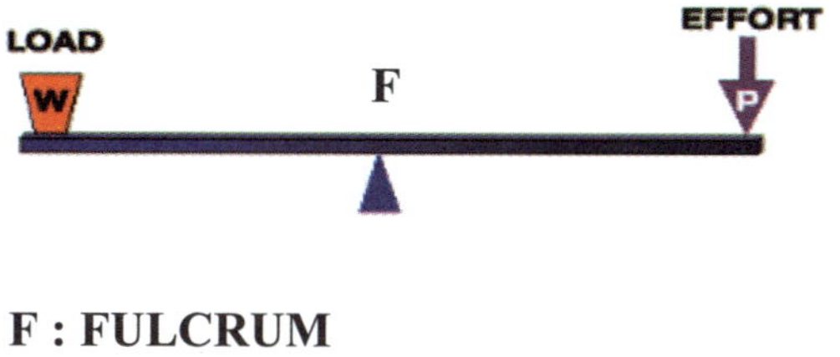

(a) a simple lever

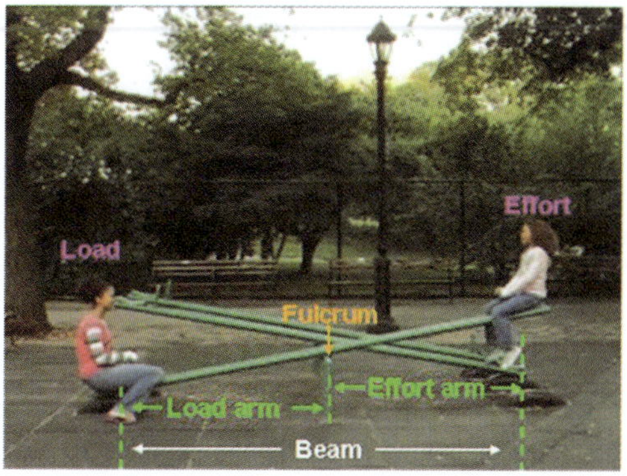

(b) principle of lever used in a seesaw

Fig.5.2 : Principle of fulcrum

Principle of Operation

The operation of a lever involves application of an **effort** at one end of the bar whilst the weight can be lifted from the other end.

Types of Lever

There are three kinds of lever depending on the positions of load, effort and fulcrum.

First type of lever

In the first type of lever, the fulcrum is positioned between the load and effort as depicted in Fig.5.3.

Examples

A common balance, pliers, seesaw, scissors, handle of a hand pump, a wheelbarrow etc.

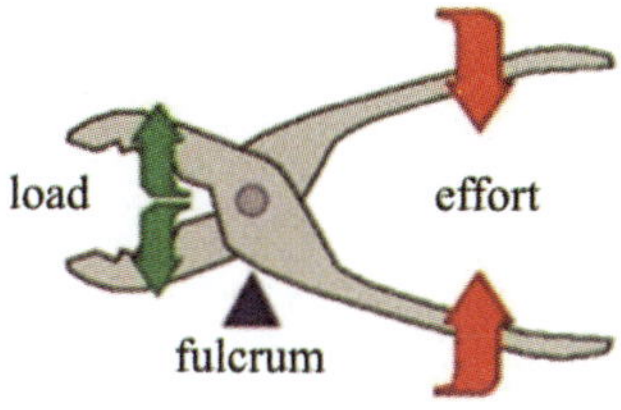

(a) common plier

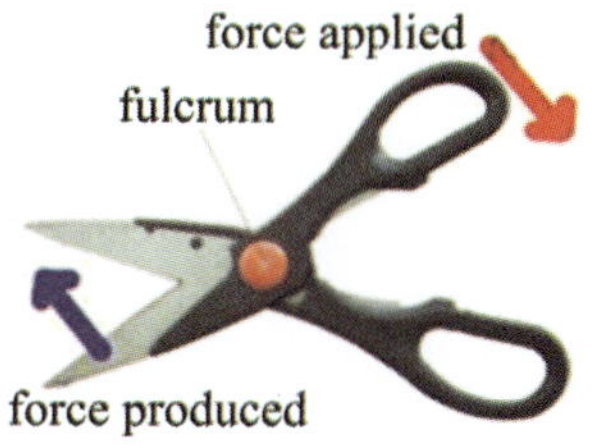

(b) a pair of scissors

(c) a wheelbarrow

Fig.5.3 : Example of lever of first type in practice

Example

1. A 4 m long beam of negligible weight is to be balanced about a point 125 cm from one end. A load of 18 kg wt. is suspended at a point 60 cm from the support on the shorter arm.

(a) A weight W is placed 250 cm from the support on the longer arm. Find W.

(b) If W = 5 kg wt., where must it be placed to balance the beam?

(c) To which class of lever does the beam belong?

The configuration of the problem is shown below

The beam is supported at 125 cm from the right end on a fulcrum as shown

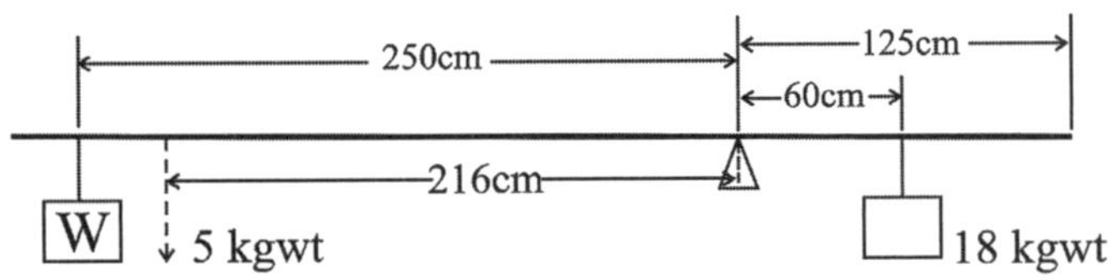

The moment on the right arm = 18 x 60 kg wt.-cm

Let the weight W is placed at a distance of d (=250 cm)cm on the left side of the beam

Then moment on the left side

$$= W \times 250 \text{ kgwt.-cm}$$

The beam will be balanced horizontally when

$$W \times 250 = 18 \times 60 \text{, giving } W = 4.32 \text{ kg wt.}$$

If W were 5 kg wt., the distance d to balance the beam will be given by

$$d = (18 \times 60)/5 \text{ or } 216 \text{ cm}$$

The lever is "class I" lever

2. A pair of scissors has its blades 15 cm long whilst its handles are 7.5 cm long. What is its mechanical advantage, MA?

For a pair of scissors, the MA is given by

$$MA = \frac{\text{length of handles}}{\text{length of blades}}$$

Substituting

$$MA = 7.5/15 \text{ or } 0.5$$

[this shows that it is always desirable to choose a pair of scissors with long handles in comparison to the blades]

3. A wheelbarrow of mass 15 kg carrying a load of 30 kg wt. with its centre of gravity at A as shown in the adjoining figure. The points B and C are the centres of the wheel and tip of the handle, respectively, such that the horizontal distance AB is 20 cm and AC is 40 cm.

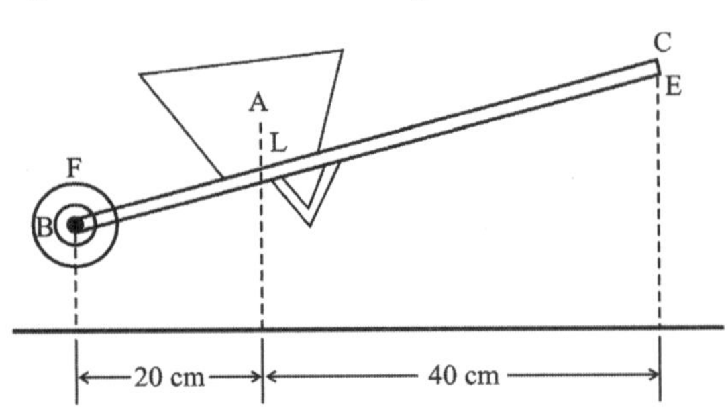

Calculate (i) the load arm, LA (ii) the effort arm, EA (iii) mechanical advantage, MA (iv) minimum effort required to keep the leg just off the ground.

(i) the load arm is simply the distance AB, thus LA = AB = 20 cm
(ii) the effort arm will be the distance BC, thus EA = BC = AB + AC = 60 cm
(iii) mechanical advantage, MA = EA/LA = 60/20 or 3
(iv) the minimum effort will be given by (total load x LA)/EA
Substituting

$$E = (15 + 30) \times 20/60$$
$$= 15 \text{ kg wt.}$$

Second type of lever

In this type of lever, the load is situated between the fulcrum and the load as shown in Fig.5.4.

Fig.5.4 : Second type of lever

Examples

A door moving about its hinges (see Fig.1.9, Chapter I, Unit 1), a nut cracker, a punching machine, a boy hitting a ball with a bat etc. as shown in Fig.5.5.

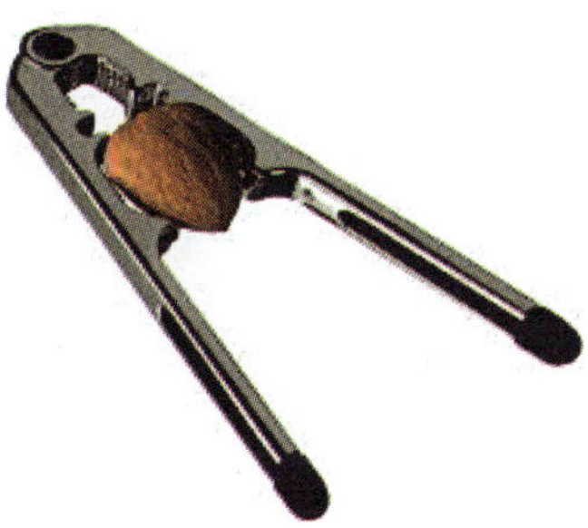

(a) a nut cracker

(b) a boy hitting a ball

Fig.5.5 : Examples of lever of second type

Example

A lever has its F (fulcrum), W (load) and P (effort) points as shown below.

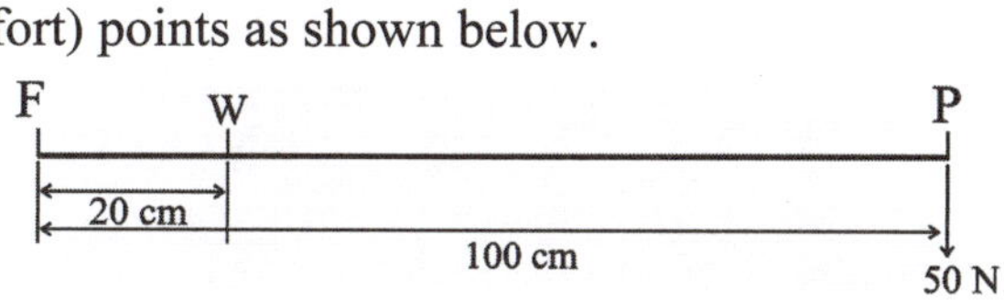

If an effort of 50 N is applied to this lever, how much load can it lift?
An effort equal to this load is applied in an

inclined plane of length 2 m and height 50 cm, what load can be pushed up on this inclined plane?

The lever is type II.

The ratio of load to effort is given by

$$\frac{\text{load}}{\text{effort}} = \frac{\text{distance of effort}}{\text{distance of load}}$$

∴ substituting

$$\frac{L}{50} = \frac{100}{20}$$

whence $L = 250 \text{ N}$

Now for an inclined plane the ratio of load to effort is given by

$$\frac{\text{load}}{\text{effort}} = \frac{\text{length}}{\text{height}}$$

∴ substituting

$$\frac{L}{250} = \frac{2.0}{0.5}$$

whence $L = 1{,}000 \text{ N}$

Third type of lever

In the third type of lever, the effort lies between the load and fulcrum as shown in Fig.5.6.

Examples

Forceps, jaws, human forearm (shown below), fire-tong.

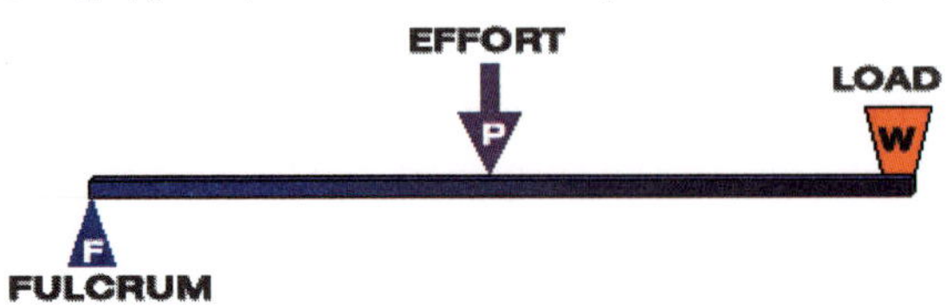

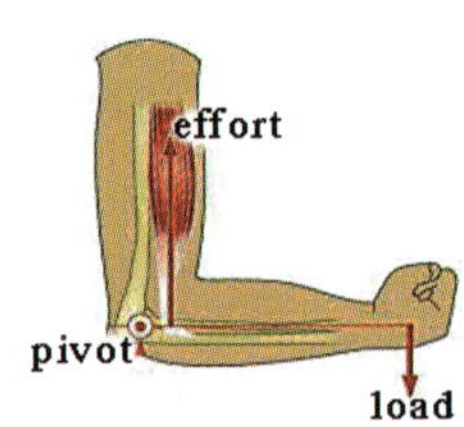

forearm as a lever of type 3

Fig.5.6 : Third type of lever

The line sketches of the three type levers are shown in Fig.5.7(a), (b) and (c) .

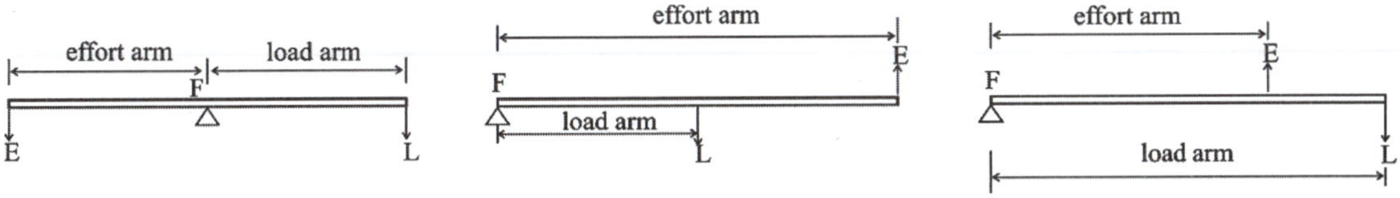

L : load E: effort F: fulcrum

(a) lever type I (b) lever type 2 (c) lever type 3

Fig.5.7 : Line sketches of the three types of lever

Examples of levers in human body

These are illustrated in Fig.5.8.

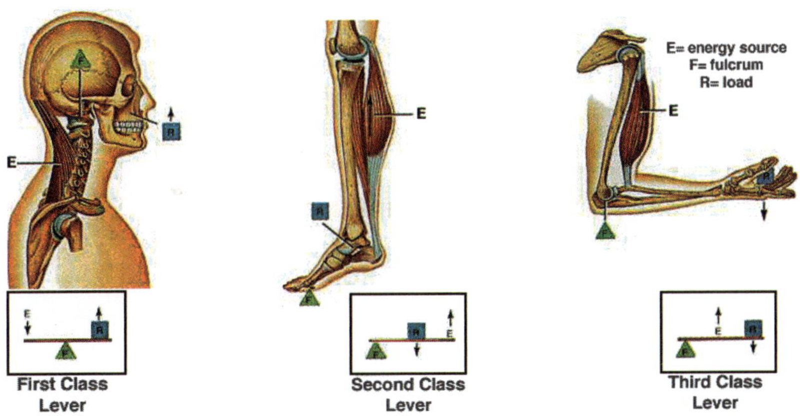

Fig.5.8 : Various types of lever in human body

Mechanical advantage of levers

The levers work on the principle of moments that come into play with the load and effort exerted about the fulcrum. Thus, referring to Fig.5.7(a), if d_E denotes the distance of effort E from the fulcrum, known as "effort arm", and d_L the distance of load L from the fulcrum, known as "load arm", then anticlockwise moment due to E about fulcrum, F = E x d_E, and clockwise moment due to L about F = L x d_L

The lever is 'balanced' when

$$E \times d_E = L \times d_L$$

Then, from the 'standard' definition of mechanical advantage, the lever of this type (type I) has the mechanical advantage given by

$$MA = \frac{d_E}{d_L}$$

$$= \frac{\text{effort arm}}{\text{load arm}}$$

It follows that longer the effort arm, or location of effort with respect to fulcrum, than the load arm, or location of load with respect to the fulcrum, greater will be the mechanical advantage. Levers are invariably 'designed' keeping this aspect in mind; observe, for example, the length of the handles compared to the position of jaws in a standard plier.

Velocity ratio

This is obtained as the ratio of velocities of movement of effort arm to that of the load arm in a given type of lever. Since a common time of movement is involved, the velocity ratio can also be given as the distance travelled by the effort arm to that travelled by the load arm. In mathematical terms, the velocity ratio would still be related to the mechanical advantage by the expression

$$\eta = \frac{MA}{VR} \times 100\%$$

when the efficiency, η, of the lever is known.

Leverage

This is a term that may be associated with the physical *dimension* of a lever offering 'mechanical' advantage in practice; for example,

(a) the active length of a screw driver to unscrew a screw of a given type/size, or

(b) a broom stick of bigger length/size that may require less effort in sweeping a floor [see Fig.5. 9].

The concept of leverage also applies to size (largely the length) of the handle of a hammer for proper, efficient driving of a nail.

Fig.5.9: The longer broom stick on the right requires less effort

PULLEY(S)

A pulley is a simple, and yet a very useful, device (or 'machine') often employed in workshops or industrial sheds to lift a heavy load from the shop floor to a desirable height for further action[1].

---

[1]Typically, this may include moving the load, or 'work piece', from one location to the other for a different machining operation etc. This is achieved by making use of an overhead crane to which the pulley may be firmly attached, the former being free to move around the entire floor area in two perpendicular directions.

In its simplest form, a typical pulley, sometimes referred to as a pulley block, comprises

• a wheel of about 6 to 8 inch in diameter (occasionally smaller) made of cast steel, or tough wood, having a semicircular groove round its periphery to receive/house an appropriate rope or chain

• a cast steel fork clamp fitted with 'bush' or ball bearings and a shaft through which is threaded the above wheel, free to rotate about the axis of the shaft or the axle.

The clamp is rigidly fitted to a firm support with the help of bolts and nuts. The rope passes round the wheel, one end of which is attached to the load (or 'job') to be lifted whilst the other end is used to apply effort.

The whole assembly is called **pulley** or pulley-block as depicted in Fig.5.10.

Fig.5.10 : A simple pulley block

A typical, simple pulley-block, also called "single-fixed type", is illustrated in Fig.5.11.

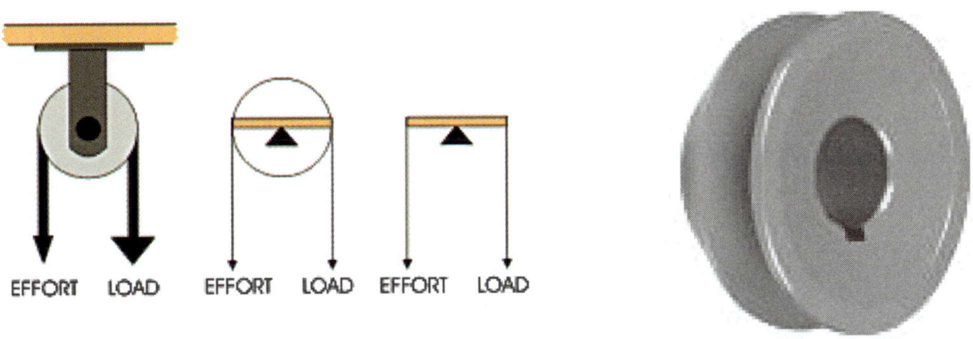

pulley block
[single-fixed type]

pulley wheel

Fig.5.11 : A simple pulley-block and an actual pulley wheel

Mechanical advantage and velocity ratio of a single-fixed type pulley

In the simple pulley, when the load L is lifted applying the effort E, and a *balanced condition is reached*, that is when there is no relative movement, the applied load is exactly equal and opposite to the load being lifted. Hence,

$$\text{Mechanical Advantage, MA} = \frac{\text{load, L}}{\text{effort, E}} = 1$$

since E = L.

Observe that there is thus no 'gain' of effort in this type of pulley.

Velocity Ratio

Since under stable condition, the distance moved by the effort in downward direction, d_E, is the same as the distance moved by the load in upward direction, d_L, it follows that

$$\text{Velocity Ratio, VR} = \frac{d_E}{d_L} = 1$$

Clearly, the efficiency of the assembly would be MA/VR = 1 or 100%.

Single Movable Pulley

A *movable* pulley has an axle in a movable block. A *single* movable pulley is supported by two parts of the same rope. The load is suspended or supported by the axle fitted to the movable pulley whilst the tension is carried by two sides of the rope, vertically up as illustrated in Fig.5.12.

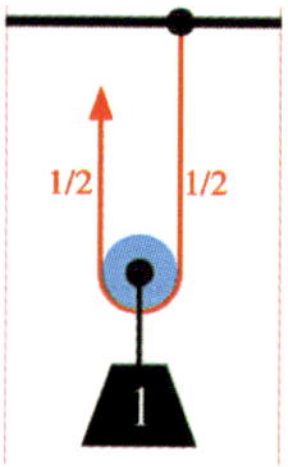

Fig.5.12 : A single movable pulley

Images of two alternative schemes are shown in Fig.5.13(a) and (b) .

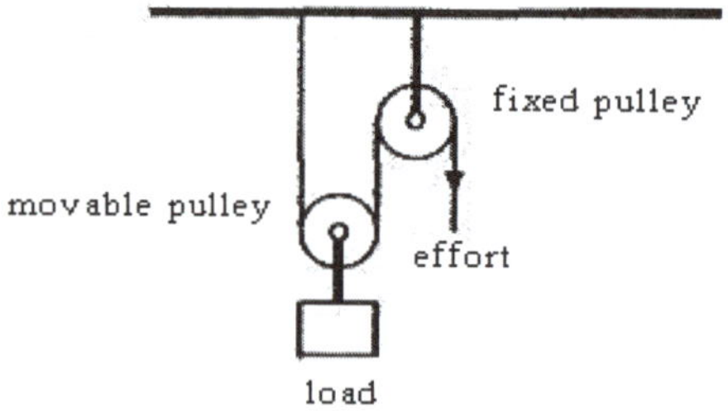

(a)

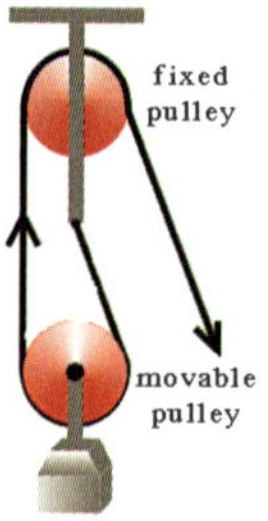

(b)

Fig.5.13: Images of alternative schemes of single movable pulley

Example

1. From the diagram shown, answer the questions

(i) What kind of pulleys are A and B?

(ii) What effort has to be applied at C to just raise the load L = 20 kg wt. Neglect the weight of pulley A and friction.

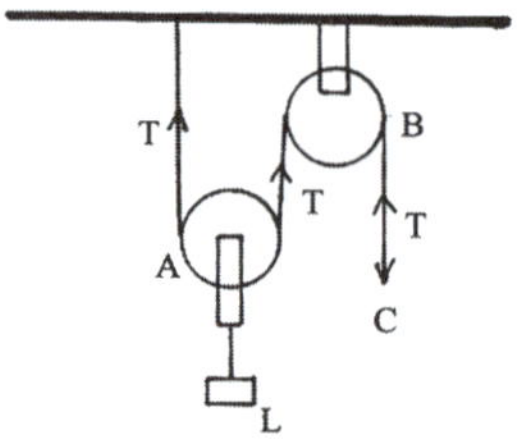

[ICSE, 2016]

(i) Pulley A: single, movable; pulley B: single, fixed
(ii) Let the distance moved by the effort in lifting the load be D_E = x m
Then distance moved by the load, D_L= x/2 m

Now $\quad$ load x D_L = effort x D_E

or $\quad$ effort = load x (D_L/D_E)

= 10 kg wt.

2. Draw a diagram to show how a single pulley can be used so as to have its MA equal to 2. [ICSE, 2014]

The system of pulley in this case would be a pulley suspended from a fixed, rigid support, with the rope going round. The arrangement is shown in the adjoining figure.

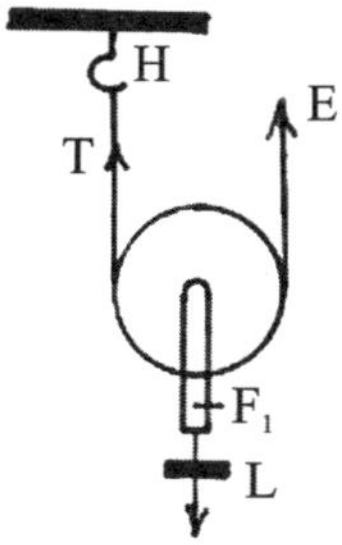

Referring to the figure

load, l = tension, T x 2 $\quad$ or $\quad$ 2 T

effort = T

∴ $\quad$ MA = load/effort = 2 T/T $\quad$ or $\quad$ 2

Mechanical advantage, velocity ratio and efficiency of single movable pulley

Mechanical Advantage

Referring to Fig.5.13(b) and disregarding the friction at the axle, and weights of the pulley wheel and rope as compared to the load, the load, L, is 'balanced' at equilibrium by tensions T and T on the two rope sides such that

L = T + T or 2 T

and $\quad$ E = T

as would be clear from Fig.5.13 (a)
Hence, for a single movable pulley,

$$\text{Mechanical Advantage, MA} = \frac{\text{load}}{\text{effort}} = \frac{2T}{T} = 2$$

or *twice* as much as in the fixed pulley system[2].

Velocity Ratio

During movement of the load, as it is lifted up by a distance d_L *each* section of the rope also rises by the distance d_L. This also means that the distance moved by effort would be d_E = 2 x d_L.

[2]Clearly, if the weights of pulley wheel etc. and friction at the axle are considered, the mechanical advantage may be somewhat less than 2.

Hence,

$$\text{VelocityRatio, VR} = \frac{\text{distance moved by the effort}}{\text{distance moved by the load}}$$

$$= \frac{2 \times d_L}{d_L} = 2$$

Efficiency

The efficiency of the system can be obtained from

$$\text{Efficiency, } \eta = \frac{MA}{VR} = 2/2 = 1 \text{ or } 100\%$$

In practice, it is not possible to achieve an efficiency of 100 % on account of friction at the axle(s) etc.

Example

1. A woman draws water from a well using a fixed pulley. The mass of bucket and water together is 6 kg. The force applied by the woman is 70 N. Find the mechanical advantage. Assume g = 10 m/s/s.

The mechanical advantage is given by

MA = load/effort

Here, the load is L = 6 x 10 or 60 N
and effort is the force applied by the woman, E = 70 N

∴ MA = L/E or 60/70 = 0.857

2. A fixed pulley is driven by a 100 kg mass falling at a rate of 8.0 m in 4.0 s. In being driven, it lifts a load of 500 kg wt. Calculate (i) the power input to the pulley taking the force of gravity on 1 kg as 10 N (1 kg wt. x g @ 10 m/s/s), (ii) if the efficiency of the pulley is 75%, find the height to which the load is raised in 4.0 s.

The load or the force applied is 100 kg wt. or 100 x 10 = 1000 N
The velocity of moving mass is distance/time or 8.0/4.0 = 2 m/s
The power input to the pulley is force x velocity
or 1000 x 2 = 2000 W

The efficiency of the pulley above is supposed to be 100%. If the efficiency were 75 %, the height to which the load would be raised will be 8.0 x 0.75 or 6.0 m.

3. A pulley system has a velocity ratio, VR, of 3 and an efficiency of 80%. Calculate (i) its MA, (ii) effort required to raise a load of 300 N.
Since efficiency of the pulley system is defined as (MA/VR) x 100 %, the MA in the present case is given by

$$\text{MA} = \text{efficiency} \times \text{VR}$$
$$= 0.80 \times 3 \text{ or } 2.4$$

Also, mechanical advantage = load/effort

$\therefore$ effort = 300/2.4 or 125 N

Block and Tackle System[3]

When heavy to very heavy loads are to be lifted, special 'combinations' of (rope and) pulleys are commonly employed so as to minimise the effort for lifting a given load. Block and tackle constitutes one such assembly. The system is also known as "compound" or "combination" of fixed and movable pulleys.

A **block** is a set of pulleys mounted on a single axle. An assembly of blocks with a rope threaded through the pulleys is called **tackle**. A **block and tackle** system 'amplifies' the tension force in the rope to lift heavy loads.

It may also be interpreted as a hoisting device in which a rope (or chain) is passed around a *pair* of blocks containing one or more pulleys. The upper block is secured overhead whilst the lower block supports the load, the effort being applied to the free end of the rope.
A block and tackle is characterized by the use of a single continuous rope to transmit a tension force around one or more pulleys to lift or move a load.

A simple schematic is shown in Fig.5.14.

Fig.5.14 : Schematic of a simple block and tackle

[3]It is interesting to refer to a broad meaning attached to the word "tackle" in usage in English. In the Oxford Dictionary , the word is stated to apply to "act of tackling in (game of) football", "trying to deal with a problem or difficulty", "grapple with (an opponent)", "confront (a person) in discussion or argument" etc.

A few images of alternative assemblies are illustrated in Fig.5.15.

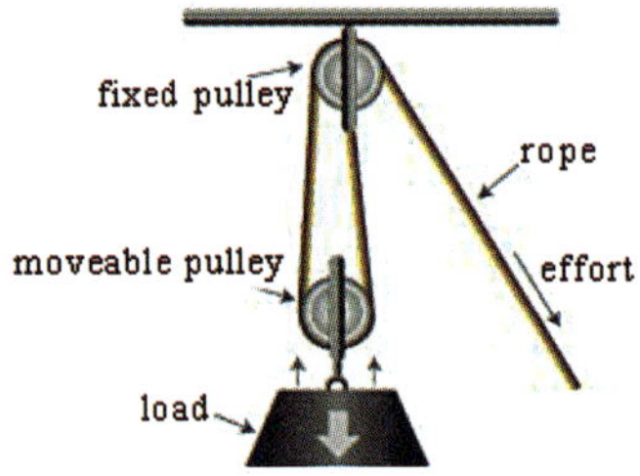

(a) a block and tackle with two pulleys

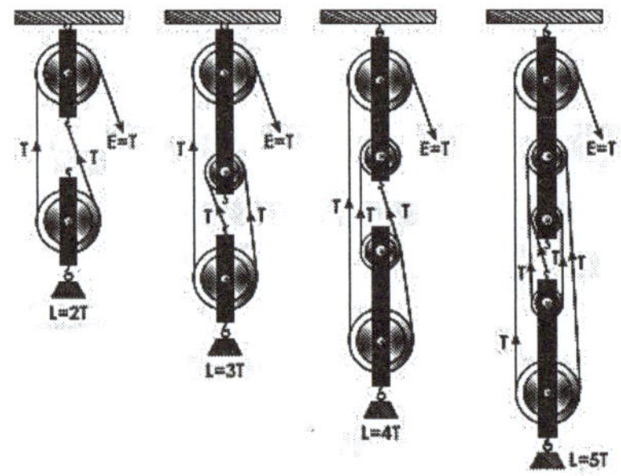

(b) block and tackle assemblies comprising 2, 3, 4 and 5 pulley blocks

Fig.5.15 : Alternative images of block and tackle

Mechanical advantage, velocity ratio and efficiency of block and tackle

Mechanical Advantage

Consider the schematic of a block and tackle shown in Fig.5.16 comprising five pulley blocks, with the rope threading the pulleys as depicted.

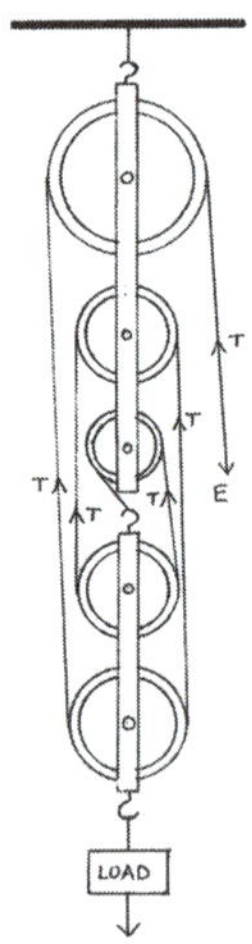

Fig.5.16 : Schematic of a block and tackle with five pulley blocks

Then, as worked out in the case of a single movable block, the load, L, is balanced by five sections or segments of the (continuous) rope, each carrying a tension, T.
Therefore,

$$L = 5 \times T \qquad \text{and} \qquad E = T$$

Hence

$$\begin{aligned} MA &= L/E \\ &= 5\,T/T \\ &= 5 \end{aligned}$$

Note that in the system shown, the last segment of the rope, S_{last}, where the effort is supposed to act, is directed *downward* and constitutes the fifth segment in the system of five pulleys.

Example

A block and tackle system has the VR equal to 3. A man can exert a pull of 200 kg wt. What is the maximum load he can raise with this pulley system if its efficiency is 60%.

Here $\qquad MA = \text{efficiency} \times VR$

$$= 0.60 \times 3 \text{ or } 1.80$$

Also, $$MA = \text{load/effort}$$

$\therefore$ $$\text{load} = 200 \times 1.80 \text{ or } 360 \text{ kg wt.}$$

The Case of Block and Tackle Comprising n Pulley Blocks

It follows that if a block and tackle comprises n pulley blocks and n rope segments threading the blocks, the mechanical advantage would also be n. In other words, the effort required to lift a load, L, will just be L/n. This demonstrates the importance of using a block and tackle system in place of a simple single movable or fixed pulley.

Velocity ratio and efficiency

Similar to the case of single movable (pulley) block, if the distance travelled by load would be d_L, the 'distance' travelled by the effort would work out to be $n \times d_L$ and hence the velocity ratio in this case would be

$$\text{Velocity Ratio, VR} = \frac{\text{distance travelled by effort}}{\text{distance travelled by the load}}$$

Or $$VR = (n \times d_L)/d_L$$
$$= n$$

same as the mechanical advantage.

The efficiency would thebe MA/VR = 1 or 100%, *disregarding any friction loss etc.*

Example

A block and tackle system has a velocity ratio (VR) of 5.

(i) Draw a neat labelled diagram of a system indicating the direction of its load and effort

ii) Rohan exerts a pull of 150 kg wt. What is the maximum load he can raise with this pulley system if its efficiency is 75 %?

[ICSE, 2015]

The diagram showing block and tackle system is drawn on the side.

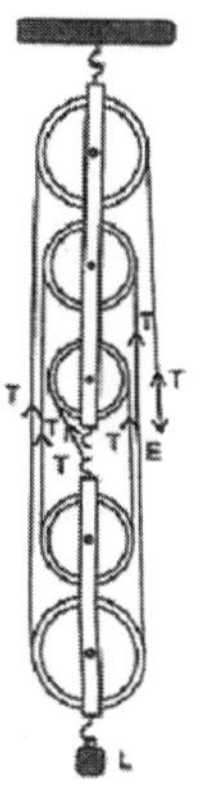

The efficiency of a block and tackle system having n pulleys is expressed as

$$\eta = \frac{MA}{VR} = \frac{\text{load, L}}{\text{effort, E} \times n}$$

With the system efficiency being 75 % (or 0.75) and an applied effort of 150 kg wt., the load that can be lifted would be

$$\text{load} = \eta \times \text{effort} \times n$$
$$= 0.75 \times 150 \times 5 \quad \text{or} \quad 562.5 \text{ kg wt.}$$

A Practical Block and Tackle System

An example of such a system is illustrated in Fig.5.17.

A double tackle has two pulleys in both the fixed and moving blocks with four rope parts or segments ($n = 4$) supporting the load (F_B) of 100N.

The mechanical advantage is 4, requiring a force of only 25N to lift the load. This arrangement demonstrates how in practice the mechanical advantage can be 'multiplied' manifold by increasing the number of pulleys both in fixed and moving blocks.

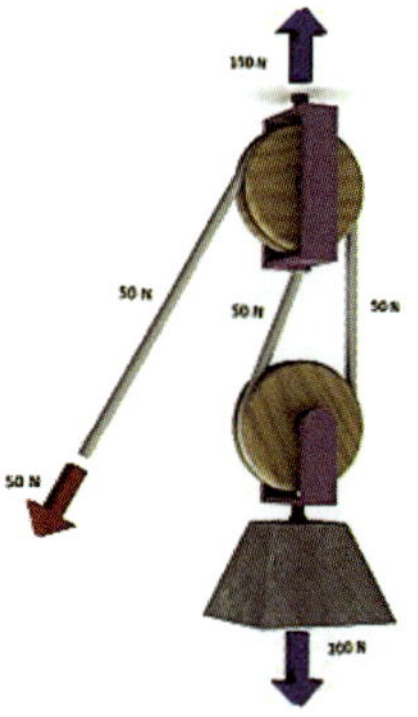

Fig.5.17 : Image of a block and tackle in practice

GEARS

A **gear** or **cog wheel** is a mechanical device or rotating 'machine' part having cut "*teeth*", or cogs, which mesh with another toothed part or gear to transmit torque. Geared devices can change the speed, torque and even direction of a power source. Gears almost always produce a change in torque, creating a mechanical advantage, through their gear ratio, and thus may be considered a simple machine. The teeth on the two meshing gears all have the same shape. Two or more meshing gears, working in a sequence, are called a gear train or a *transmission.* A gear can mesh with a linear toothed part, called a rack, thereby producing translation instead of rotation[4].

The gears in a transmission are analogous to the wheels in a crossed, belt pulley system. An advantage of gears is that the teeth of a gear prevent slippage.
When two gears mesh, if one gear is bigger than the other, a mechanical advantage is produced, with the rotational speeds, and the torques, of the two gears differing in proportion to their diameters[5].

---

[4]An important component in mechanics or mechanical energy transfer, a gear has much usage in English to express many 'actions' or idioms. To wit., in the Oxford dictionary
Gear: *n.* **1** (often in *pl.*) **a** set of toothed wheels that work together, esp. those connecting the engine of a vehicle to the road wheels.
2 *colloq***.** Clothing; adjust to or adapt to; make ready or prepared to etc.

[5]In transmissions with multiple gear ratios—such as bicycles, motorcycles, and cars—the term "gear" as in "first gear" refers to a *gear ratio* rather than an actual physical gear.

An example of commonly used gears is shown in Fig.5.18.
A train of multiple "reducing" gears employed in a microwave oven is shown in Fig.5.19.

Fig.5.18 : A pair of meshed gears of different diameters

Fig.5.19 : Multiple reducer gears in a microwave oven (rule for scale)

Driving and Driven Gear

It is common to have each gear carrying a shaft that is connected to a mechanism leading to rotation of the gears. Thus, the gears which is driving one of the two gears is called "driving" gear; the other that is driven is called "driven" gear.

Types of Gear

Some of the commonly employed gears to suit various requirements are discussed below.

Bevel Gear

A bevel gear, depicted in Fig.5.20, is shaped like a right circular cone with most of its tip cut off. When two bevel gears mesh, their imaginary vertices must occupy the same point. Their shaft axes also intersect at this point, forming an arbitrary non-straight angle between the shafts. The angle between the shafts can be anything except zero or 180 degrees. Bevel gears with equal numbers of teeth and shaft axes at 90 degrees are called *miter gears*.

Fig.5.20 : Bevel gear

Worm Gear

A worm gear, shown in Fig.5.21, is a species of helical gear, but its helix angle is usually somewhat large (close to 90 degrees) and its body is usually fairly long in the axial direction. These attributes give it screw like qualities. The distinction between a worm and a helical gear is that at least one tooth persists for a full rotation

Fig.5.21 : Worm gear

around the helix. If this occurs, it is a 'worm'; if not, it is 'helical gear'. A worm may have as few as one tooth. If that tooth persists for several turns around the helix, the worm appears, superficially, to have more than one tooth, but what one in fact sees is the same tooth reappearing at intervals along the length of the worm. The usual screw nomenclature applies: a one-toothed worm is called *single thread* or *single start*; a worm with more than one tooth is called *multiple thread* or *multiple start*. The helix angle of a worm is not usually specified. Instead, the lead angle, which is equal to 90 degrees minus the helix angle, is given.

In a worm-and-gear set, the worm can always drive the gear. However, if the gear attempts to drive the worm, it may or may not succeed. Particularly if the lead angle is small, the gear's teeth may simply lock against the worm's teeth, because the force component circumferential to the worm is not sufficient to overcome friction

Rack and Pinion Gear

A rack and pinion gear assembly is characterised with a *linear* movement. It comprises a flat or 'linear' part called "rack", AB; the other being a small rotational gear called "pinion", CD, that meshes with the teeth in the rack. These type of gears are used in converting circular motion to linear and vice versa.
Rotational motion applied to the pinion causes the rack to move relative to the pinion, thereby translating the rotational motion of the pinion into linear motion. A simple rack and pinion assembly is shown in Fig.5.22.

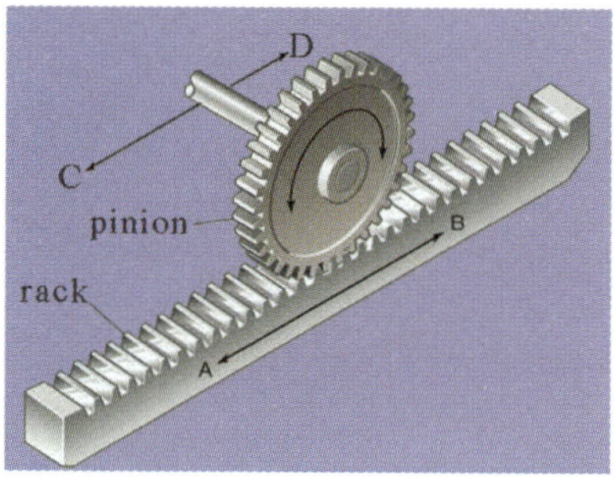

Fig.5.22 : A rack and pinion gear assembly

'Internal" Gears

These are meant for special applications in machines. Two common combinations are shown in Fig.5.23.

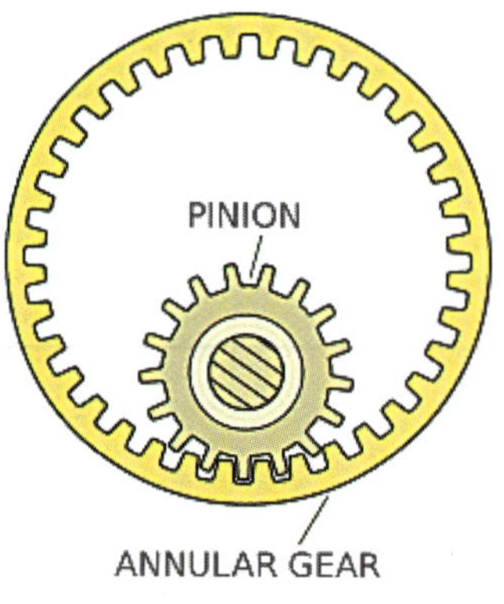

Fig.5.23 : Examples of internal gears

Typical applications of gears

- Bicycle gears

An interesting example of use of gears (or rather toothed wheels or sprockets) connected through a specially designed chain, without normal meshing, is in a bicycle where the driving torque is transferred to the rear wheel by the chain as the rider pedals the front wheel. The arrangement is illustrated in Fig.5.24.

Fig.5.24 : Gearing in a bicyle

- "Gear box" in a car

This provides an ingenious arrangement of meshed gears which are made to slide suitably by means of a lever operated by the driver of a car so as to transfer the engine output, or developed torque, to the rear axle of the car. This is essential to match the required torque at the rear axle at starting of the car, maximum owing to static friction at the wheels, and later at stages of gradual increase of the car speed such that in the last stage, or "top" gear, the full developed power or torque of the engine is directly transferred to the rear axle and the car can reach its maximum speed.

- Gear mechanism in a wall clock

A typical old-fashioned wall clock[6] employs a rather complicated arrangement of co-planer gears of numerous radii and having different number of teeth, meshing appropriately to form a "gear train". The energy for rather slow movement or rotation of gears is primarily derived from a wound helical spring that is wound periodically, say once a week. The relative movement of the gears is transferred to two concentrically mounted shafts carrying "hour" and "minute" hands, making these rotate appropriately over a graduated dial, marked 1 to 12 (o'clock). The wall is fitted with a pendulum at the lower end, oscillating periodically and synchronising its movement with the movement of the gear train.

---

[6]Often nick-named as "grandfather's clock".

A similarly wound spring and related movement is used for the chime of hour and half-hour notes.

A view of a typical wall clock is shown in Fig.5.25(a) whilst an idea of the internal gear mechanism is given in Fig.5.25(b).

(a) a wall clock

(b) gearing inside the clock

Fig.5.25 : A wall clock and its internal gear mechanism

Analysis of a gear pair or principle of operation

Input-output Torque or Power Relationship

The schematic of a gear train or pair that is commonly used in numerous applications is shown in Fig.5.26 with an "input" gear on the left and an "output" gear on the right. For the purpose of this analysis, assume the input gear is attached to a motor and is thus the driving gear whilst the output or the driven gear is attached to a shaft on a machine that performs a desired function.

Observe that whereas the input gear is rotating counter-clockwise with an angular velocity of ω_{in}, the output gear is rotating clockwise with an angular velocity of ω_{out}. An input torque, T_{in}, is applied by the motor onto the input gear, and an opposing output torque, T_{out} appears on the output gear.

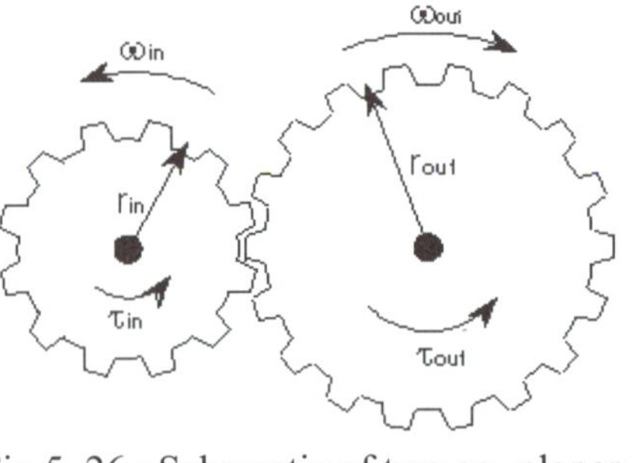

Fig.5. 26 : Schematic of two co- planer meshed gears

Assuming that there are no frictional losses, the input and output *power* can be set equal to each other in terms of torque and angular velocity on input and output side as:

$$P_{in} = \tau_{in}\omega_{in}$$
$$P_{out} = \tau_{out}\omega_{out}$$
$$\tau_{in}\omega_{in} = \tau_{out}\omega_{out}$$

This is the basic equation or relationship of power or torque transfer between the two meshing gears *that are co-planer.*

Ratio of Teeth and Radii

In a pair of meshed gears, the number of teeth on each gear are related to their respective radii. Let the number of gears on one of the gears, say the driving gear, be N_a and that on the other, the driven gear be N_b. Then the linear distance traversed circumferentially at the location of meshing for each tooth must be same for the two gears; again assuming no slip. This can be expressed as

$$2\pi r_{in}/N_a = 2\pi r_{out}/N_b$$

where r_{in} and r_{out} are the respective radii of the two gears at the common line of meshing or movement as shown in Fig.5.26.

It follows from the above expression that

$$\frac{r_{in}}{r_{out}} = \frac{N_a}{N_b}$$

If the input or driving gear makes n_a revolutions per second and the driven gear makes n_b revolutions per second then the *linear* speeds of the two gears must also match.

Or

$$2\pi r_{in}n_a = 2\pi r_{out}n_b$$

giving

$$\boxed{\frac{r_{in}}{r_{out}} = \frac{n_b}{n_a}}$$

defining the relationship between the number of teeth, radii and revolutions per second of respective gears.

Gear Ratio

The ratio of number of teeth on the output or driven gear to those on the input or driving gear is defined as "gear ratio" of the combination.

Or

$$\text{gear ratio}\,(R) = \frac{N_b}{N_a}$$

When $N_b < N_a$, the gearing is called "reduction gearing".

Velocity Ratio

The velocity ratio of gearing is defined as

$$\text{velocity ratio} = \frac{\text{frequency of driving gear, rps}}{\text{frequency of driven gear, rps}}$$

Referring to above expressions, the velocity ratio can be defined by any of the following ratios

$$\text{Velocity Ratio, VR} = \frac{n_a}{n_b} = \frac{r_{out}}{r_{in}} = \frac{N_b}{N_a}$$

Mechanical Advantage

This is defined as the ratio of output torque to the input torque.
As derived before $\qquad \tau_{in}\omega_{in} = \tau_{out}\omega_{out}$

Therefore

$$\text{Mechanical Advantage, MA} = \frac{\tau_{out}}{\tau_{in}} = \frac{\omega_{in}}{\omega_{out}} = \frac{\eta_a}{\eta_b}$$

or equal to velocity ratio, in ideal case.

Uses of gears or gearing

From the various expressions or ratios derived earlier, and the types of gear described, the gears can be used for

- Change of Torque

This is a common use in most vehicles used for transport on the road such as cars, motor cycles and trucks etc. that invariably employ a "gear box" or arrangement of three to five gears to change the torque used by the vehicle vis-à-vis the developed torque by the engine depending on whether the vehicle is driving on a level road or going up an hill. Clearly, the driving torque required would be relatively more in the latter case.

- Change of Speed

Change of rotational or linear speed can be achieved by choosing gears of different ratios. Thus, a driven gear having lesser number of teeth than the driving gear would result in reduced speed and *vice-versa*[7]

- Change of Direction of Rotation

The two meshed gears result in opposite direction of rotation as illustrated in Fig.5.26 where the input or driving gear is seen to rotate in an anticlockwise direction whilst the output or driven gear rotates in clockwise direction. This automatically ensuing action can be put to use for changing direction of rotation of desired object or member of a machine, a common example being a lathe machine or a planer in a machine shop.

- Change of Direction of *Axis* of Rotation

Recall the worm and bevel gears shown in Figs.5.21. The directions of rotation of gear pairs in these are mutually perpendicular to each other. This can be applied in practice to mechanisms requiring change in mutual direction of rotation in addition to changes in relative torques and speeds[8].

---

[7]Clearly, this would automatically result in the corresponding change of torque from one to the other gear; a common feature in all driven vehicles using a gear box or gear-changing mechanism.

[8]A common application is in what is called “differential” in a car, located towards the rear of the car chassis and is essential to change the ‘axial’ rotation of the main shaft to the rear wheel axles at right angles.

EXERCISES

1. In the lever system shown, if the beam length is 1 m, calculate (i) the value of effort E, (ii) the mechanical advantage.

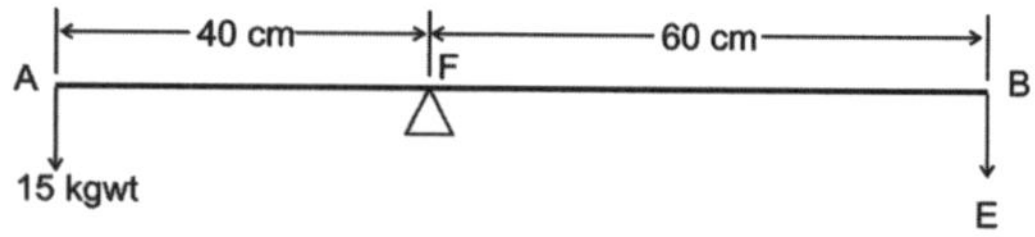

[10 kg wt.; 1.5

2. A nut is opened by a wrench of length 20 cm. If the least force required is 2 N, find the moment of force needed to loosen the nut.

[ICSE, 2015]
[0.4 Nm

3. A boy of mass 30 kg is sitting at a distance of 2 m from the middle of a see-saw. Where should a boy of mass 40 kg sit so as to balance the see-saw?

[ICSE, 2012]
[1.5 m

4. A man can open a nut by applying a force of 150 N by using a lever handle of length 0.4 m. What should be the length of the handle if he is able to open it by applying a force of 60 N?

[ICSE, 2011]
[1 m

5. The diagram shows a lever in use.

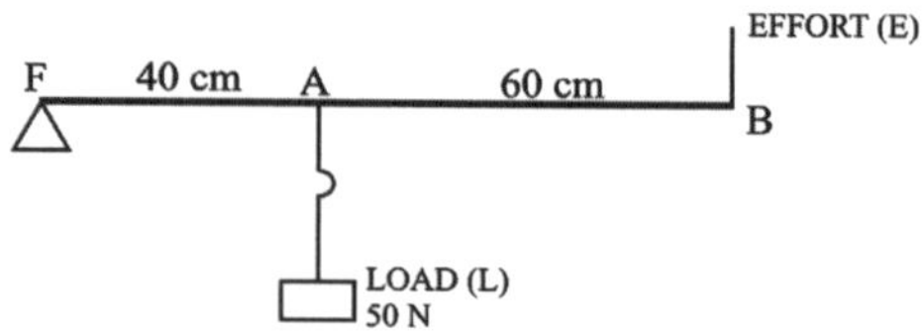

(i) To which class of lever does it belong?
(ii) If FA = 40 cm, AB = 60 cm, then find the mechanical advantage of the lever.

[ICSE, 2011]
[class II lever; 2.5

6. A uniform metre scale can be balanced at the 70.0 cm mark when a mass of 0.05 kg is hung from the 94.0 cm mark.
(i) Draw a diagram of the arrangement
(ii) Find the mass of the metre scale.

[ICSE, 2011]

[(i)

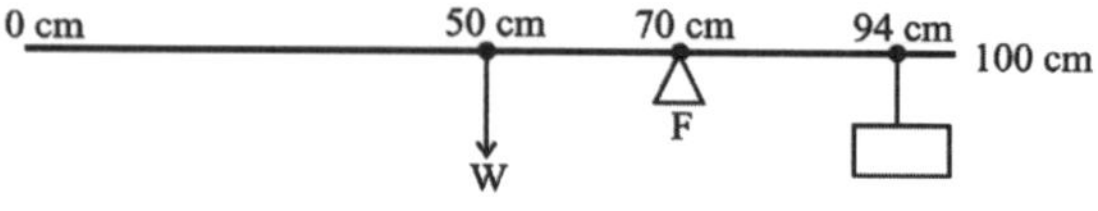

[(ii) mass of the metre scale = 0.06 kg

7. A block and tackle pulley system has a velocity ratio 3. Draw a labelled diagram of this system. In the diagram, indicate clearly the points of application and directions of the load and effort.

[ICSE, 2007]

The diagram of the arrangement is shown below.

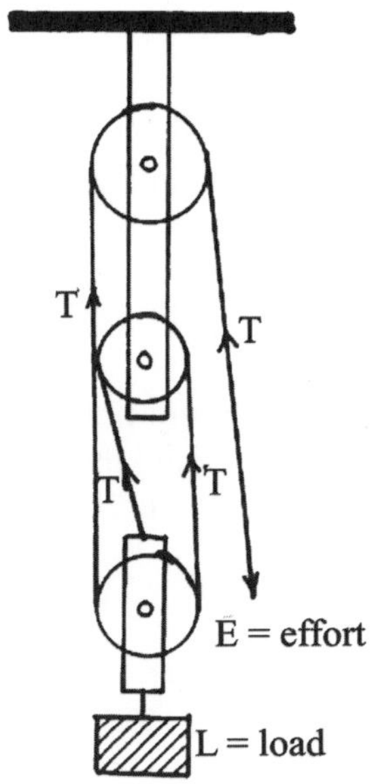

VI Principle of Conservation of Energy

Principle of conservation of energy.
Statement of the principle of conservation of energy: theoretical verification that $U + K =$ constant for a free falling body; application of this law to simple pendulum (qualitative only); [simple numerical problems].

The Law of Conservation of Energy

The law states

Energy can be neither created nor destroyed.

However, it can be transformed from one form to another, that is within a system the amount of energy is constant.

Thus, when the water at the top of the waterfall plunges over the edge, potential energy is converted to kinetic energy[1]. The energy of an air molecule moving in wind can be converted to rotational energy when moving the rotor of a windmill which then can be converted into electricity by a wind-turbine generator. In almost all of these processes some energy is also converted into heat energy – a kind of loss.

Energy conservation during free-fall

Consider a mass m which is falling vertically under the influence of gravity. The velocities, acceleration due to gravity and distance (of fall) are related by the equation

$$v^2 = u^2 - 2\,g\,s$$

as discussed for the translational motion.

[1]This forms the basis of generating electricity by using the kinetic energy to drive hydro-turbines, coupled to alternators or generators in a hydro power plant, such as Nagarjuna Sagar plant.

Suppose that the mass falls from a height h_1 to h_2; its initial velocity is u and final velocity is v. It follows that the net vertical displacement of the mass is

$$s = h_2 - h_1$$

Moreover, $u = v_1$ and $v = v_2$, being the velocities at h_1 and h_2, respectively.

Then, the previous expression can be rearranged as

$$\frac{1}{2}mv_1^2 + mgh_1 = \frac{1}{2}mv_2^2 + mgh_2$$

The above equation clearly represents a conservation law, of some description, since the left-hand side only contains quantities evaluated at the initial height, whereas the right-hand side only contains quantities evaluated at the final height.

Or, the *total* energy at height h_1 (and velocity v_1) is equal to *total* energy at height h_2 (and velocity v_2).

In order to clarify the meaning of the above equation, consider the *kinetic energy* of the mass,

$$K = \frac{1}{2}mv^2$$

and the *gravitational potential energy* of the mass,

$$U = mgh$$

by virtue of its *motion* and *position*, respectively.

It follows that the equation in question can be written

$$E = K + U = \text{constant}$$

That is, at any instant during the free fall, the sum of kinetic and (gravitational) potential energy would be a constant.

This is known as the **law of conservation of energy**.

The expression for potential energy shows that this energy is proportional to height of the object, that is,

$$PE \propto h$$

or $$PE = k\,h, \quad k \text{ a constant } (= mg \text{ here})$$

This shows that the graph of PE with respect to height would be straight line having a slope of k as depicted in Fig.6.1. The magnitude of PE for a height h_1 is given by $PE = k\,h_1$ (or mgh_1)as marked in the figure.

The relationship of K , the kinetic energy, and U, the potential energy, can also be depicted graphically as in Fig.6.2 where KE is represented by a straight line with a *negative* slope, shown by a dotted line in the figure. The total energy would be given by the point of intersection of the two lines, each being equal to ½ PE, or ½ mgh.

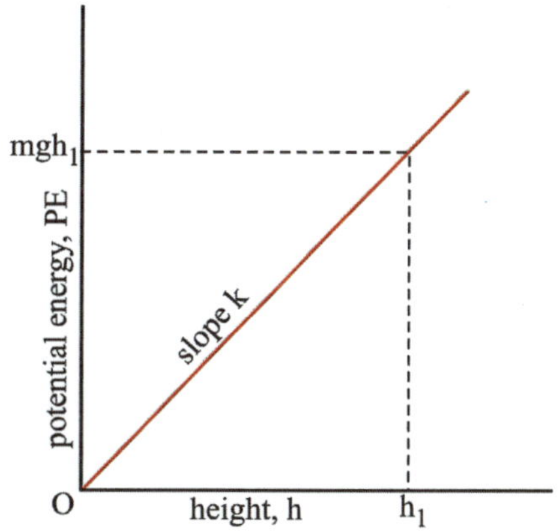

Fig.6.1 : Potential energy vs. height

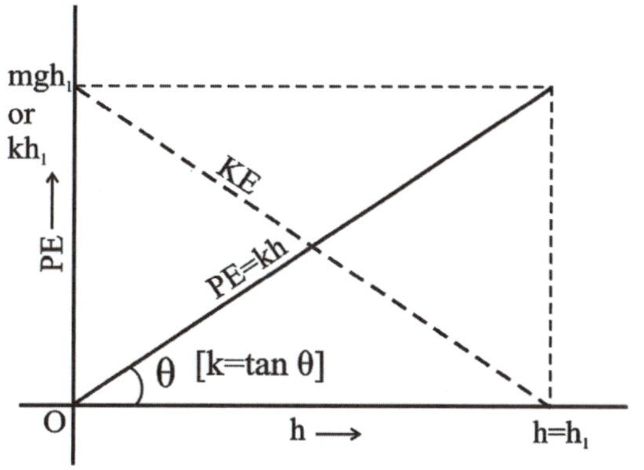

Fig.6.2 : Graphical representation of potential and kinetic energy

The Motion of a Pendulum

A simple pendulum comprises a small mass, m (also known as pendulum bob), attached to the end of a string or fine wire of length, L and a 'small' mass. The other end of the string is fixed to a 'rigid' support, point P as depicted in Fig.6.3. If the mass is displaced slightly it oscillates to and fro about the point P. The swing of a pendulum follows a *simple harmonic motion* if the swing is of a small angle and represents a classic example of alternate conversion of kinetic energy into potential energy and *vice versa*, depending on the position of the pendulum at any instant.

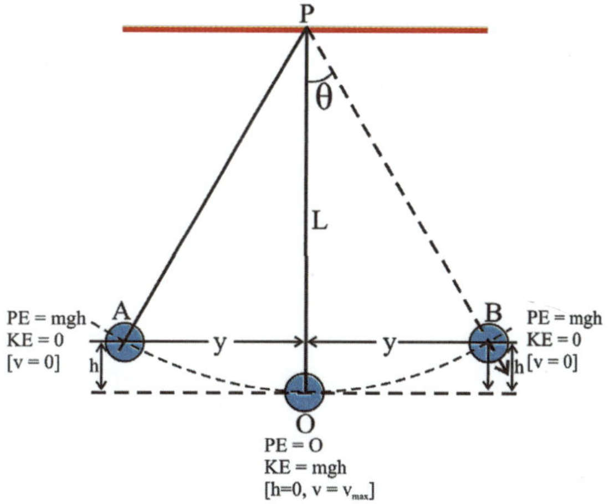

Fig.6.3 : A simple pendulum

Referring to the figure, suppose that the oscillating mass is at point B at some instant where OB is y and angle OPB = θ. Then at B, the force pulling the bob towards O is directed along the tangent at B and is equal to mg sin θ. The tension T in the string has no component in this direction since PB is perpendicular to the tangent at B.

Now, since force = mass x acceleration,

$$-mg \sin \theta = ma$$

where a is the acceleration along the arc OB.
[the negative sign indicates that the force is towards O whilst the displacement y is measured along the arc from O in the opposite direction]

When θ is *small*, sin θ = θ in radians[2] and also = y/L. Hence

$$- mg\, \theta = - ma\, (y/L)$$

Or

$$a = - (a/L)\, y$$

That is, the acceleration is proportional to the distance y from a fixed point – a characteristic of simple harmonic motion[3].
With reference to a datum as shown in the figure when the bob reaches point B at a height h, it acquires a potential energy = mgh where its velocity is zero (it stops at the farthest point of swing) and hence its kinetic energy is zero. As it swings in the reverse direction and reaches point O, its PE is zero since h = 0, but the KE is maximum (given by mgh). Similarly, at point A the PE is mgh and KE = 0 and so on[4].

A wall clock employing a pendulum is shown in Fig.6.4.

The concepts of kinetic energy vs. potential energy are fully discussed in Chapter IV, Unit1.

Fig.6.4 : Pendulum in a wall clock

Common Occurrences of Law of Conservation of Energy

The law of conservation of energy can be seen to occur in some of these everyday occurrences or instances of energy transference:

---

[2]For example,

θ, rad [from 180 degrees = π rad (or 3.1415927 rad)]	0.1744	0.3488	0.5232	0.6976
sin θ	0.1736	0.342	0.5	0.6428

The values of sin θ are very closely comparable to θ up to θ=30°. The swing of a typical pendulum in practice is within 10°.

[3]An important aspect of simple harmonic motion (SHM) is the variation of acceleration vs. displacement that would be *periodic* when plotted.

[4]This follows from the discussion of energy conservation during free fall of a body.

- Water can produce electricity
  Water, stored in huge reservoirs beside a dam or otherwise, is made to run down through pipes (called penstock), converting its potential energy to kinetic energy. This energy is then used to rotate the turbine of a generator to produce electricity. In this process, the potential energy of water in a dam is turned into kinetic energy which is then converted into electric energy.

- A boy runs across the room and bumps into his brother, pushing him to the floor. The kinetic energy he possessed because of his running is transferred to her brother, causing him to move or fall down

- When a moving car hits a parked car and causes the parked car to move, kinetic energy is transferred from the moving car to the parked car.

- A cat sitting on the highest branch of a tree has potential energy by virtue of his height. If he falls off the branch and falls to the ground, his potential energy is converted into kinetic energy.

- A fly ball hits a window in a house, shattering the glass. The energy from the ball is transferred to the glass, making it shatter into pieces and fly in various directions.

- Two football players collide on the field and both fly backwards. Energy is transferred from each player to the other, sending them in the opposite direction from which they had been running.

- Fingers hitting piano keys transfer energy from the player's hand to the keys.

- When a car hits the road sign, the sign falls over. Energy is transferred from the moving car to the stationary sign, causing the sign to move. No energy is lost in the transfer.

- When a bowling ball knocks over pins that had been standing still, energy is transferred from the ball to the pins. No energy is lost.

These law-of-conservation-of-energy examples show how commonplace this physics concept is in everyday life.

See also Fig.4.1, Chapter IV, Unit 1.

Example

1. A stone of mass 64.0 g is thrown vertically upward from the ground with an initial velocity of 20.0 m/s. Calculate the potential energy acquired by the stone at the maximum height attained by it.

[ICSE, 2008]

The maximum *kinetic* energy possessed by the stone as it leaves the ground when height is zero is

$$\frac{1}{2} \times 0.064 \times 20^2 \text{ or } 12.8 \text{ J}$$

By the law of conservation of energy, this would also be maximum potential energy acquired by the stone as it comes to rest at the maximum height above the ground.

Alternatively, deducing the maximum height acquired by the stone upon coming to rest, using the expression $2\,g\,s = v^2 - u^2$
whereas the height or distance is s
g the gravitational acceleration
v the final velocity, and
u the initial velocity
Here, assume $g = 10$ m/s/s
$v = 0.0$
$u = 20$ m/s

∴ substituting, the distance or height is found to be, $s = 20$ m
Then $PE = m\,g\,h(\text{or } s)$ or $0.064 \times 10 \times 20 = 12.8$ J

2. A ball of mass 50 g is thrown vertically upwards with an initial velocity of 20 m/s. Calculate
(a) its initial kinetic energy
(b) maximum height reached by the ball, neglecting air friction
Assume $g = 10$ m/s/s.

The kinetic energy is given by $KE = \frac{1}{2}\,m\,u^2$
Substituting

$$KE = \frac{1}{2} \times 0.05 \times 20^2$$
$$= 10 \text{ J}$$

If the height reached by the ball is h m when it comes to rest (the maximum height attained), then by law of conservation of energy
the potential energy at height h = initial kinetic energy at ground
or

$$m\,g\,h = \frac{1}{2}\,m\,u^2$$

∴

$$h = u^2/2\,g$$
$$= 20^2/2 \times 10 \quad \text{or } 20 \text{ m}$$

3. A simple pendulum when in simple harmonic motion attains its maximum height of 5 cm in extreme position. If the mass of the pendulum bob is 50 g, calculate
(a) total energy of the pendulum at any instant
(b) velocity of the bob at its mean position

Assuming $g = 10$m/s/s.
(a) Total energy of the pendulum at any instant = its potential energy at maximum height (say)

$= m\,g\,h$

or $= 0.05 \times 10 \times 0.05$

$= 0.025$ J

(b) (maximum) kinetic energy at the mean position = maximum potential energy, by the law of conservation of energy

$\therefore\ \frac{1}{2} m v^2 = m\,g\,h$

or $v = \sqrt{(2\,g\,h)}$

$= \sqrt{2 \times 10 \times 0.05}$ or 1 m/s

4. The diagram shown in the figure exhibits a ski jump. A skier weighing 60 kg takes off at point A at a height of 75 m and following skiing reaches point B at a height of 15 m. Calculate
(a) the change in his potential energy between points A and B
(b) the velocity with which the skier arrives at B if 75% of his KE acquired as in (a) is only available. Assume g = 10 m/s/s.

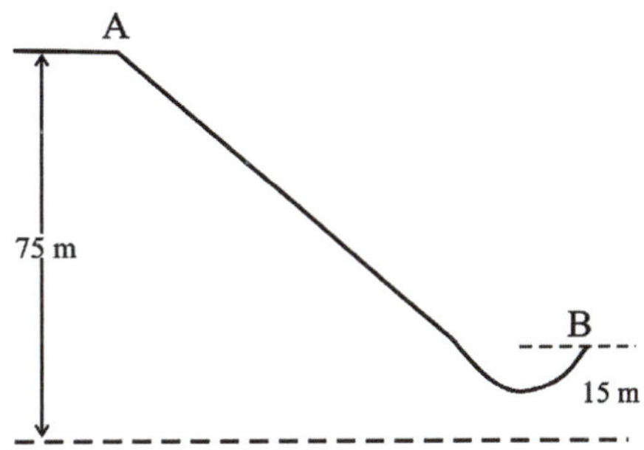

(a) The change in potential energy would be proportional to difference of heights

$\therefore$ change in PE $= 60 \times 10 \times (75 - 15)$

$= 36{,}000$ or 3.6×10^4 J

(b) By the law of conservation of energy, this PE is converted to KE at point B
The actual kinetic energy available at B $= 36{,}000 \times 0.75 = \frac{1}{2} m v^2$

$\therefore\ \frac{1}{2} \times 60 \times v^2 = 36{,}000 \times 0.75$

whence $v = 30$ m/s

5. A roller coaster car starts from rest at point A (as marked in the figure) on a hill at a height of 40 m. What is the speed of the car at the bottom of the hill? What is the speed of the car when it reaches point B as marked? Assume g = 9.8 m/s/s.

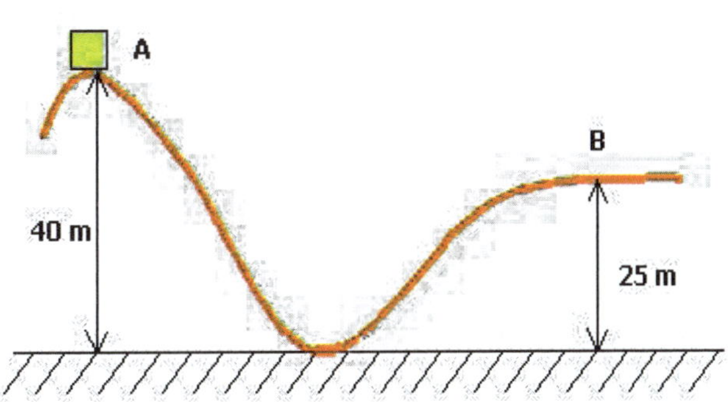

The energy at A is purely gravitational potential energy since the car starts from rest. Therefore

$E_A = Mgh_A = M \times (9.80 \text{ m/s/s}) \times (40 \text{ m}) = (392\ M)$ J, where M is the mass of the car.

The energy at the bottom of the hill is purely kinetic, no potential energy since h = 0. Therefore

$$E_{bottom} = ½ Mv^2$$

Using conservation of energy,

$$½Mv^2 = (392\ M)\ J$$

or $$v^2 = 2(392\ )$$

or $$v = \sqrt{(2x392)}$$

giving $$v = 28\ m/s$$

The energy at B is a combination of potential and kinetic energy Therefore

$$E_B = ½\ Mv^2 + Mgh_B = ½\ Mv^2 + (245\ M)$$

$$E_B = E_A$$

$$½\ Mv^2 + (245\ M) = (392\ M) \quad \text{where} \quad v^2 = 2(392 - 245)\ \ (m/s)^2$$

or $v^2 = 294\ \ (m/s)^2$, giving

$$v = 17.1\ m/s$$

6. The water to a hydro-station is supplied from a reservoir with water level being 50 m above the turbine level. Assuming that only 40% of the energy stored is available for power generation, calculate the mass of water required to flow through the turbine to generate a power output of 1 MW per second. Assume g = 10 m/s/s

Let the mass of water required be M kg.
Then the potential energy of the stored water is m g h

or $$M \times 10 \times 50\ \ J$$

Only 40 % of this energy is available

∴ (kinetic) energy available for power generation $= 0.4 \times M \times 500$

$= 200\ M\ \ J$

The required generation in each second is 1 MW or 1 MWs, that is 10^6 J

∴ $$200\ M = 10^6$$

giving $$M = 5{,}000\ kg$$

EXERCISES

1. A ball of mass 200 g is thrown vertically upwards with an initial velocity at ground of 20 m/s. What will be its maximum potential energy as it goes up?

[40 J

2. The bob of a simple pendulum attains a velocity of 5 m/s when at its mean position. What will be its maximum height at the extreme vertical position if 60% of its energy is lost in overcoming air friction? Assume g to be 10 m/s/s.

[0.5 m

2. Light

Unit 2: Topics covered

Light

(i) Refraction of light through a glass block and a triangular prism - qualitative treatment of simple applications such as real and apparent depth of objects in water and apparent bending of sticks in water. Applications of refraction of light.

Partial reflection and refraction due to change in medium. Laws of refraction; the effect on speed (V), wavelength (λ) and frequency (f) due to refraction of light; conditions for a light ray to pass undeviated. Values of speed of light (c) in vacuum, air, water and glass; refractive index μ =c/V, V = f λ. Values of μ for common substances such as water, glass and diamond; experimental verification, refraction through glass block; lateral displacement,· multiple images in thick glass plate/mirror; refraction through a glass prism simple applications: real and apparent depth of objects in water; (Simple numerical problems and approximate ray diagrams required).

(ii) Total internal reflection: Critical angle; examples in triangular glass prisms; comparison with reflection from a plane mirror (qualitative only). Applications of total internal reflection.

Transmission of light from a denser medium (glass/water) to a rarer medium (air) at different angles of incidence; critical angle (C) μ = 1/sinC. Essential conditions for total internal reflection. Total internal reflection in a triangular glass prism; ray diagram, different cases – angle of prism (60^0,60^0,60^0), (60^0,30^0,90^0), (45^0,45^0,90^0) use of right angle prism to; obtain δ=90^0 and 180^0 (ray diagram); comparison of total internal reflection from a prism and reflection from a plane mirror.

(iii) Lenses (converging and diverging) including characteristics of the images formed (using ray diagrams only); magnifying glass; location of images using ray diagrams. and thereby determining magnification.

Types of lenses (converging and diverging), convex and concave action of a lens as a set of prisms; technical terms; centre of curvature, radii of curvature, principal axis, foci, focal plane and focal length,; detailed study of refraction of light in spherical' lenses through ray diagrams; formation of images - principal rays or construction rays; location of images from ray diagram for various positions of a small linear object on the principal axis; characteristics of images. Sign convention and direct numerical problems using the lens formula are included. (derivation of required)

Scale drawing or graphical representation of ray diagrams not required.

*Power of a lens (concave and convex) - [simple direct numerical problems]: magnifying glass or simple microscope: location of image and magnification from ray diagram only [formula and numerical problems **not** included]. Applications of lenses.*

(iv) Using a triangular prism to produce a visible spectrum from white light; Electromagnetic spectrum. Scattering of light.

Deviation produced by a triangular prism; dependence on colour (wavelength) of light; dispersion 'and spectrum; electromagnetic spectrum: broad classification (names only arranged in order of increasing wavelength); properties common to all electromagnetic radiations; properties and uses of infrared and ultraviolet radiation. Simple. application of scattering of light e.g. blue colour of the sky.

I Refraction of Light

Refraction of light through a glass block and a triangular prism – qualitative treatment of simple applications such as real and apparent depth of objects in water and apparent bending of sticks in water. Applications of refraction of light.
Partial reflection and refraction due to change in medium; laws of refraction; the effect on speed (V), wave length (λ) *and frequency (f) due to refraction of light; conditions for a light ray to pass undeviated; values of speed of light (c) in vacuum, air, water and glass; refraction index* $\mu = c/V$, $V = f\lambda$ *; values of for common substances such as water, glass and diamond; experimental verification; refraction through glass block; lateral displacement; multiple images in thick glass plate/mirror; refraction through a glass prism; simple applications; real and apparent depth of objects in water; apparent bending of a stick under water; [simple numerical problems and approximate ray diagrams required].*

What is Light?

As discussed previously, light is a form of energy[1]. The characteristic feature of light is that it is also an electromagnetic wave or radiation, comprising a narrow and distinct spectrum of

- wavelength

and

- corresponding frequency

From within a very broad range of electromagnetic waves, light constitutes a narrow band as depicted in Fig.1.1(a) and (b). Observe how different colours of 'normal' white light relate to wavelength and hence corresponding frequency of radiation.

[1] This is evident from the fact that plants and vegetables grow when they absorb sunlight – process known as photosynthesis. Also, electrons are emitted by certain metals or materials when light is incident on them as in a photovoltaic cell (PVC).

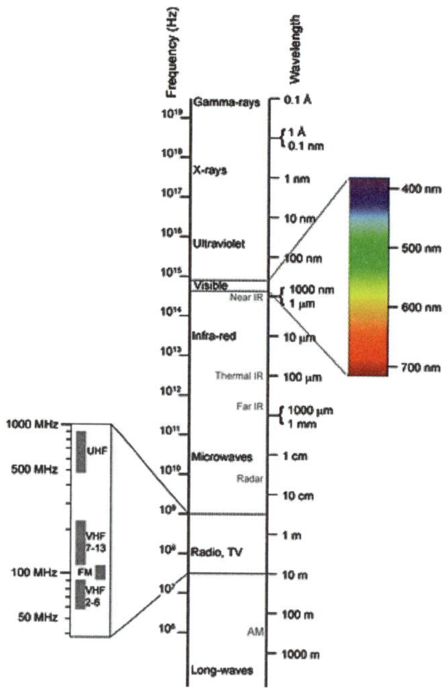

(a) broad range of electromagnetic waves

frequency Hz.	wave-length A/U*	ultra-violet ↑
7.5 x 10^{14}	4,000	violet
		blue
6.0 x 10^{14}	5,000	green
5.0 x 10^{14}	6,000	yellow
		orange
4.3 x 10^{14}	7,000	red
3.75 x 14^{14}	8,000	↓ infra-red (heat)

*1 AU equals 10^{-10} m

(b) wavelength and colour

Fig.1.1 : Electromagnetic waves or radiation and spectrum of light

Note that the values of wavelength and frequency of any wave pertain to vacuum, or very approximately in air. For any of the waves, the product of wavelength and frequency is a constant c, the speed of light being 3×10^8 m/s (very approximately).

Ray of Light

A *ray* of light is the *direction* along which the light energy or light waves travel[2]. A *beam* is a collection of rays and more of physical significance.

Types of beam of light

In general, a beam of light can be

- a ***parallel*** beam

Such a beam comprises rays of light which 'run' parallel to each other for large distances, apparently 'meeting' at "infinity"; for example, light coming from the sun or light emitted

[2]Although rays are represented in diagrams, very commonly, by straight lines, in practice a ray has a finite width or thickness.

from a searchlight, typically from a light house along the sea shore

- a ***divergent*** beam

For example that emitted by a lamp when the light 'spreads' in all direction

- a ***convergent*** beam

Typically from a source of light as in a projection 'lantern'

These are shown in Fig.1.2.

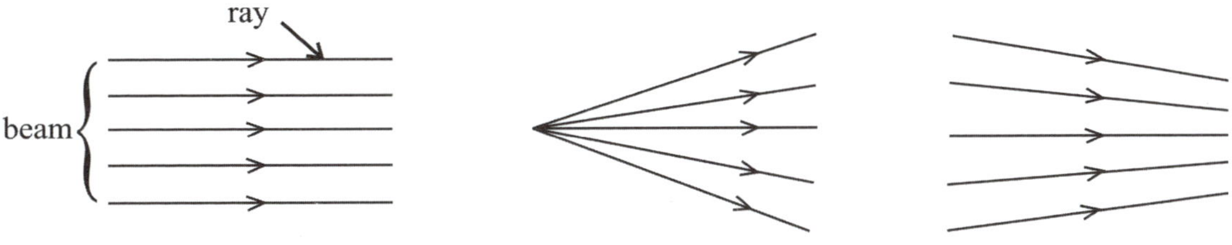

(a) parallel beam (b) divergent beam (c) convergent beam

[also relate to actions of (i) convergent or convex (ii) divergent or concave lenses]

Fig.1.2 : Types of light beam

Refraction of Light

Refraction is the bending of a light wave when it enters a medium where its speed is different.

In **refraction of light** when the latter passes from a fast medium to a slow medium, the light ray 'bends' ray toward the normal to the boundary between the two media. This is shown conceptually in Fig.1.3(a) whilst actual bending of a typical light ray from one medium to another is depicted in Fig.1.3(b), indicating the incident and refracted rays.

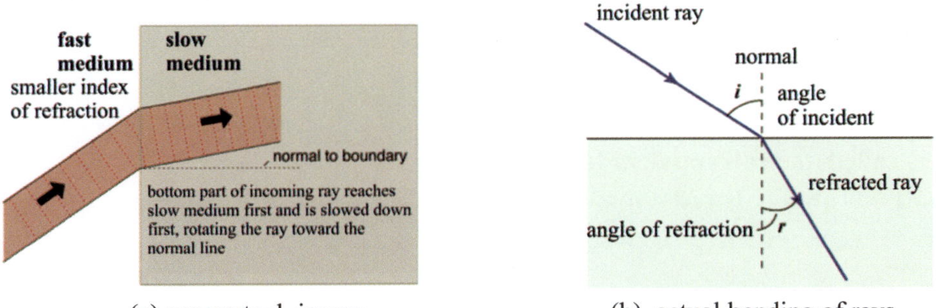

(a) conceptual image (b) actual bending of rays

Fig.1.3 : Phenomenon of refraction

An interesting example of effect of refraction 'seen' through a tumbler filled with water is illustrated in Fig.1.4.

Fig.1.4 : Refraction in a glass of water; the image is flipped

Causes of Refraction

The refraction of light occurs because light travels with different speeds in different media. If the speed of light in vacuum (or very nearly even in air) is taken as reference or datum, the speed would change, usually decreasing, when directed to pass through any other substance, *denser than air*, and the rays of light would change course, that is, resulting in refraction.

Laws of refraction

Laws of refraction of light relate to describe the relationship between the angles of incidence and refraction when referring to light (or other) waves passing through a boundary between two different *isotropic* media, such as water, glass, or air.

The laws can be defined as

- The incident ray, the refracted ray and the normal to the surface at the point of incidence all lie in one plane
- For any two given pair of media, the ratio of the sine of the angle of incidence to the sine of the angle of refraction is a constant

This law is also called Snell's law after the scientist Willebrod Snellius[3] who first formulated it.

The law is graphically illustrated in Fig.1.5, reproduced from Fig.1.3(b).

Thus, referring to the figure,

$$\frac{\sin i}{\sin r} = \text{a constant} = \mu$$

where μ is called the **refractive index** of the *second* medium with respect to the *first* medium.

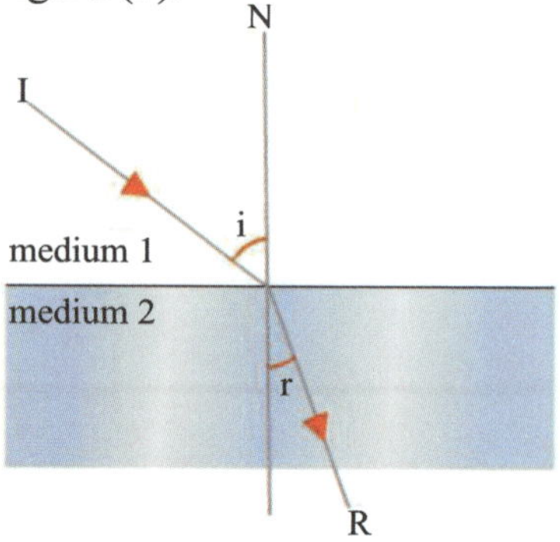

Fig.1.5 : Snell's law of refraction

[3]Willebrord Snellius, a Dutch astronomer and mathematician: 1530-1626

Factors Affecting the Refractive Index

As stated before, the phenomenon of refraction takes place because of change of the speed of light as it travels from one optical medium to another.

Thus, the refractive index can be defined in terms of the speed of light in the two media.

Accordingly, the refractive index of glass with respect to air, being 1.50, is given by the relation

$$_{air}\mu_{glass} = \frac{\text{speed of light in air}}{\text{speed of light in glass}} = \frac{c_{air}}{v_{glass}}$$

In general, if a ray of light is passing from medium 1 to medium 2, then

$$_1\mu_2 = \frac{\text{speed of light in medium 1}}{\text{speed of light in medium 2}}$$

If medium 1 is air or vacuum, the refractive index of medium 2 is referred to as the *absolute* refractive index.

Apart from speed of light, the refractive index of a medium depends on the following factors:

- the nature of the medium, defined by its 'optical density[4]
- the colour or wavelength of the incident light
- temperature of the medium: with an increase of temperature, the speed of light through the medium increases so that the value of refractive index decreases.

Values of Refractive Index (μ) for Various Materials

material	refractive index, μ
Vacuum	1.000000 (by definition)
Air (at STP)	1.000277
CO_2	1.001
H_2	1.000152
Water	1.33
Ice	1.39
Crown glass (pure)	1.50[3]
Diamond	2.419

[4]A glass that is not absolutely clear (and clean) will have a refractive index higher than 1.50.

Observe that, as reflected by the values of μ in the above table, the harder a substance or material as compared to vacuum or even air, the greater the value of μ. In terms of physical property of light, this would mean a steeper bending of light towards the 'normal' for a given angle of incidence.

Experimental Verification of Refraction of Light

The course of traverse of light through a glass plate can be observed experimentally by using simple means to verify the two principles of refraction of light. The details of steps to be adopted are given in Appendix A.

The principle of reversibility of light

The principle of reversibility of light states that the paths of light rays are reversible. This holds true in the case of refraction and, indeed, for rays passing through any optical system. Thus, referring to Fig.1.5,

$${}_a\mu_g = \frac{\sin i}{\sin r}$$

and by principle of reversibility of light

$${}_g\mu_a = \frac{\sin r}{\sin i}$$

Hence

$${}_g\mu_a = \frac{1}{{}_a\mu_g}$$

Conditions for a light to pass undeviated

"Undeviated" traverse of light does not mean that there is no refraction. However, there are cases when the ray of light following refraction may be considered to be undeviated such as

- when the angle of incidence at the boundary of two media is zero;
- when the refractive index of medium 2 is same as the refractive index of medium 1;
- when the incident ray and the emergent ray are parallel to each other;
- for 'thin' lenses, the refraction (as defined) inside the lens is negligible and therefore the ray passing through the optic centre follows a straight line.

Speed of Light in Different Media

The speed of light in *vacuum*, commonly denoted by letter ***c***, is a universal physical constant, important in many areas of physics. Its precise value is **299792458 m/s** (approximated to 3.00×10^8 m/s); the length of the metre is defined from this constant and the international standard for time.

The speed of light in other media is related to their refractive indexes and speed of light in air (or vacuum). For example,

- the refractive index for water being 1.33, the speed of light in water is given by c/1.33 or **2.25×10^8** m/s;

- the refractive index for glass being 1.5, the speed of light in glass would be c/1.5 or **2.0×10^8**m/s, very approximately.

Refraction of Light Through a Glass Block

A glass block may be understood as a flat glass sheet having considerable thickness, for example about 5 to 10 mm or more, comparable to its length and width, with the light falling normally on the surface.

When a ray of light enters a glass block at an angle other than the normal, it changes speed, wavelength and direction as shown in Fig.1.6.

In traversing from air to a more dense medium (glass), light bends towards the normal as explained before. This means that $i > r$. As the ray emerges from the glass block, it bends away from the normal, the change in angle of the light ray is the same as it enters and leaves the glass. How much the light bends depends on its colour, that is its wavelength. If the incident ray had continued without changing its direction, shown dotted in Fig.1.6, the emergent ray would be parallel to it.

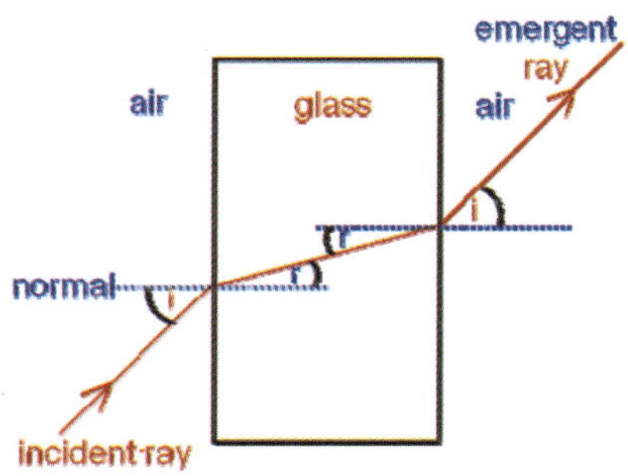

Fig.1.6 : Deviation of a ray through a glass slab

Lateral Displacement

The lateral diffraction or displacement of light is the distance a ray of light moves when passing through a denser medium, such as glass. When light enters glass, the beam is bent, but upon exiting the glass into the air again, it returns to its previous angle, resulting in a lateral movement of the light beam.

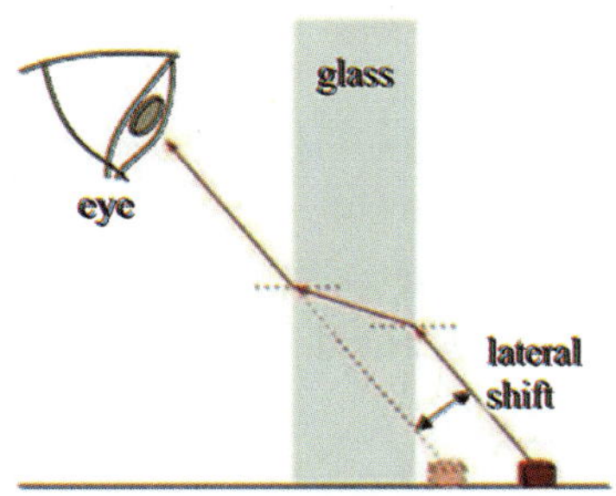

Fig.1.7 : Lateral displacement of a beam

Referring to Fig.1.7, the perpendicular distance between the direction of the incident ray and that of emergent ray is called **lateral displacement** or **shift** as shown.

Example

A ray of monochromatic light is incident from air on a glass slab:
(i) Draw a labelled ray diagram showing the change in the path of the ray till it emerges from the glass slab
(ii) Name the two rays that are parallel to each other
(iii) Mark the lateral displacement in the diagram

[ICSE, 2010]

(i) A labelled diagram of the ray, assuming an arbitrary angle of incidence on the glass surface, is drawn in the adjoining figure giving all details
(ii) The two parallel rays are named
(iii) The lateral displacement is marked

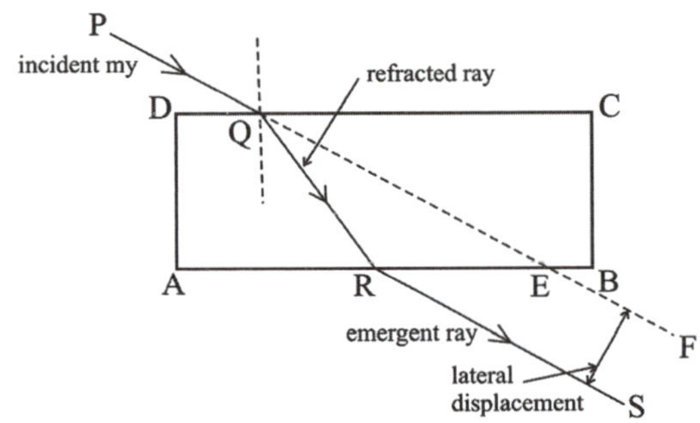

Factors affecting lateral displacement

The factors that may result in a given lateral displacement are

- the thickness of the refracting material (glass in the above case)
- the refractive index of the material
- angle of incidence of the ray of light

Multiple Images in Thick Glass Mirror

If an object is placed in front of a thick glass mirror and is viewed *obliquely*, a number of images are seen. Out of these images, the second image is the brightest whilst others are of diminishing brightness.

Referring to Fig.1.8, when a ray of light falls on the surface of the glass plate, a small part of light, about 4 %, is reflected forming a faint *virtual* image at I', whilst a larger part of the light, nearly 96 %, is refracted inside the glass. The ray which now strikes the mirror, is strongly reflected back by the mirror. The image I formed is therefore the brightest.

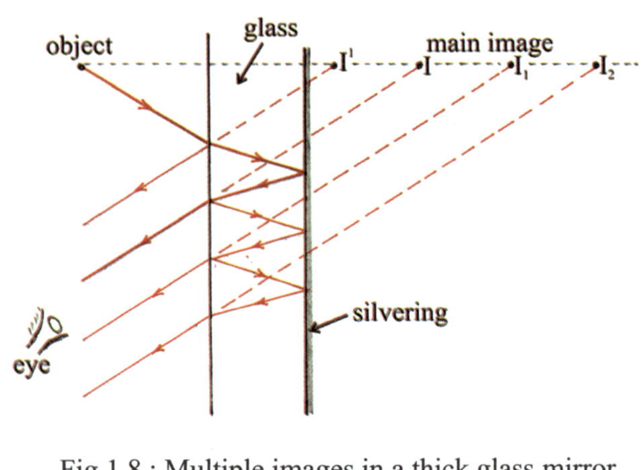

Fig.1.8 : Multiple images in a thick glass mirror

Reflection of Light

Reflection is the change in direction of a *wave front* at an interface between two different media so that the wave front *returns* into the medium from which it originated. Common examples include the reflection of light, sound and water waves. Mirrors exhibit (specular) reflection for light waves. Two examples of reflection are given in Fig.1.9.

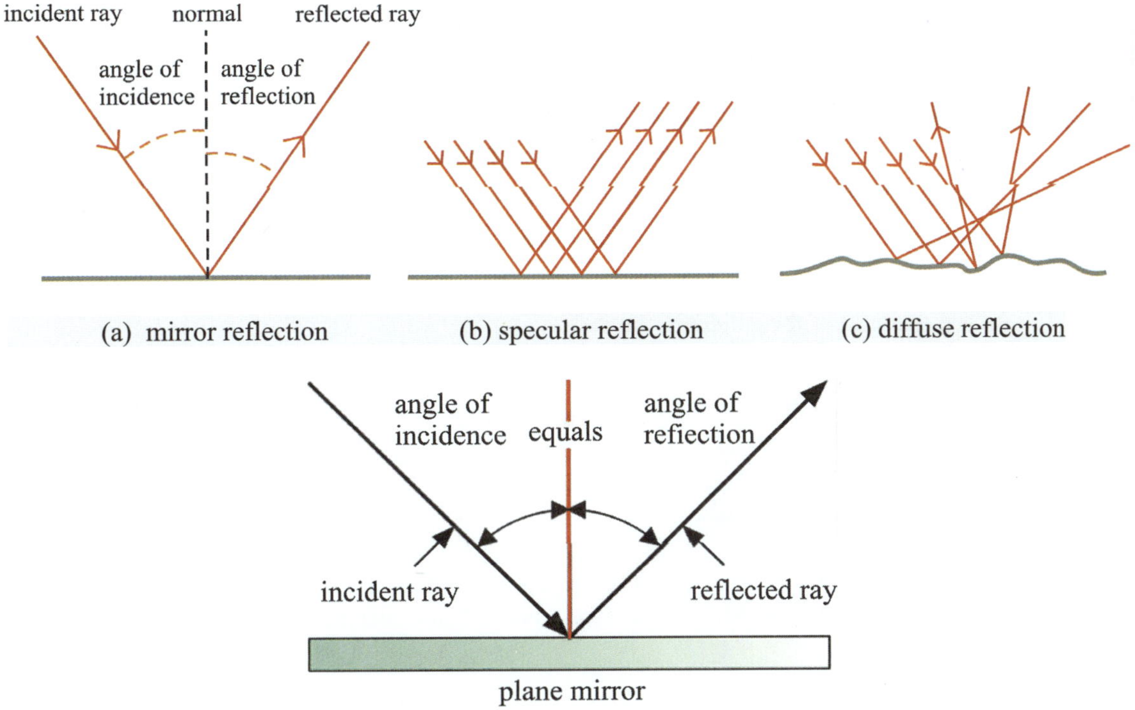

Fig.1.9 : Examples of reflection of light rays

The Law of Reflection

Objects can be seen by the light they emit or, more often, by the light they reflect. Reflected light obeys the law of reflection, that is, the *angle of reflection equals the angle of incidence* as depicted in Fig.1.10, where $\theta_i = \theta_r$.

For objects such as mirrors, with surfaces so smooth that any hills or valleys on the surface are smaller than the wavelength of light, the law of reflection applies on a large scale. All the light travelling in one direction and reflecting from the mirror is reflected in one direction; reflection from such objects is known as *specular reflection* as illustrated in Fig.1.9(b).

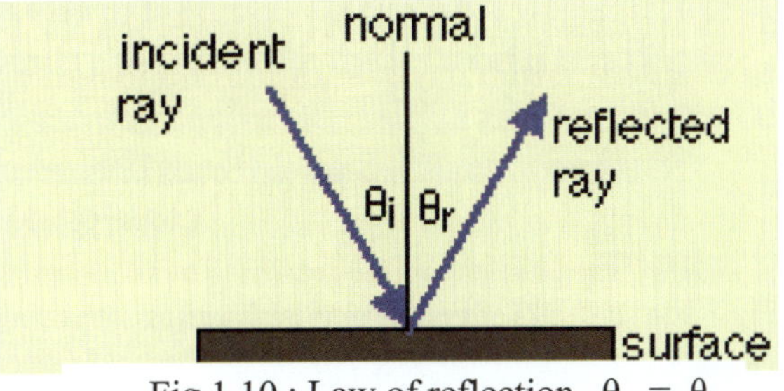

Fig.1.10 : Law of reflection, $\theta_r = \theta_i$

Diffused Reflection

Most objects exhibit diffuse reflection, with light being reflected in all directions. All objects obey the law of reflection on a microscopic level, but if the irregularities on the surface of an object are larger than the wavelength of light, which is usually the case, the light reflects off in all directions. See Fig. 1.9(c).

Plane Mirrors

A **plane mirror** is a mirror with a flat surface; being used by most people every day. Images produced by plane mirrors have a number of properties:

1. the image produced is upright
2. the image is the same size as the object (i.e., the magnification M = 1)
3. the image is the same distance from the mirror as the object appears to be (i.e., the image distance = the object distance)
4. the image is a *virtual* image, as opposed to a real image, because the light rays do not actually pass through the image. This also implies that an image could not be focused on a screen placed at the location where the image is.

Partial Reflection and Refraction

Partial Reflection and Refraction occurs when a wave is travelling between two mediums. Some of the wave is reflected back and the rest is refracted through into the other medium as illustrated in Fig.1.11.

An example of this is the sun and the lake on a summer day. The sun will reflect part of its light off the water and the rest will refract into the water. See Fig.1.12.

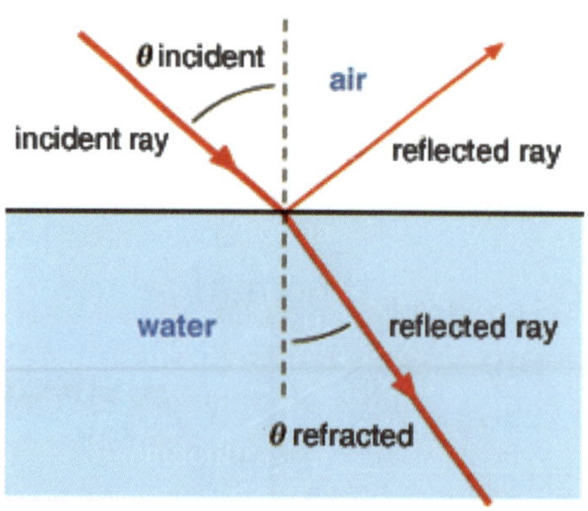

Fig.1.11: Partial reflection and refraction

Fig.1.12 : The Sun over a lake resulting in partial reflection and refraction

Refraction Through a Glass Prism

In optics, a **prism** is a transparent optical element with flat, polished surfaces that refract light. At least two of the flat surfaces must have an angle between them, with the base making up third flat surface of square or rectangular shape and in colloquial use "prism" usually refers to this type . The exact angles between the surfaces depend on the application. Some types of optical prism are not, in fact, in the shape of geometric prisms. Prisms can be made from any material that is transparent to the wavelengths for which they are designed, usually light. Typical materials in use are (mostly) glass, occasionally plastic with obvious disadvantage for prolonged use.

A commonly used triangular glass prism is shown in Fig.1.13.

(a) an actual prism

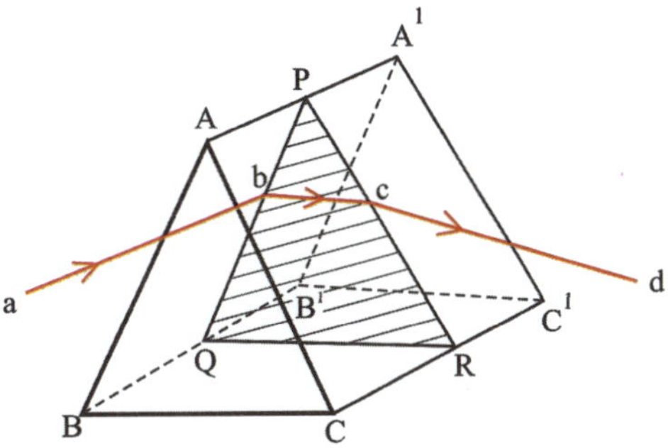

(b) sketch showing path of a light ray, abcd

Fig.1.13 : A triangular glass prism

Dispersion of light

A *dispersive* prism can be used to break light up into its constituent spectral colours[5]. Furthermore, prisms can be used to reflect light, or to split light into its components. Two images of the phenomenon are illustrated in Fig.1.14.

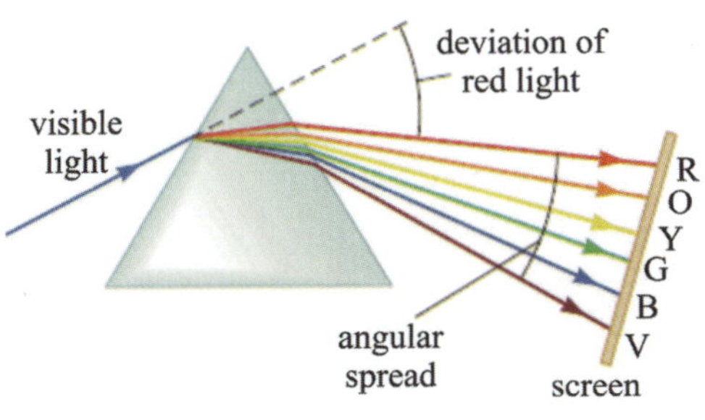

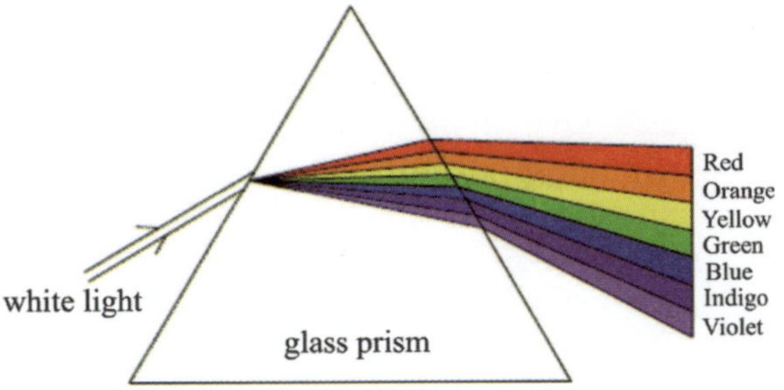

Fig.1.14 : Dispersion of light into constituent colours

[5]Recall that 'white' sun light comprises seven colours as reflected in a rainbow.

Principle of Dispersion of Light

It is well-known that light changes speed as it moves from one medium to another (for example, from air into the glass of the prism). This speed change causes the light to be refracted and to enter the new medium at a different angle. The degree of bending of the path of light depends on the angle that the incident beam of light makes with the surface, and on the ratio between the refractive indices of the two media. The refractive index of glass varies with the *wavelength* or *colour* of the light that is involved in refraction; this causes light of different colours to be refracted differently and to leave the prism at different angles. This phenomenon is known as **dispersion.**

Prisms are sometimes used for the internal reflection at the surfaces rather than for dispersion. If light inside the prism hits one of the surfaces at a sufficiently steep angle, *total internal reflection* occurs and *all* of the light is reflected. This makes a prism a useful substitute for a mirror in some situations.

Primary Colours

Primary colours are the colours of light which when mixed in *equal proportion* result in white light (or 'colour'). These are *red, green* and *blue.* These colours cannot be formed by mixing other colours.

Traverse of Light Through a Triangular Prism

An example of traverse of light through a typical "triangular" prism *in general*, showing all details, is depicted in Fig.1.15.

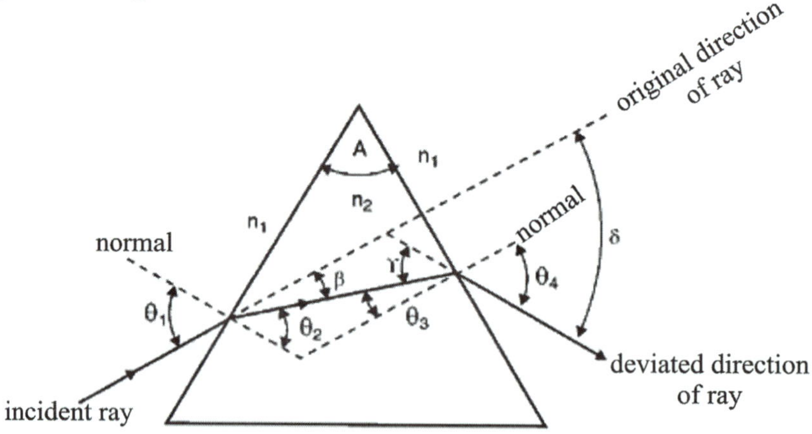

Fig.1.15 : Refraction of light through a triangular prism

Note the angles of refracted light on the two inclined sides of the prism with respect to the incident light on the left surface.

Example

A ray of light is incident on an equilateral glass prism placed in minimum deviation position. Draw and show the path of the refracted ray and the emergent ray.
How are angle of incidence and angle of emergence related to each other in this position of the prism?

[ICSE, 2008]

The complete ray diagram is drawn below.

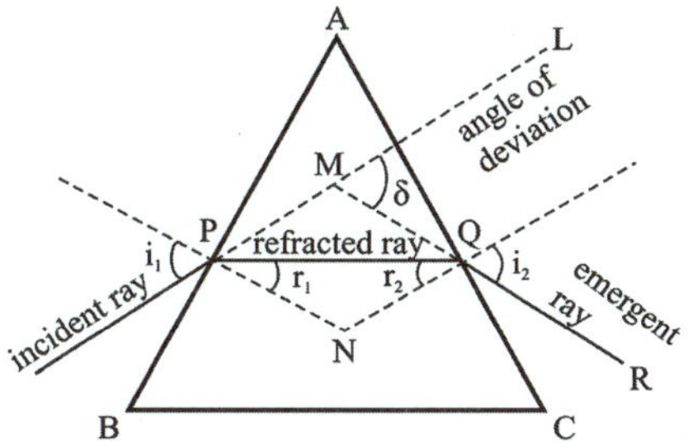

In minimum deviation position $\angle i_1 = \angle i_2$, that is angle of incidence = angle of emergence as marked in the figure.

Experiment to determine refraction of light through a prism

An experimental verification of refraction of light through a triangular prism can be carried out using a prism, a sheet of paper and a few pins.

The details of the experiment are given in Appendix B.

Applications of Refraction

Real and apparent depth of objects in water

The effect of refraction is to cause *apparent* positional change in the objects when they are immersed in water and viewed by human eye from above the water surface. The objects seem to be closer to the surface than they are in reality, the phenomenon being known as **apparent depth.** An example of apparent depth in practice and the refraction of light rays that might cause it is shown in Fig.1.16 .

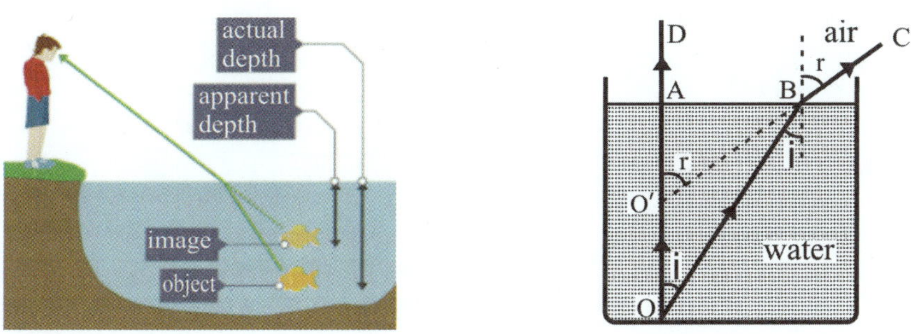

Fig.1.16 : Apparent depth and refraction of light

Explanation of Cause of Apparent Depth

This can be explained by reference to the "ray diagram" shown in Fig.1.17.

- In the diagram, A is the actual position of the object under water
- The light rays coming from the object travel to the water surface and bend away from the normal
- Because light travels in a straight line, when these rays are extended back they seem to come from position B
- This is where the object appears to be.
- The depth at which the object appears to be below the surface is thus the **apparent depth**
- The depth at which the object actually is, is termed the **real depth**.

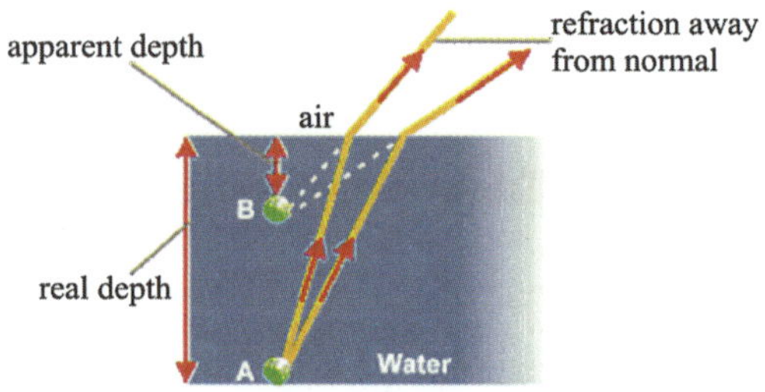

Fig.1.17 : Ray diagram to explain apparent depth

In this case, the refractive index of water is related to the real and apparent depth of the given object by

$$\text{refractive index} = \frac{\text{real depth}}{\text{apparent depth}}$$

Apparent bending of a stick under water

This phenomenon also occurs due to refraction of light. When a stick is immersed in water, it is being inserted from a rarer medium – air – to a denser medium, that is water. Due to refraction when the rays of light pass from rarer medium to denser medium they move towards the normal. So the part of stick immersed in water appears to be broken towards the normal drawn at the interface. Hence a stick appears short as it appears raised up in water. See Fig.1.18.

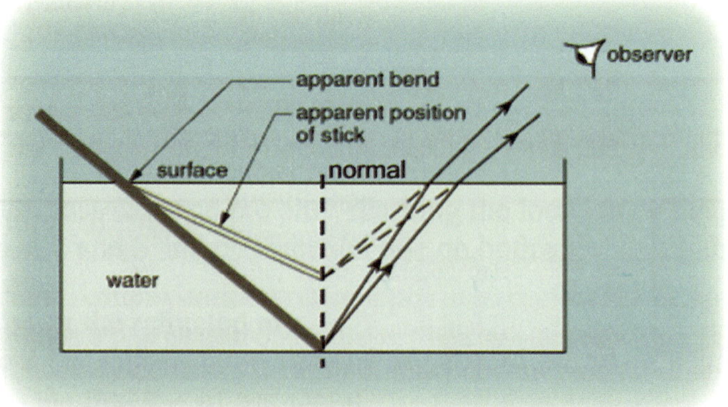

Fig.1.18 : A stick appearing bent in water

Example

1. A pencil is inserted into a glass container having water. Draw a ray diagram showing how the eye sees the pencil.

[ICSE, 2015]

The ray diagram is drawn below

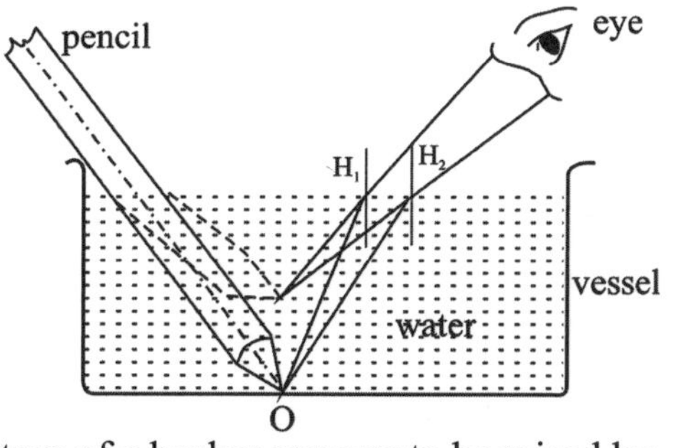

2. A coin placed at the bottom of a beaker appears to be raised by 4.0 cm. If the refractive index of water is 4/3, find the depth of water in the beaker.

[ICSE, 2013]

Let the depth of water, or the real depth of the coin be x cm
Since the coin appears to be raised by 4 cm, the apparent depth of the coin will be (x – 4) cm
Then

$$\frac{4}{3} = \frac{x}{x-4}$$

or $$4x - 16 = 3x$$

Solving $$x = 16$$

$\therefore$ real depth of water $= 16$ cm

3. A rectangular glass block of refractive index 1.5 is kept above a postage stamp . The image of the stamp is observed to get raised by 8.0 mm. What is the thickness of the glass block?
Let the thickness of glass block, or the “real depth”, be x mm
It is known that

$$\text{refractive index of medium, } \mu = \frac{\text{real depth}}{\text{apparent depth}}$$

Here $\mu_g = 1.5$ and apparent depth of the stamp $= x - 8$
therefore

$$1.5 = \frac{x}{x-8}$$

whence $$x = 24 \text{ mm}$$

Appendix A

Experimental verification of refraction of light

Refer to Fig.1.19.

Follow the steps listed below

- Place a rectangular glass slab on the white sheet of paper placed centrally on a drawing board
- Trace the boundary ABCD of the glass slab
- Remove the glass slab and draw a normal N_1N_2 at O
- Draw a straight line IO, inclined at an angle say 30° with the normal. IO is the incident ray
- Fix two pins P and Q on the incident ray IO
- Place the glass slab covering its boundary ABCD
- Looking from the other side of the glass slab, fix two other pins R and S such that P, Q, R and S appear to lie on the same straight line

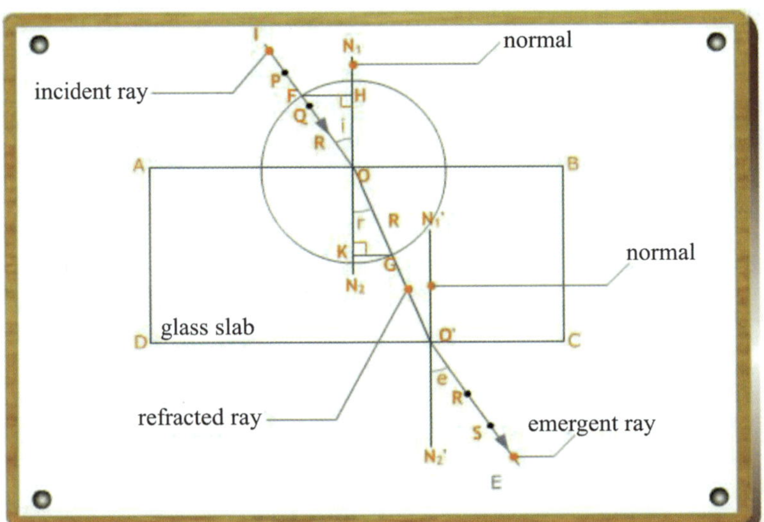

Fig.1.19 : Experimental verification of principles of refraction

- Remove the glass slab and the pins, marking the location of pin points P, Q, R and S
- Join the pins R and S and produce the line on both sides being the ray O'E as the emergent ray
- Join OO' being the refracted ray
- The incident ray, the refracted ray and the normal are all coplanar
- This proves the first law of refraction

To prove the second law of refraction

- With O as centre, draw a circle of a convenient radius 'R' in such a way that it cuts the incident and the refracted rays at F and G, respectively
- From F and G draw perpendiculars to the normal N_1N_2
- Then, FHO and GKO are right-angled triangles

$$\sin i = \frac{FH}{OF}$$

$$\sin r = \frac{GK}{OG}$$

$$\mu = \frac{\sin i}{\sin r} = \frac{FH}{OF} \times \frac{OG}{GK}$$

But OG = OF = R

$$\mu = \frac{FH}{OG} \times \frac{OG}{GK}$$

$$\mu = \frac{FH}{GK}$$

- Measure and record the lengths FH and GK
- Repeat the experiment for different values of angle of incidence, i
- Find the value of $\frac{FH}{GK}$ for different values of i
- In each case it is found that the ratio $\frac{FH}{GK}$ is the same, that is, the ratio of $\frac{FH}{GK}$ is a constant
- This verifies the second law of refraction or Snell's law

Appendix B

A sheet of white paper is fixed on a drawing board using drawing pins. A glass prism is placed on it in such a way that it rests on its triangular base. The outline of the prism is traced on the paper. A straight line PE is drawn inclined to one of the refracting surfaces, AB, of the prism. Two pins are fixed at points P and Q, on the line PE as shown. The images of the pins, fixed at P and Q are looked through the other face AC of the prism. Two more pins are fixed at points R and S, such that the pins at R and S and the images of the pins at P and Q lie on the same straight line.

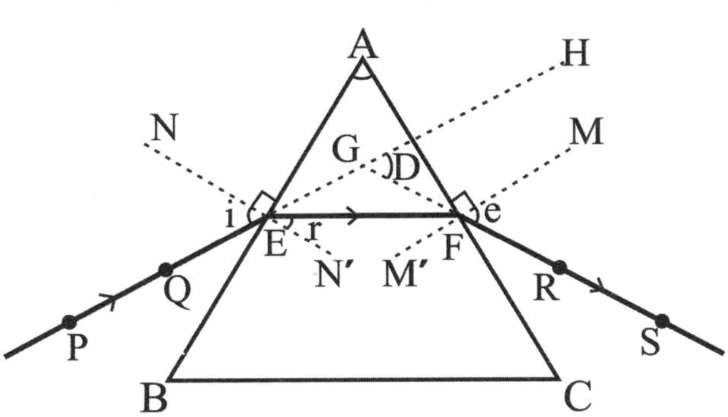

The pins and the glass prism are removed. The points P and Q are joined and the line is extended till it meets the nearest boundary of the prism. Similarly, the points R and S are joined and produced back. The lines should meet the boundary of the prism at E and F, respectively. E and F are joined. Perpendiculars are drawn to the refracting surfaces AB and AC of the prism at points E and F respectively. The angle of incidence ($\angle i$), the angle of refraction ($\angle r$) and the angle of emergence ($\angle s$) marked as marked.

It is seen that a ray of light enters from air to glass at the first surface AB. This light ray on refraction bends towards the normal. At the second surface AC, the light ray enters air from glass and bents away from normal. These rays obey the laws of refraction of light. But emergent ray is not parallel to the incident ray.

The angle between the direction of the incident ray and that of the emergent ray is called angle of deviation, δ.

EXERCISES

1. A ray of light strikes the surface of a rectangular glass block such that the angle of incidence is (i) 0°; (ii) 42°. Sketch a diagram to show the approximate path taken by the ray in each case as it passes through the glass block and emerges from it.

[ICSE, 2009]

The ray diagrams for the two cases are drawn below.

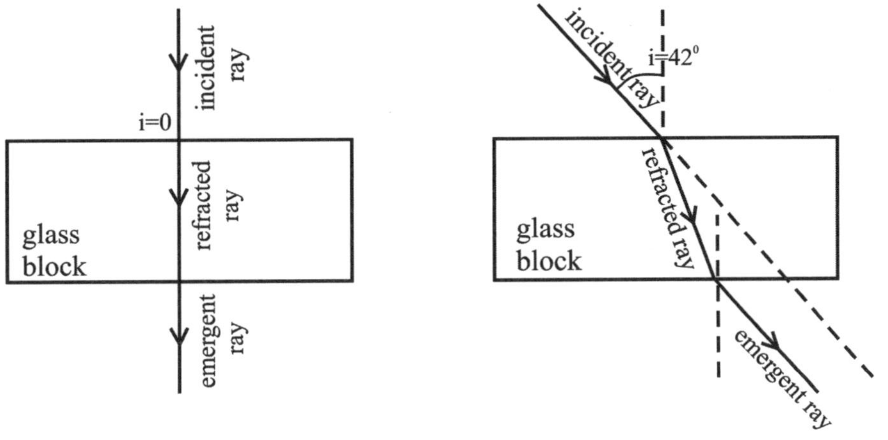

II Total Internal Reflection

Total internal reflection: critical angle; examples in triangular glass prisms; comparison with reflection from a plane mirror (qualitative only); applications of total internal reflection.
Transmission of light from a denser medium (glass/water) to a rarer medium (air) at different angles of incidence; critical angle (C) $\mu = 1/\sin C$; essential conditions for total internal reflection; total internal reflection in a triangular glass prism; ray diagram, different cases – angles of prism ($60^o, 60^o, 60^o$), ($60^o, 30^o, 90^o$), ($45^o, 45^o, 90^o$); use of right angle prism to obtain $\delta = 90^o$ and 180° (ray diagram); comparison of total internal reflection from a prism and reflection from a plane mirror.

Transmission of Light from a Denser to Rarer Medium

Refraction from a Denser Medium to Rarer Medium

When the light *emerges* from a dense medium, say water, into a rarer medium such as air, much depends on the angle at which the light strikes the surface, or the interface/boundary of the two media. Recall that when light is incident on water/air boundary, it would be refracted *away* from the (common) normal as it emerges for a given angle of incidence. What would happen if this angle increases from a small angle, say zero, to progressively larger angle(s)?

To understand this, consider the following "thought experiment".

Suppose that a laser beam is submerged in a tank of water and pointed upwards towards water-air boundary. Then suppose that the angle at which the beam is directed upwards is slowly altered, beginning with small angles of incidence and proceeding towards larger angles of incidence. What would be observed in such an experiment? According to the principles of behaviour of light at the boundary, both refraction *and* reflection would occur at the interface with relative brightness of the refracted and reflected light, depending on the angle of incidence. At angle of incidence close to 0 degrees, most of the light is transmitted across the boundary and very little of it is reflected.
As the angle is increased to greater values, there would be less refraction and more reflection. Finally, it would be observed that the angles of the reflection and refraction are not equal. Also, since the light waves would refract away from the normal, the angle of

refraction would be greater than the angle of incidence. And if this were the case, the angle of refraction would also be greater than the angle of reflection since the angles of reflection and incidence are the same.

As the angle of incidence is increased, the angle of refraction would eventually reach a 90-degree angle. These principles are depicted in the diagram given in Fig.2.1.

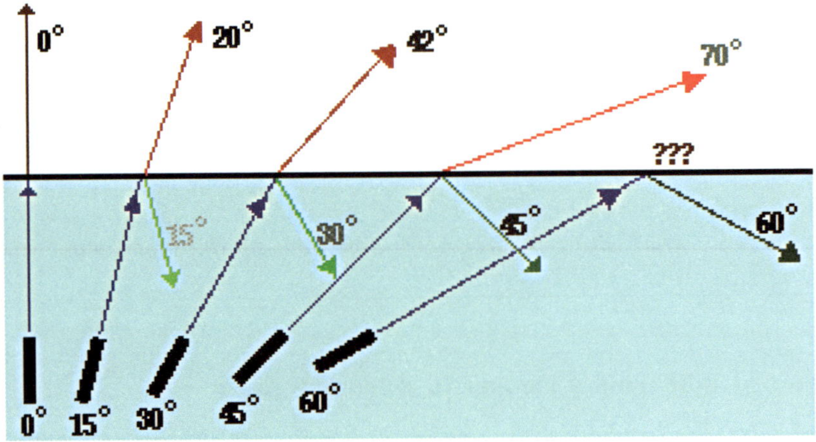

Note that

as the angle of incidence increases progressively from 0° to greater angles, it is observed that

- there is less refraction; the refracted ray becomes dimmer
- there is more reflection; the reflected ray becomes relatively brighter
- the angle of *refraction* approaches 90° until the refracted ray is no longer seen

Fig.2.1 : Refraction and reflection of light at water-air boundary at increasing angles of incidence

The maximum possible angle of refraction is 90°. If it were possible that the angle of refraction were greater than 90°, then the refracted ray would 'lie' on the incident side of the medium and that is just not possible.

Critical angle

So in the case of the assumed laser beam in the water, there is some specific value for the angle of incidence that yields an angle of refraction of 90-degrees. This particular angle of incidence is known as **critical angle** for the given media boundary and can be deduced from the knowledge of refractive indexes of the two media. Thus, for the water-air boundary, with $\mu_{(water)}$ being 1.33, $\mu_{(air)}$ equal to 1.00 and $\theta_r = 90^\circ$ the critical angle in this case is obtained as 48.6°.

TOTAL INTERNAL REFLECTION, TIR

Any angle of incidence that is greater than 48.6 degrees would not result in refraction. Instead, when the angles of incidence is greater than 48.6 degrees (the critical angle), all of the energy (the *total* energy) carried by the incident wave of light to the boundary stays within the water (*internal* to the original medium) and undergoes total reflection off the boundary. This phenomenon is known as **total internal reflection.**

Clearly, critical angle and total internal reflection for a given boundary of a denser medium to a rarer medium are inter-related. This is illustrated in Fig.2.2.

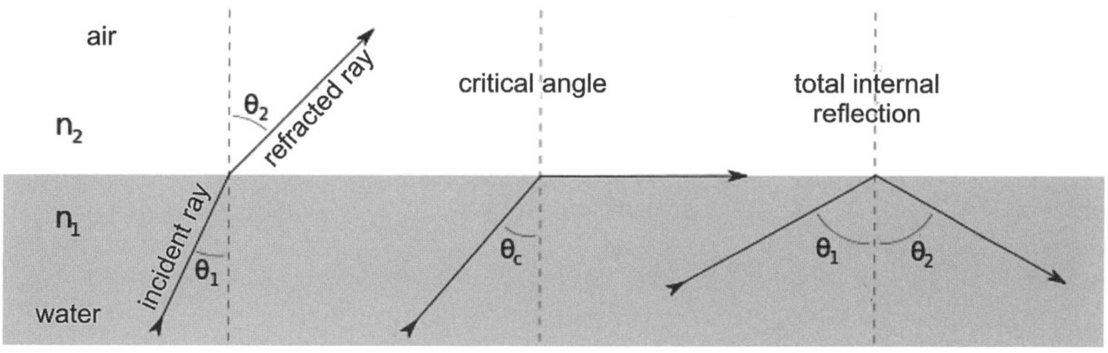

Fig.2.2 : Critical angle and total internal reflection

Example

A ray of light enters a glass slab PQRS as shown in the diagram. The critical angle of glass is 42°. Complete the diagram by drawing the path of the ray till it emerges from the glass slab, marking the various angles where required.

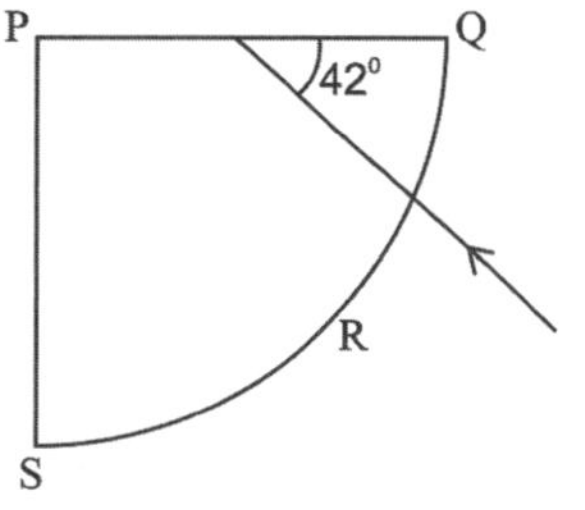

[ICSE, 2010]

The diagram showing path of the ray is drawn in the adjoining figure. Notwithstanding the curved portion of the glass slab, QRS, the angle of incidence of the ray on the flat surface PQ at O is 48° which is greater than the critical angle for glass. This leads to total internal reflection of the ray. The ray is reflected by angle 48° as shown. It is incident on surface PS at 42° at T, and is refracted along TS.

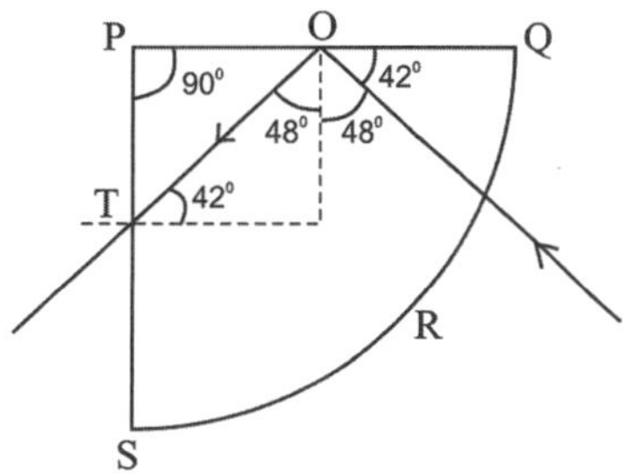

Derivation of Expression for Critical Angle

Referring to Fig.2.3 and following the laws of refraction (Snell's law), in general, in terms of quantities pertaining to incident and refracted (or transmitted) rays

$$n_i \text{ (or } \mu_i) \sin \theta_i = n_t \text{ (or } \mu_t) \sin \theta_t$$

When θ_i is increased gradually such that the angle θ_t reaches the limiting value of 90°, the angle θ_i is termed the critical angle for the given boundary.
Then

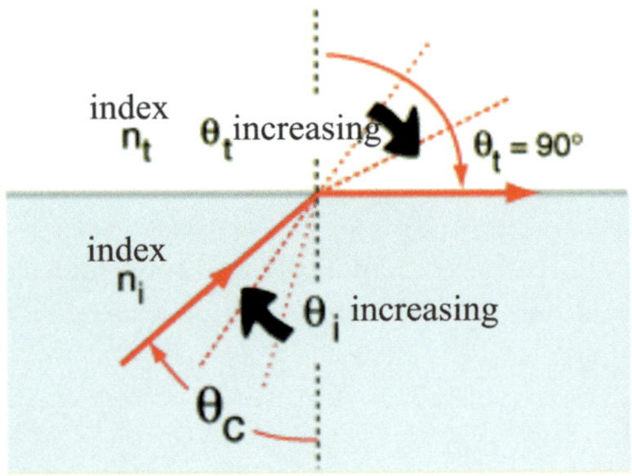

Fig.2.3 : Refraction of light and critical angle

$$\mu_i \sin \theta_c = (\mu_t) \sin 90^o$$

or

$$\mu_i \sin \theta_c = (\mu_t) \times 1$$

$$\sin \theta_c = (\mu_t / \mu_i)$$

whence

$$\theta_c = \sin^{-1} (\mu_t / \mu_i)$$

and is dependent on the ratio of two refractive indexes.

When the "rarer" medium is air, μ_t is very nearly 1.0 and hence

$$\theta_c = \sin^{-1} (1/ \mu_i)$$

or

$$\mu_i = 1/ \sin \theta_c$$

Thus, knowing μ_i , the refractive index of "dense" medium (water in the above discussion), the critical angle, θ_c, for the medium can be obtained and *vice-versa*.

When the dense medium is glass and rarer medium is air, the critical angle for the glass-air boundary, with μ_i for glass being 1.5, would be given by

$$\theta_c = \sin^{-1} (1/1.5)$$

$$= 41.8^o \text{ , or } 42^o \text{ very approximately}$$

If the dense medium is glass and rarer medium water, with the refractive indexes being 1.5 and 1.33 respectively, then using the general expression the critical angle for glass for this combination can be deduced to be

$$\theta_c = \sin^{-1} (1.33/1.5)$$

$$= 62.45^o$$

Applications of Total Internal Reflection

- **Sparkle of diamonds**

Critical Angle for Diamond-air Boundary

The refractive index for diamond is 2.42. Therefore using the above expression

$$\theta_c = \sin^{-1}(1.0/2.42)$$

$$= 24.4^o$$

This is rather extremely small. This peculiarity about the diamond-air boundary plays an important role in the brilliance of a diamond gemstone. Having a small critical angle, light has the tendency to become "trapped" inside of a diamond once it enters. A light ray will typically undergo total internal reflection *several times* before finally refracting out of the diamond. Because of the extremely small critical angle, most rays approach the diamond at angles of incidence greater than the critical angle. This gives diamond a tendency to sparkle. The effect can be enhanced by the cutting of a diamond gemstone with a strategically planned shape. The diagrams in Fig.2.4 below depict the total internal reflection within a diamond gemstone with a strategic and a non-strategic cut.

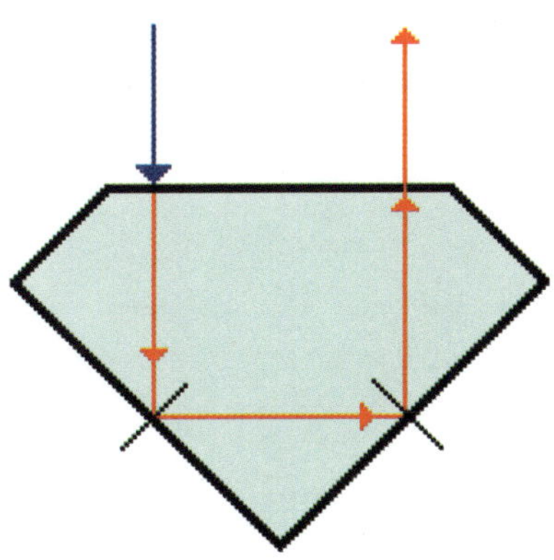

light entering the top surface undergoes TIR a couple of times before exiting surface again

(a) total internal reflection (TIR) in a diamond with strategic cuts

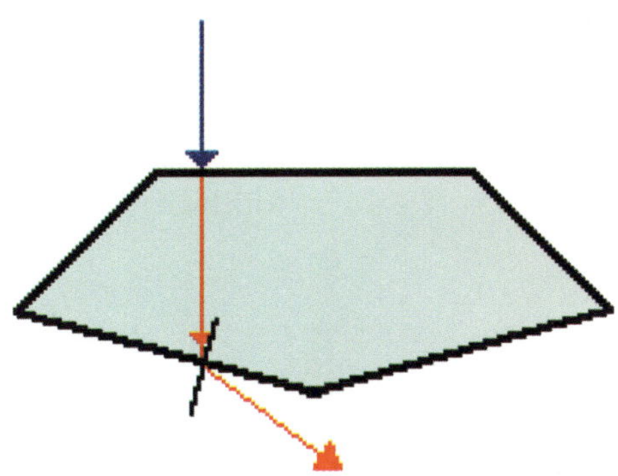

light entering similarly exits the lower left boundary since its angle of incidence is less than the critical angle

(b) TIR without strategic cuts

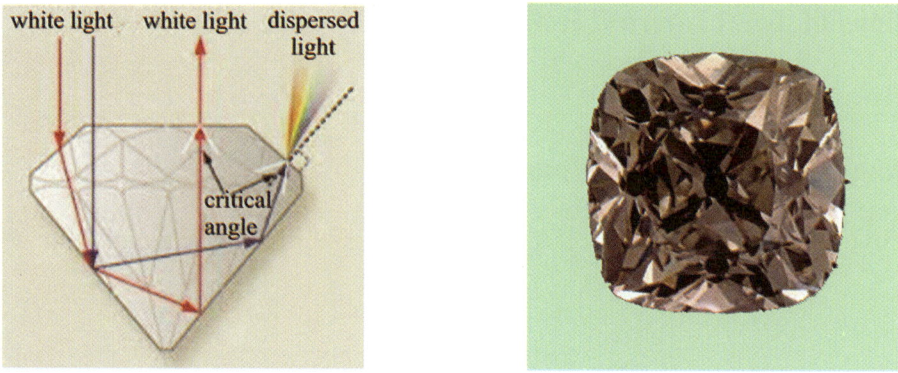

Fig.2. 4 : Total internal reflection in diamonds and importance of "diamond cut"

• Fibre optics or optical fibres

An important application of TIR is in the manufacture of fibre optics now universally in use for high to very-high speed transmission of data and information in communication networks and medical equipment. An illustration of transmission of light using a fibre optic is given in Fig.2.5.

Fig.2.5 : Travel of light through fibre optic

The phenomenon of total internal reflection inside an optical fibre is depicted in Fig.2.6.

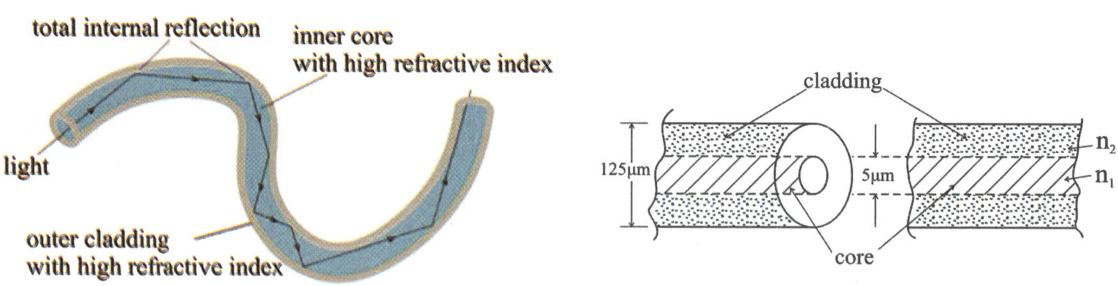

(a) a fibre optic and travel of light ray

(b) schematic of construction

Fig.2.6 : Total internal reflection in fibre optic

Principle of Operation

The principle of operation of total internal reflection *inside the glass core* of an optical fibre can be explained by reference to Fig.2.7.

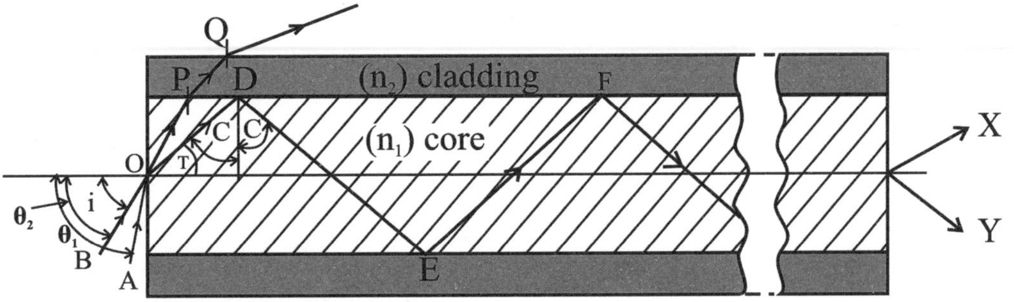

Fig.2.7 : Light path inside the core of fibre optic due to TIR or multiple total reflections

With a 'large' angle of incidence, θ_1, a ray of light AO entering one end of the glass core at O is refracted into the core along OP and then refracted along PQ in the cladding. At Q, the fibre surface, the ray passes into the air. In this case, only a very small amount of light, due to reflection, passes along the fibre.

With a 'smaller' angle of incidence, θ_2, however, a ray such as BO is refracted in the core along OD and meets the boundary between the core and cladding *at their critical angle, C.* In this case, μ sin i is constant

$$\mu_1 \sin C = \mu_2 \sin 90^\circ = \mu_2$$

where μ_1 is the core refractive index and μ_2 that of the slightly smaller cladding refractive index. If $\mu_1 = 1.5$, the refractive index of glass (material for the core), and $\mu_2 = 1.48$ (say).

$$\sin C = \frac{\mu_2}{\mu_1} = \frac{1.48}{1.5} = 0.986$$

so that $C = 80.6^\circ$.

The ray OD is now *totally* reflected at D along DE where it again strikes the core-cladding boundary at the critical angle. At E, therefore, it is totally reflected along EF.

In this way, by total reflection, a ray of light entering one end of a fibre (core) can travel along the fibre by *multiple reflections* with a fairly high light intensity. At the other end of the fibre the ray emerges in a direction X (odd number of multiple reflections) or a direction Y (even number of multiple reflections).

• Mirage in desert

A phenomenon in which total internal reflection plays the key role is mirage in a desert. Mirage is an optical illusion which occurs usually in deserts on hot summer days. On such a day, temperature of air near the earth is maximum and hence is rarer or lighter. The upper layers of air, which are relatively cool, are denser. A ray of light from the top of a tree travels from denser to rarer and bend away from the normal. At a particular layer, I_c, if the angle of incidence is greater than the critical angle applicable to the boundary so-formed, total internal reflection occurs. To a far away observer, this ray, that is IB, i.e., appears to be coming from I, that is, mirror image of O. Thus inverted image of tree creates an optical illusion of reflection from a pond of water.

This is illustrated in Fig.2.8.

Conditions for Total Internal Reflection

The two essential conditions necessary for total internal reflections are

- The index of refraction must decrease across the boundary in the direction of light refraction
- The angle of incidence of the light ray must exceed the critical angle of the interface

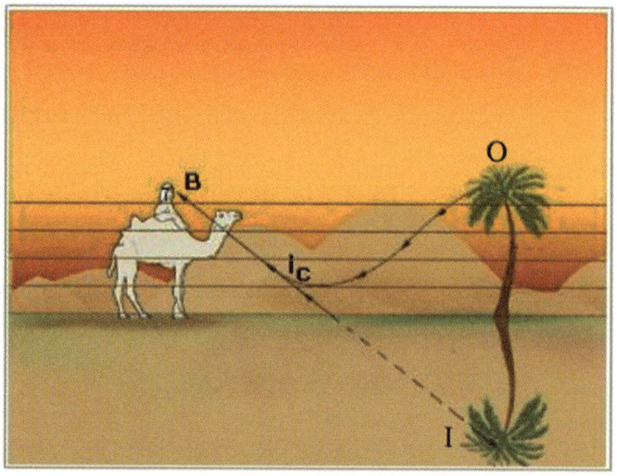

Fig.2.8 : Phenomenon of mirage

Total Internal Reflection in a Triangular Glass Prism

Total internal reflection, TIR, in (triangular) prisms depends on

- the material of the prism; this is usually taken to be glass unless specified otherwise and therefore refractive index of glass would apply

- the angle of incidence of light on any of the surfaces

- the three angles of the prism, viz., whether
  an equilateral or 60°, 60°, 60° prism
  an isosceles prism having a given 'apex' angle and proportionately two equal 'base' angles, for example, 75°, 30°, 75° (apex angle being 30°)
  a 'right-angled' prism having angles such as 90°, 45°, 45° or 90°, 30°, 60°[1]

Equilateral prism: 60°, 60°, 60°

This is the commonest example of a triangular prism in which any of the three angles would form the apex and the base angles would be 60° each. The general diagram of refraction of light through such a prism was discussed earlier. See Fig.1.15, Chapter I.

---

[1]Note that in all these case, reference is being made to the *"cross section"* of the prism, irrespective o the 'axial' length in a direction normal to the plane of the cross section.

Total Internal Reflection

The condition leading to total internal deflection in the prism, as an example, is illustrated in Fig.2.9.

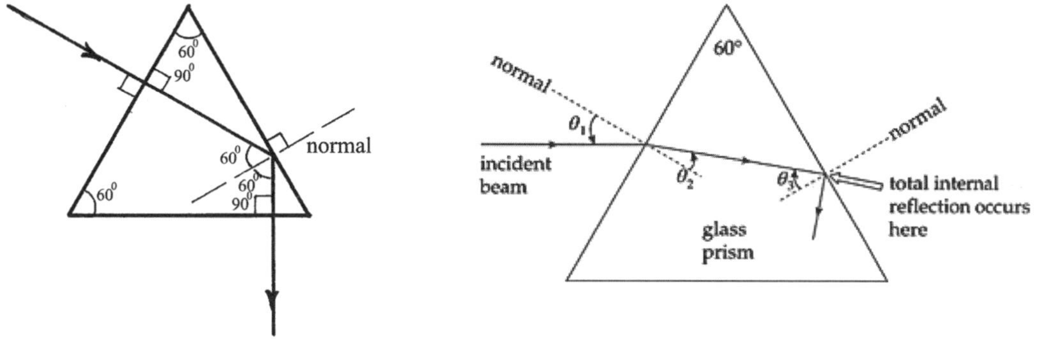

(a) light entering LHS surface at right angles

(b) light entering at a general angle

Fig.2.9 : Total internal reflection in a 60°, 60°, 60° prism

Referring to Fig.2.9(a), observe that light enters (any of) the surface at 90°, reaches the opposite inclined surface at an angle of 30° with the surface, the angle with the normal being 60°. It gets totally reflected and emerges from the 'base' surface *at right angles*. Thus, all the surfaces of the prism are involved. The location of emerging beam would naturally depend on the point of incident ray on the first surface, being arbitrarily shown in the figure. When referred to the angle of incidence on side AC, it is seen to be 60° which is greater than the critical angle (= 42°); hence total internal reflection.

Clearly, a ray entering the side AB at any other angle less than 90° would also lead to total internal reflection provided the angle of incidence internally on the opposite surface, AC, would be greater than the critical angle.

Right-angled prism: 60°, 30°, 90°

This is one special case of a prism in which one angle is 90° whilst the other angles are 60° and 30°. Any of the sides of the triangle ABC can form the 'base'.

Total Internal Reflection

An example of total internal reflection in such a prism is illustrated in Fig.2.10.
The triangle ABC depicts the cross-section of the prism. A ray of light, PQ, enters side AB at right angle and after entering the prism, strikes the opposite side AC at point R as shown, making an angle of 60° with the normal, MN. This angle

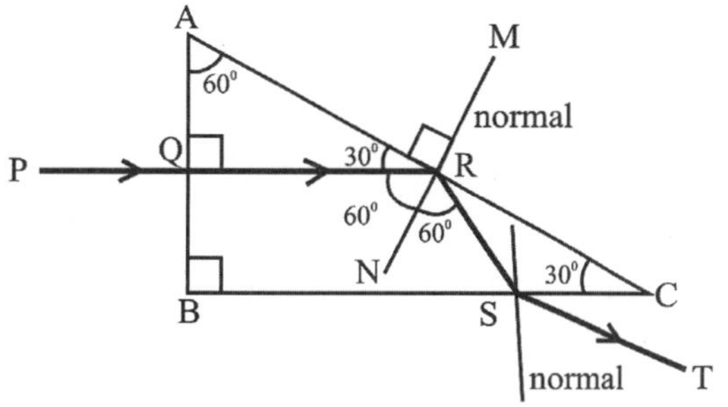

Fig.2.10 : Total internal reflection in a 60^0, 30^0, 90^0 prism

being appreciably greater than the critical angle for glass, the ray undergoes total internal reflection and emerges through the (base) side BC as indicated.

Right-angled prism: 45°, 45°, 90°

This type of prism is of immense practical importance in the making of numerous optical devices and instruments. This follows from the property of the prism to bend the light at 90° or even 180°, that is *reverse* the direction of light.

Total Internal Reflection

Fig.2.11 shows the cross section of such a prism in which two light rays A and B are incident on one of the 'flat' surfaces at right angles. On entering the prism, the rays strike the hypotenuse side of the prism where they undergo total internal reflection, the angle of incidence of each ray being 45° which is just greater than the critical angle (= 42°). Both the rays emerge through the other flat side of the prism, having changed their direction by 90° with respect to the incident rays; change of direction by 180° is shown in Fig.2.11(b).

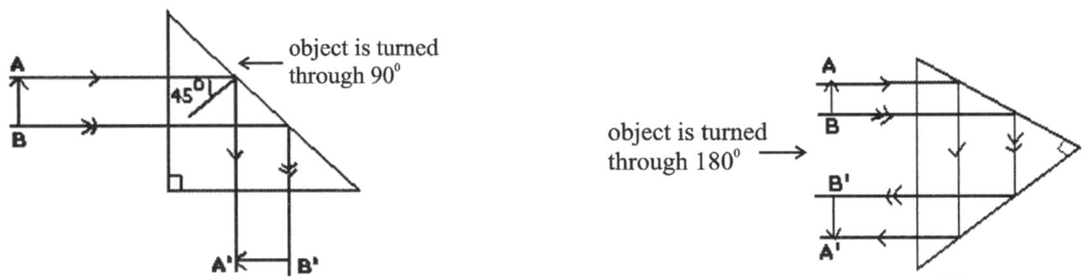

Fig.2.11 : Total internal reflection in a 45°, 45°, 90° prism

An important application of such prisms is in a periscope.
Also, the principle of total internal reflection is used in a binocular.

An isosceles prism

A special case of total internal reflection is shown in Fig.2.12 in an isosceles prism having an apex of 50° and base angles of 65° each. The ray of light is incident on side AB at an angle i and is refracted into the prism at an angle of 75° with respect to side AB and strikes the opposite side at an angle of 45° with respect to normal, and being reflected by the same angle to emerge from the base as shown.

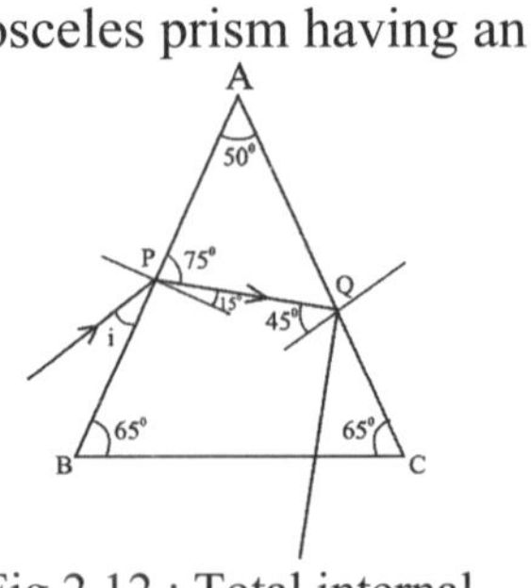

Fig.2.12 : Total internal reflection in an isosceles prism

Angle of Deviation in a Prism

The angle of deviation (angle δ) through a triangular prism is defined as the angle between the incident ray and the emerging ray. This is illustrated in Fig.2.13. It can be shown that when the angle of incidence i is equal to the angle of refraction r' for the emerging ray, the angle of deviation is at a minimum. [See also Fig.1.15, Chapter I].

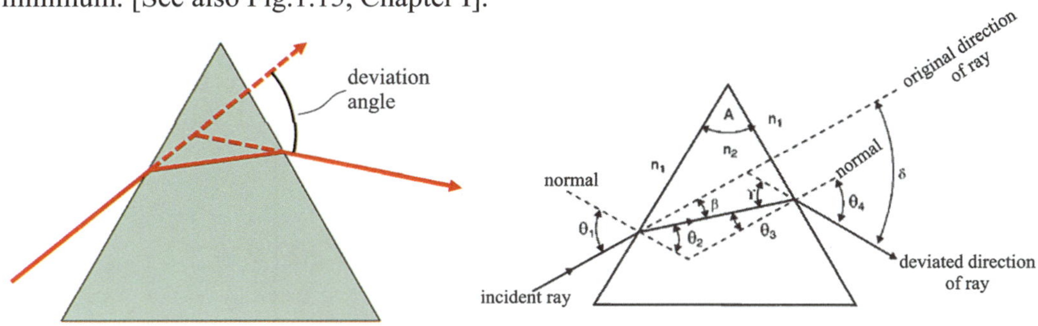

Fig.2.13 : Angle of deviation in a prism

Minimum Angle of Deviation

When the angle of incidence i is equal to the angle of refraction r' for the emerging ray, the angle of deviation is at a minimum.

Note that a 45°, 45°, 90° prism can result in a 90° or 180° deviation depending on the surface of prism being used for the incident rays of light. [See Fig.2.11].

Comparison of total internal reflection from a prism and reflection from a plane mirror

Total Internal Reflection From a Prism	Reflection From a Plane Mirror
■ It takes place when light passes from a denser medium to a rarer medium and the angle of incidence is greater than the critical angle for the interface	It takes place when light is incident on a plane mirror (surface) from *any* medium at *any* angle of incidence
The entire light is reflected	Only part of light is reflected whilst the rest is refracted and absorbed
■ There is no loss of energy	The energy of reflected ray is less than that of the incident. ray
■ The image is much brighter and the brightness is permanent	The image is less bright and brightness gradually decreases

The two phenomena are depicted in Fig.2.14.

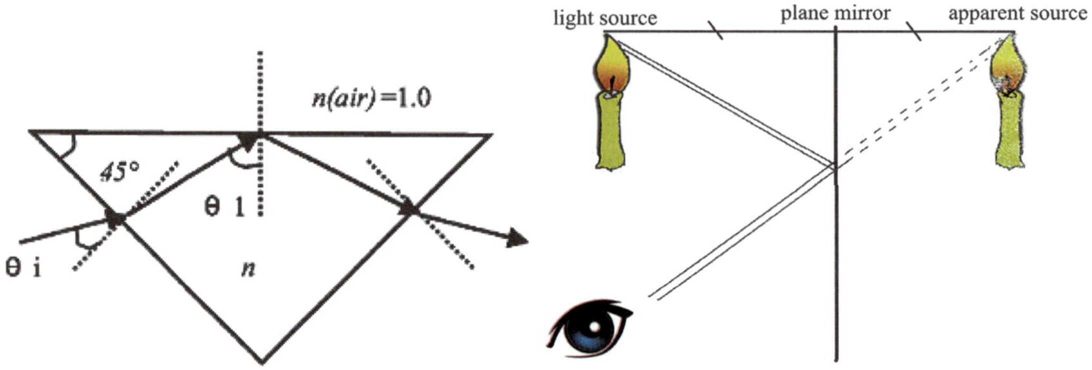

(a) a prism (b) a plane mirror

Fig.2.14 : Total internal reflection from a prism and reflection from a mirror

III Lenses

Lenses (converging and diverging) including characteristics of the images formed (using ray diagrams only); magnifying glass; location of images using ray diagrams and thereby determining magnification.
Types of lens (converging and diverging), convex and concave, action of a lens as a set of prisms; technical terms; centre of curvature, radii of curvature, principle axis, foci, focal plane and focal length; detailed study of refraction of light in spherical lenses through ray diagrams; formation of images –principal rays or construction rays; location of images from ray diagrams for various positions of a small linear object on the principal axis; characteristics of images; sign convention and direct numerical problems using the lens formula are included (derivation of formula not required).

LENSES

What is a lens?[1]

A **lens** is a transmissive optical device that affects the focus of a light beam through refraction. A simple lens consists of a single piece of material, while a compound lens consists of several simple lenses (*elements*), usually along a common axis. Lenses are made from transparent materials such as glass, ground and polished to a desired shape. A lens can focus light to form an image, unlike a prism, which refracts light without focusing.

Types of Lens

There are mainly two types of lens

- converging lens
- diverging lens

[1]The word *lens* comes from the Latin name of the lentil – a pulse - because a double-convex lens is lentil-shaped. The genus of the lentil plant is *Lens*, and the most commonly eaten species is *Lens culinaris*.

These are shown in Fig.3.1.

A ***converging lens*** is a lens that converges rays of light, usually travelling parallel to its principal axis – a 'line' passing through its mid-point and at right angles to its vertical axis – the rays appearing to be meeting at a point called *focal point* of the lens. [See Fig. 3.1(a)]. Converging lenses can be identified by their shape; they are relatively thick across their middle and thin at their upper and lower edges. In general, a converging lens is identified as a **convex** lens.

A ***diverging lens*** is a lens that diverges rays of light that are usually traveling parallel to its principal axis. Diverging lenses can also be identified by their shape; they are relatively thin across their middle and thick at their upper and lower edges. A diverging lens is generally identified as a **concave** lens.
[See also Fig.1.2, Chapter I].

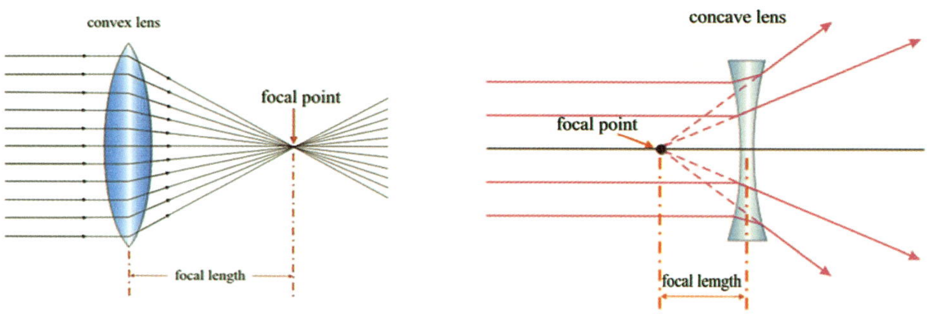

(a) a convex lens (b) a concave lens

Fig.3.1 : Types of lens

Both convex and concave lenses may be imagined to be figures of intersection or position of two spheres of *equal* radii, with their centres positioned so as to 'define' a convex or concave lens as depicted in Fig.3.2.

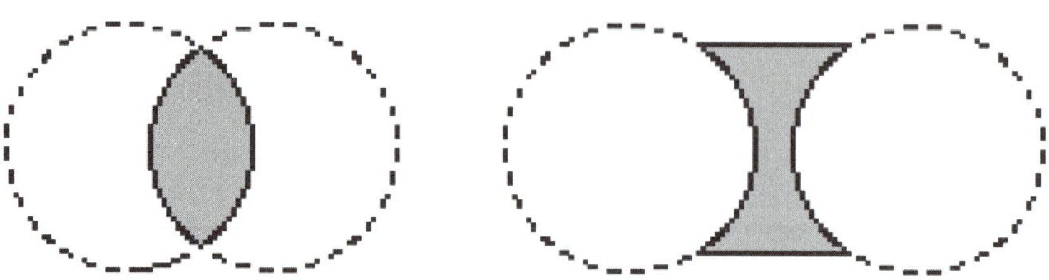

(a) a double convex lens (b) a double concave lens

Fig.3.2 : Convex and concave lens as intersection of curved parts of two circles

Variations

A familiar convex lens is also known as a *double* convex lens that is symmetrical across both its horizontal and vertical axis. Each of the lens' two faces can be thought of as originally being part of a sphere. The fact that a double convex lens is thicker across its middle is an indicator that it will converge rays of light that travel parallel to its principal axis.

Likewise, a usual concave lens is known a *double* concave lens which is also symmetrical across both its horizontal and vertical axis. As for a double convex lens, the two faces of a double concave lens can be thought of as originally being part of a sphere. The fact that a double concave lens is thinner across its middle is an indicator that it will diverge rays of light that travel parallel to its principal axis.

Other common examples of convex and concave lens used in practice are shown in Fig.3.3.

Fig.3.3 : Images of commonly used convex and concave lenses

Action of a Lens as a Set of Prisms

A lens – convex or concave - can be considered to comprise a set of 'prisms' – theoretically infinite in number - or pieces of constituting material, say glass, through which the light rays are deviated as shown in Fig.3.4.

Observe that, taken individually, each piece may not exactly resemble a conventional, commonly identified prism. However, in terms of action of light rays when refracted at the 'entry' and 'exit' surface, yields the overall effect as in a converging, that is convex, or diverging or concave lens as depicted by the ray diagrams in Fig.3.4.

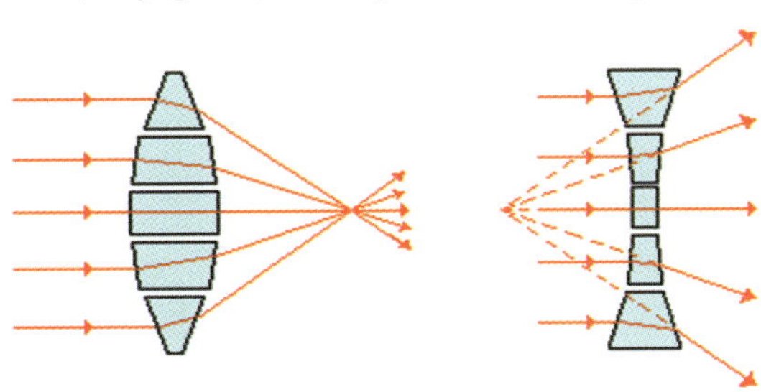

Fig.3.4 : Convex and concave lenses comprising set of prisms

The rays which pass through the centre of the lens are not deviated, while the ones passing through near the edge of the lens will be deviated the most.

Technical Terms Associated With Lenses

Centre of Curvature

Centre of curvature of a lens is defined as the centre of the *sphere* of which the lens is a part. For example, referring to Fig.3.5, C = centre of curvature of the lens carved out of the sphere S.

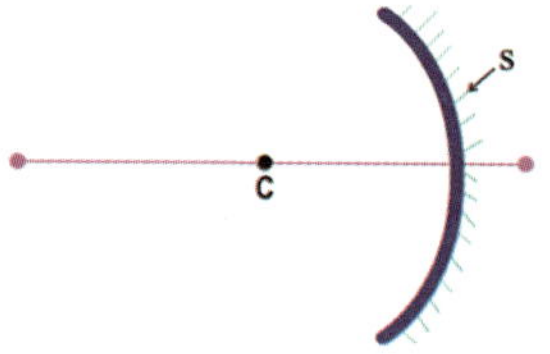

Fig.3.5 : Centre of curvature of a lens

Radii of Curvature

Radius of curvature of a lens is the radius of sphere of which the lens (or mirror) is a part as depicted in Fig.3.6..

When a lens is so devised as to be 'part' of two different spheres, it would have two radii of curvature as shown in Fig.3.6 for a convex lens; this may apply similarly for a concave lens of two different radii.

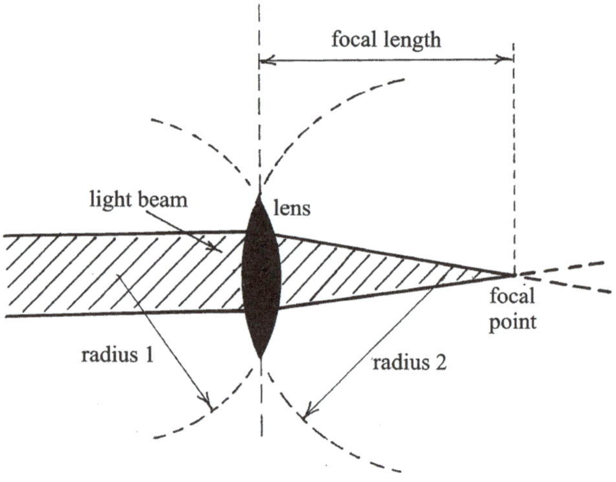

Fig.3.6 : Radii of curvature of a lens

Principal Axis

Principal axis of a lens is a line that passes through the centre of the lens, perpendicular to the lens surface or rather its vertical axis. Lines X-Y in the diagrams of Fig.3.7 illustrate the principal axes of the convex and concave lenses of the usual type.

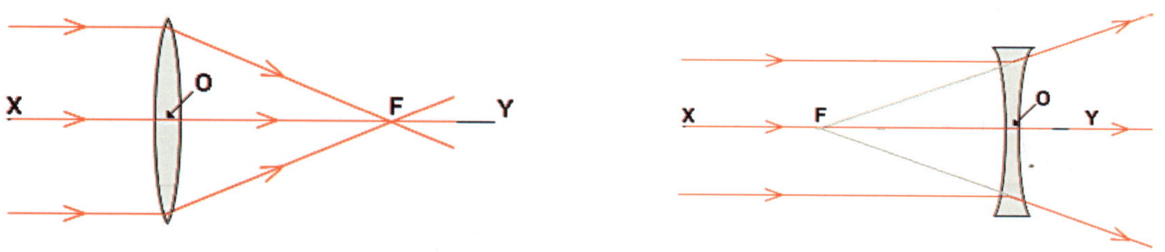

Fig.3.7 : Diagrams showing principal axes of lenses

Foci of Lenses

A lens has two *principal* foci.

Since light may pass through a lens in either direction, there will be two principal foci equidistant from the optical centre, one on either side of the lens. These are denoted by the symbols F and F' or F_1 and F_2, Apart from their use in locating the principal axis of a lens, the centres of curvature of the faces of a lens are not of particular importance in elementary work.

First principal focus[2]

It is a point on the principal axis of the lens such that the rays of light starting from it (convex lens) or appearing to meet at the point (concave lens) after refraction from the two surfaces of the lens become parallel to the principal axis of the lens.

Second principal focus

It is a point on the principal axis of the lens such that the rays of light parallel to the principal axis of the lens after refraction from both the surfaces of the lens pass through this point (convex lens) or appear to be coming from this point.
These are illustrated in Fig.3.8(a) and (b).

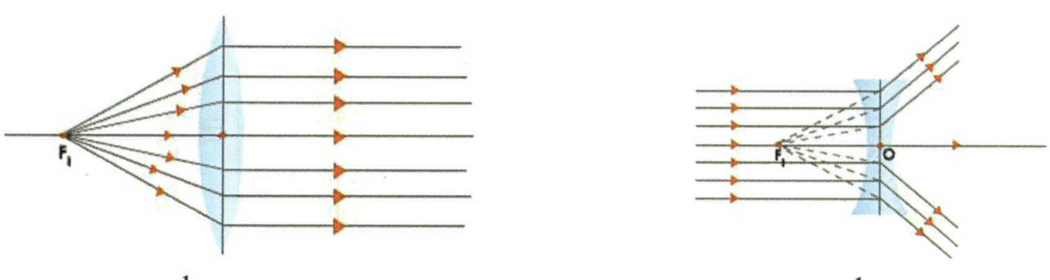

convex lens concave lens

(a) first principal focus of lens

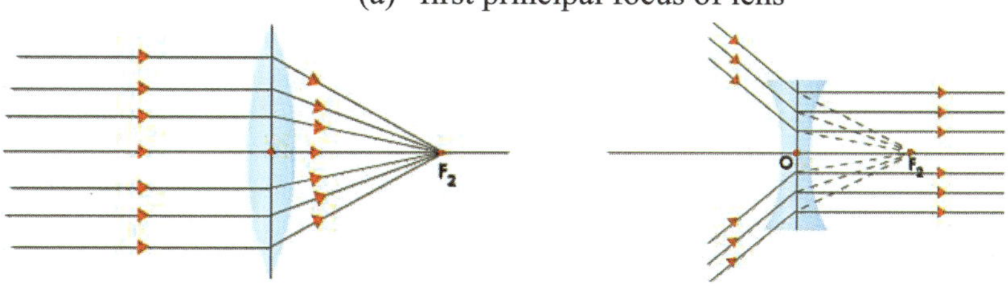

convex lens convex lens

(b) second principal focus of lens

Fig.3.8 : Principal foci of convex and concave lenses

[2]The word "focus" is also used idiomatically in general English usage meaning "to concentrate on" or "converging on" etc.

Note that the concept of foci would particularly apply in the case a of a lens – convex or concave – when the lens is not symmetrical about its vertical axis or is a 'compound' lens comprising a combination of 'basic' lenses; both F_1 and F_2 being same for a 'normal' lens.

Focal Plane and Focal Length

Focal plane

Focal plane of a lens is defined as the (imaginary) plane that would pass through the principal focus of the lens and is perpendicular to the principal axis. This is illustrated in the images shown in Fig.3.9.

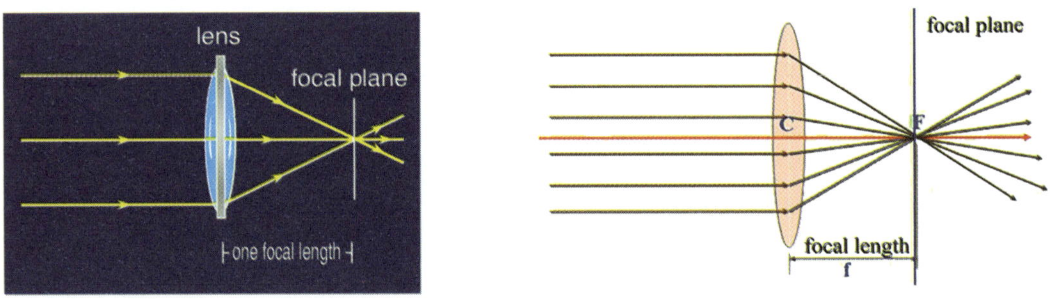

Fig.3.9 : Focal plane of a lens

The importance of the focal plane of a lens – or a combination of them, comprising a compound lens – is in the design and construction of a photographic camera where the film is positioned at the focal plane.

Focal length

This forms one of the most important parameters of lenses. This is the distance between the optical centre (or the vertical axis) and the *principal* focus of a lens. A graphical illustration of focal length of a convex lens and a concave lens is given in Fig.3.10. See also Fig.3.8.

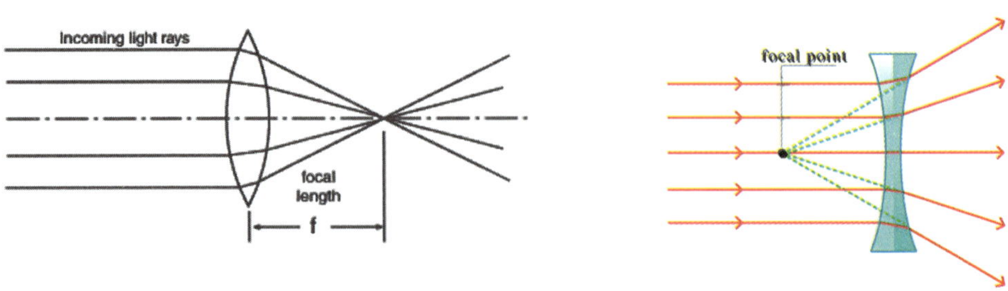

(a) convex lens (b) concave lens

Fig.3.10 : Focal length of a lens

An easy way to understand the 'physics' of focal length is to visualise light rays from the sun, essentially parallel in nature as shown in Fig.3.10, incident on the lens at right angles to the focal plane or parallel to the principal axis when the rays are seen to converge at a point, or the focus, on the other side in the case of a convex lens or appear to converge on the same side in the case of a concave lens. It is usually denoted by f.

The heat contained in the sun rays when focused at the focal point of a convex lens can be intense enough to burn a piece of paper as depicted in Fig.3.11.

Fig.3.11 : Burning action of a convex lens

Refraction of Light in Spherical Lenses[3]

Rules of refraction for spherical lenses

- A ray from the object parallel to the principal axis after refraction passes through the (second) principal focus. F_2, (in a convex lens) or appears to diverge (in a concave lens) from the first principal focus, F_1.
- Any light ray that passes through F_1 (for a convex lens) or appears to meet at it (for a concave lens) emerges parallel to the principal axis after refraction.
- A ray of light passing through the optical centre of the lens emerges without any deviation after refraction.

  A graphical illustration of these rules for convex and concave lenses are given in Fig.3.12(a) and (b).

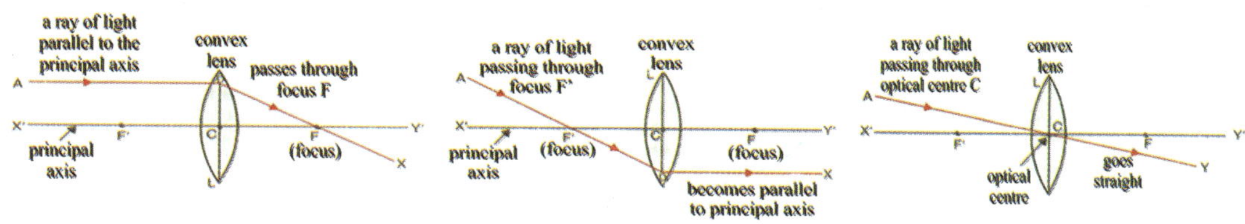

(a) refraction of light through a convex lens

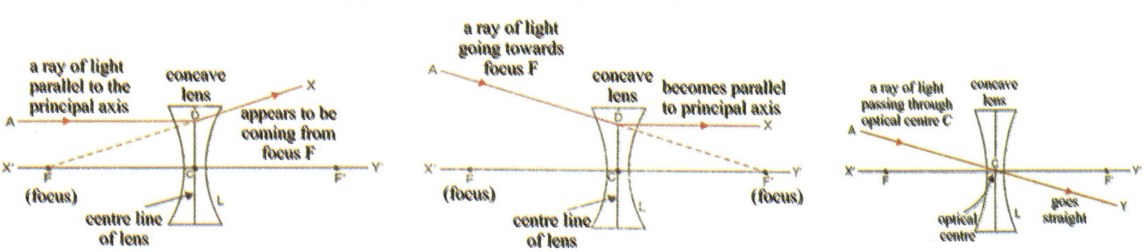

(b) refraction of light through a concave lens

Fig.3.12 : Graphical depiction of rules of refraction through lenses

[3]Note that in all the cases of refraction through the lens(es) considered here implicitly assume a ***thin*** lens – almost "paper thin" - so that the light ray passes through the lens in a straight line up to the vertical axis, esp. above the principal axis, and deviate to pass through as described in each case.

Lens equations

These relate the positions of an object and its image to focal length of a lens.

The basic equations applicable to *'thin'* lens(es) following refraction about image formation are

$$\frac{1}{u}+\frac{1}{v}=\frac{1}{f}$$

and

magnification $M = v/u$ or h_2/h_1.
[in terms of 'heights' h_1 and h_2 of u and v, respectively]

where

f - focal length of the lens
[positive in the case of a convex lens; negative in the case of a concave lens]

u - distance of the object from the optical centre

v - distance of the image from the optical centre
[is positive if the image is a real image and formed on the right side of the optical centre of the lens; is negative if the image is virtual image and formed on the side of the object]

Sign convention for spherical lenses

This is explained by reference to Fig.3.13.

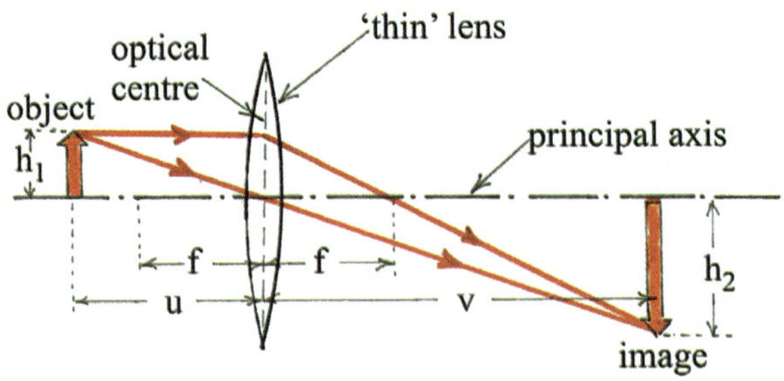

Fig.3.13 : Image for sign convention

For example,

let u = 30 cm; f = 20 cm. Then, using the lens equation, v = 60 cm

1. All figures are drawn with light assumed to be travelling from left to right

2. All distances are measured from a reference surface; for example, the optical centre. Distances to the left of the surface are reckoned negative.
3. Distances measured in the direction of the incident rays are reckoned positive; those in the direction opposite to the direction of incident rays are taken to be negative. Thus, distance of a real object is negative.
4. Heights measured above the principal axis are assumed to be positive; those measured below the principal axis are taken to be negative.
5. The focal length of a *convex* lens is assumed to be positive.
6. The focal length of a *concave* lens is negative.
7. The distance of the object is negative, but the distance of a *real* image is positive. Accordingly, the distance of a *virtual* image is negative.
   Height h_1 of the object is positive.
   Height h_2 of a virtual image is positive, but height h_2 of a real image is negative.
8. Magnification

$$M = \frac{h_2}{h_1}$$

   is positive for a virtual image and negative for a real image.

9. Angles measured clockwise from the principle axis are reckoned negative.

Characteristics of the images formed according to the position of the object

Convex Lens

When the object is placed *at the principal focus* (F_1) in front of a convex lens, the image formed is characterised by the following

- the image is *real*[4]
- the image is *inverted*
- the image is (highly) *enlarged*
- the image is formed at infinity on the right side of the lens

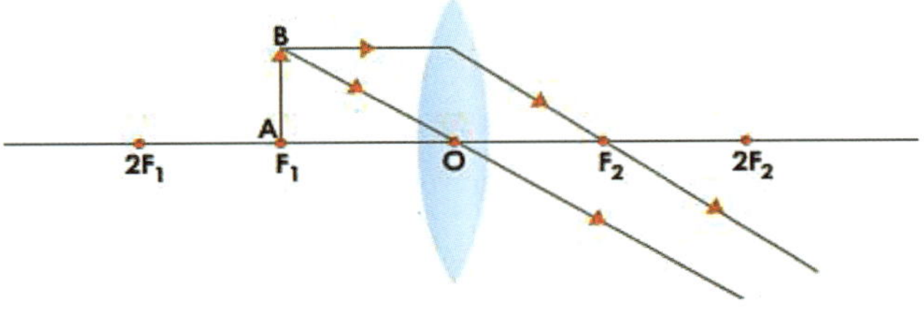

When the object is placed *between the centre and the principle focus* (F_1) in front of a convex lens, the image has the following features

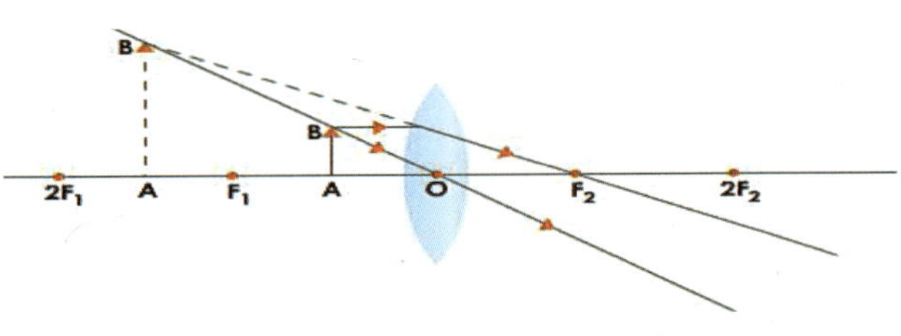

- the image is v*irtual*

[4]In practical terms, a REAL image is one that can be 'projected' on a screen of some kind; for example, a photographic film in a camera.

- the image is *erect*
- the image is *enlarged*
- the image is formed in front of the lens

When the object is placed between the principal focus (F_1) and $2xF_1$ in front of a convex lens, the image formed is characterised by

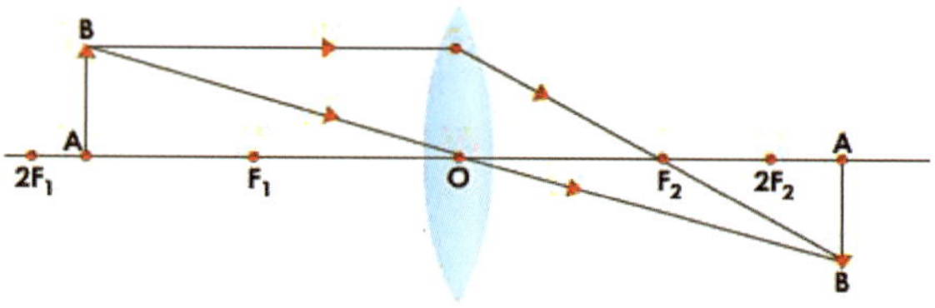

- the image is *real*
- the image is *inverted*
- the image is *enlarged*
- the image is formed *beyond* the $2F_2$ on the right side of the lens

When the object is placed *at* the $2xF_1$ in front of a convex lens, the image has the following properties

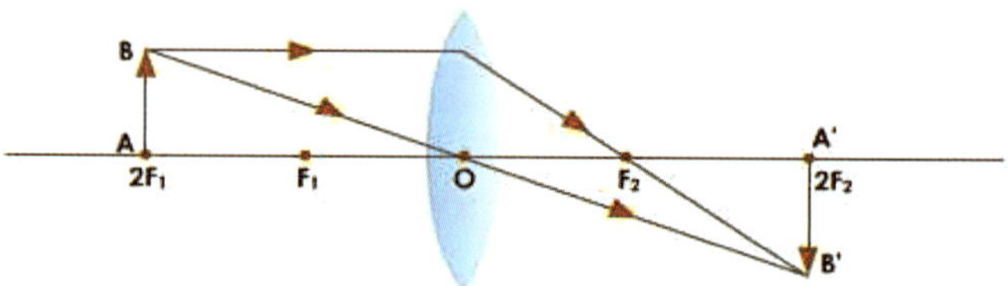

- the image is *real*
- the image is *inverted*
- the image is of the same size as the object
- the image is formed at the $2F_2$ on the right side of the lens

When the object is 'placed' at infinity, the image formed in the case of a convex lens is characterised by

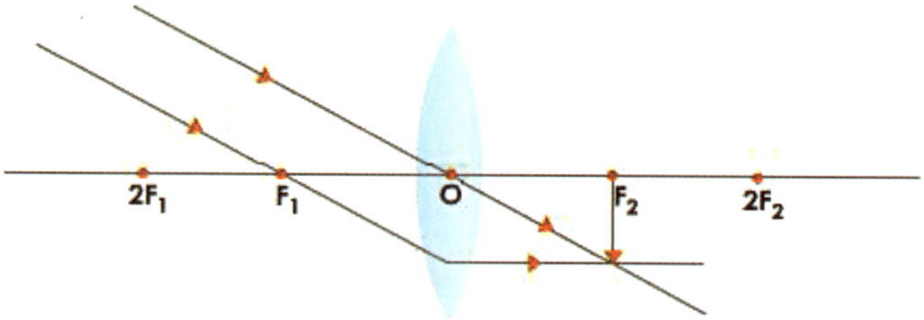

- the image is *real*
- the image is *inverted*
- the image is *highly diminished*
- the image is formed at F_2 on the right side of the lens

Example

Where should an object be placed so that a real and inverted image of the same size as the object is obtained using a convex lens?
Draw a ray diagram to show the formation of the image as specified above.

[ICSE, 2015]

The object should be placed at 2 x focal length of the convex lens. The ray diagram is as shown.

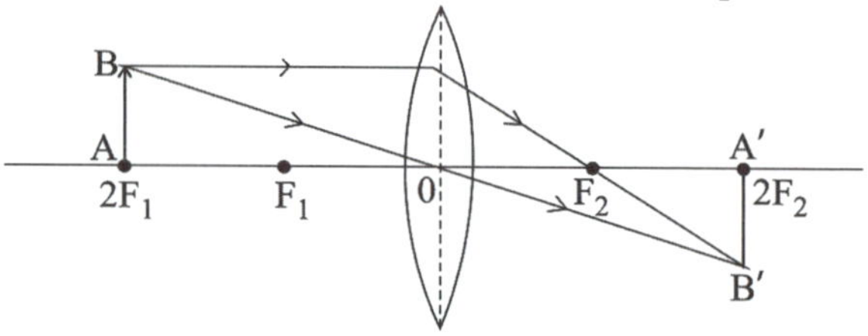

Concave Lens

When the object is supposedly at infinity in the case of a concave lens, the image formed has the following features

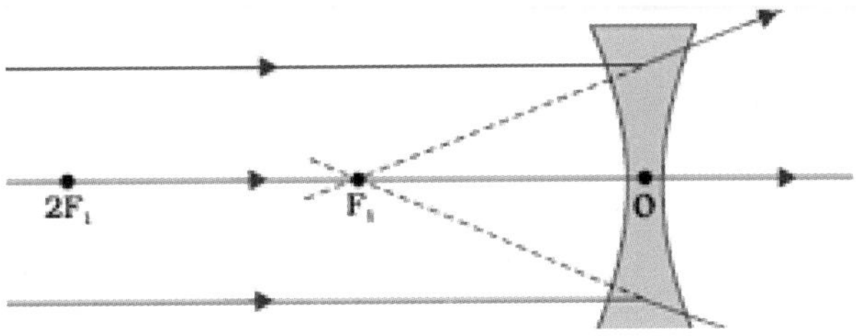

- the image is *virtual*
- the image is *'erect'*
- the image is *highly 'diminished'*
- the image is formed at the principal focus F_1 *in front of the lens*

When the object is *between* infinity and the optical centre of the lens, the image formed has the following properties

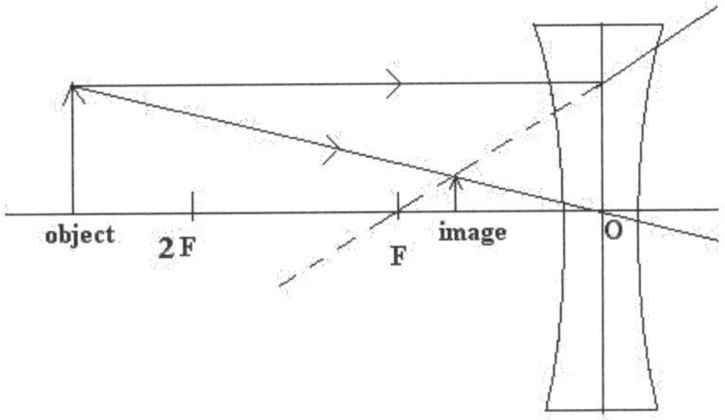

- the image is *virtual*
- the image is *'erect'*
- the image is *diminished*
- the image is formed between the principal focus (F_1) and the optical centre of the lens[5]

Example

1. An object is placed 12 cm from a converging (or convex) lens of focal length 18 cm. Find the position of the image.

Since the lens is converging, f = + 18 cm. Also, the object is real and therefore u = + 12 cm.
[note that the object is located between the first principal focal point and the optical centre of the lens]
Substituting in the lens equation

$$\frac{1}{v} + \frac{1}{+12} = \frac{1}{+18}$$

whence

$$v = -36 \text{ cm}$$

Since v is negative in sign, the image is *virtual,* being 36 cm away from the optical centre and on same side as the object[6]. The magnification, M = v/u = - 36/ 12 or -3.
[The negative sign implies that the image is *upright*]

---

[5]Observe that a convex lens results in a real image, in general, whilst a concave lens produces a virtual image
[6]As discussed later, this represents the case of image formation in a magnifying lens.

2. A beam of light converging to a point 10 cm *behind* a converging lens, is incident on lens. Find the position of the point image if the lens has a focal length of 40 cm.

The configuration is depicted in the diagram of Fig.3.14.

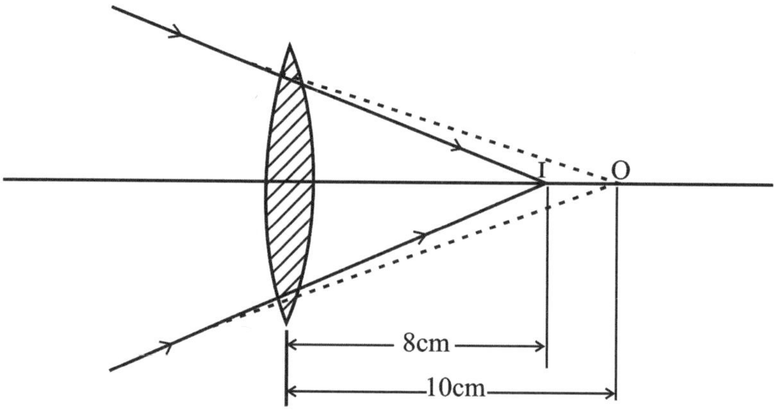

Fig.3.14 : Lens configuration for the problem

If the incident ray converges to the point O, then O is a *virtual* image. Thus u = - 10 cm. Also, f = + 40 cm, since the lens is converging.
Substituting in the lens equation,

$$\frac{1}{v} + \frac{1}{-10} = \frac{1}{+40}$$

from which v is obtained as v = 8.
Since v is positive in sign the image is *real* and it is 8 cm from the lens, indicated by I in the figure.
If the beam of light formed an object of finite size at O, and a real image of this object at I, then

$$\text{magnification} = v/u = (-)\ 8/10 \text{ or } -0.8$$

So the image is *smaller* than the object.

Divergent lens

Suppose a beam converges to a point 10 cm behind a *diverging* lens of focal length 40 cm, so f = - 40 cm. Then u = - 10 cm – a virtual object – giving v = 13.3 cm indicating a real image being formed 13.3 cm *behind* the lens.

3. A luminous object and a screen are placed on an optical bench and a convex lens is placed between them to form a sharp image of the object on the screen as shown in Fig.3.14, the linear magnification of the image being found to be 2.5. The lens is now moved 30 cm nearer the screen and a sharp image is again formed. Calculate the focal length of the lens.

Referring to Fig.3.15, let O and I represent the object and screen positions, respectively, and L_1, L_2 are the two positions of the lens.

At the instance, when the lens is at position L_1, let $OL_1 = u_1$ and $L_1I = v_1$. Then a magnified image of the object would be formed on the screen; the magnification being given to be 2.5.

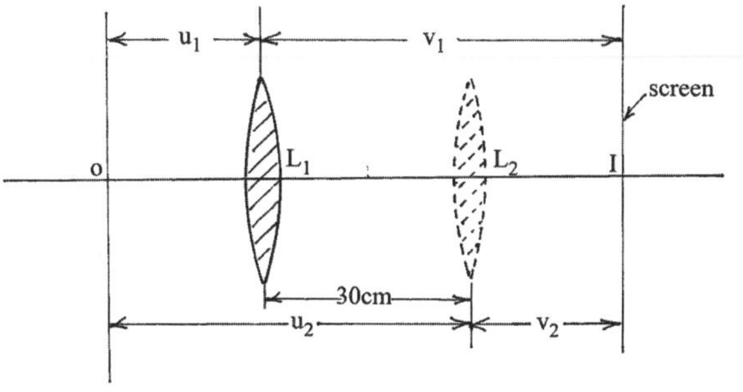

Fig.3.15 : Arrangement of lens and object

At the instance when the lens is at position L_2, let $OL_2 = u_2$ and $L_2I = v_2$
Then the following equations apply

$$u_1 + v_1 = u_2 + v_2$$

$$u_2 = v_2 + 30$$

$$\frac{v_1}{u_1} = 2.5 \, (\text{magnification})$$

$$\frac{1}{u_1} + \frac{1}{v_1} = \frac{1}{u_2} + \frac{1}{v_2} \quad (\text{same focal length in two cases})$$

The solution of the above set of equations would yield

$$u_1 = 20 \text{ cm}; \; v_1 = 50 \text{ cm}; \; u_2 = 50 \text{ cm and } v_2 = 20 \text{ cm}$$

Using the equation for focal length

$$\frac{1}{f} + \frac{1}{u} = \frac{1}{v}$$

and substituting the values of u_1 and v_1, or u_2 and v_2, and simplifying

$$f = \frac{50+20}{1 \times 20} \quad \text{or } 14.3 \text{ cm}$$

Referring to the figure again, and comparing the values of u_1, v_1, u_2 and v2, it follows that the position of object and image, that is distances OL_1 and L_2I, *are interchangeable in the case in question.*

4. A slide projector has a converging lens of focal length 20 cm. The projector is used to magnify the area of a slide, 5 cm^2, to an area of 0.8 m^2 on a screen.
Calculate the distance of the slide from the projector lens.

The ratio *area* of image/area of object (the slide) = $0.8 \text{ m}^2/5 \text{ cm}^2$
$= 8000/5$ or 1600

Then, *linear* magnification applicable to the lens = square root of *area* ratio = 40
Therefore $v/u = 40$ or $v = 40\,u$

Substituting in the lens equation with both u and v being real and positive

$$\frac{1}{40u} + \frac{1}{u} = \frac{1}{+20}$$

yielding

$$u = 20.5 \text{ cm}$$

Ray diagrams: step-by-step method[7]

Convex Lens

As an example to illustrate the method, the description below pertains to drawing a ray diagram for an object located *beyond* the 2F point of a (double) convex lens.

1. Pick a point on the top of the object and draw three incident rays travelling towards the lens as shown.

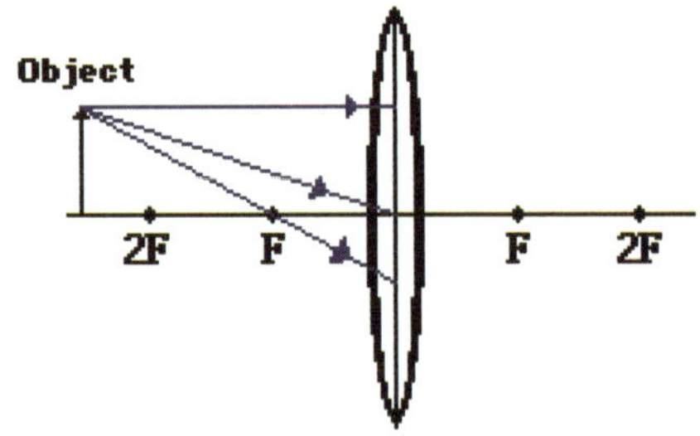

2. Once these incident rays strike the lens, refract them according to the *three rules of refraction* for converging lenses.

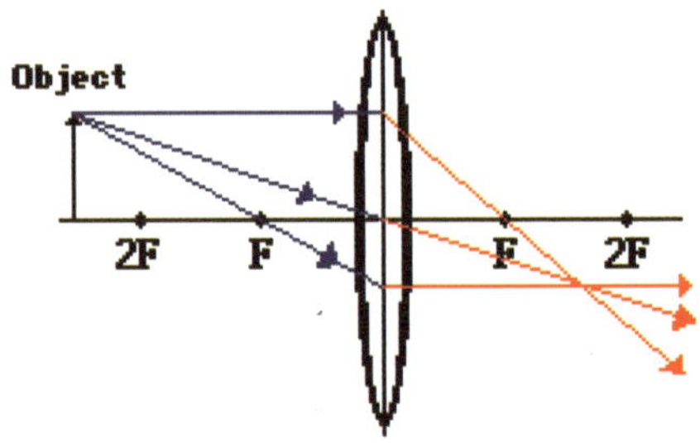

3. Mark the image of the top of the object. This point is located where the three rays intersect each other, point M in the adjoining figure. This point denotes the point where all light from the top of the object would intersect upon refracting through the lens. Of course, the rest of the object has an image as well and it can be found by applying the same three steps to another chosen point.

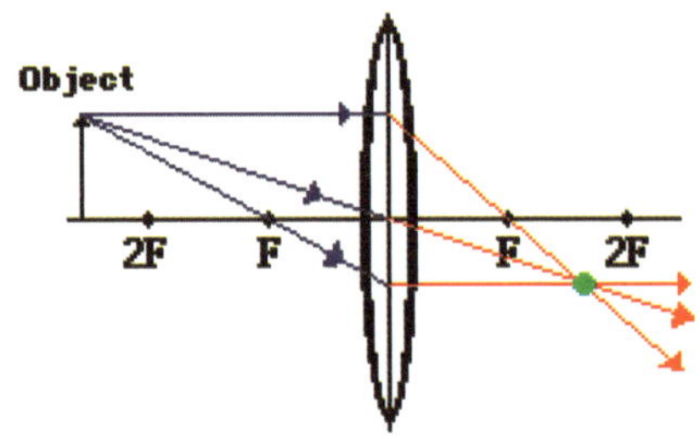

4. Repeat the process for the bottom of the object. The ray diagram if drawn properly would lead to determine the location, size, orientation, and type of image that is formed by the double convex lens. Typically, this requires determining where the image of the upper and lower extreme of the object is located and then tracing the entire image. After completing the first three steps, only the image

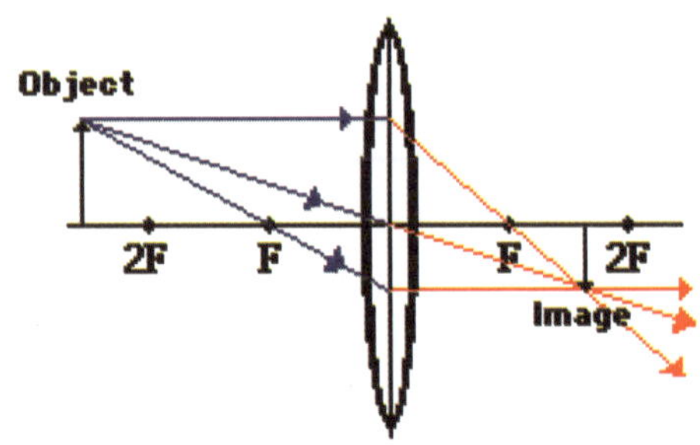

---

[7]Recall the three standard aspects of refracted rays through a (*thin*) lens discussed earlier.

location of the top extreme of the object has been found. Thus, the process must be repeated for the point on the bottom of the object. If the bottom of the object lies upon the principal axis (as it does in this example where the object is simply represented by an arrow), then the image of this point will also lie upon the principal axis, as indicated, and be the same distance from the optical centre.

Concave Lens

The method is similar to that described for the convex lens

1. Pick a point on the top of the object and draw three incident rays travelling towards the lens as indicated in the adjoining figure.

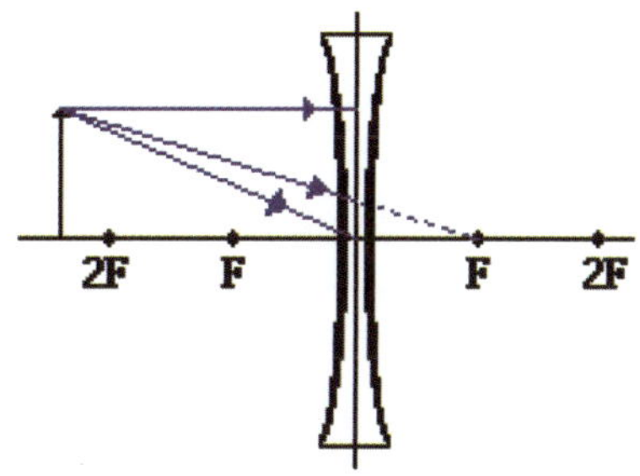

2. Once these incident rays strike the lens, refract them according to the *three rules of refraction* for double concave lenses.

The ray that travels towards the focal point will refract through the lens and travel parallel to the principal axis. The ray that travelled parallel to the principal axis on the way to the lens will refract and travel forward in a direction such that when extended back it would pass through the focal point, F, on the object's side of the lens. The ray that travelled to the optical centre of the lens will continue to travel in the same direction without deviation.

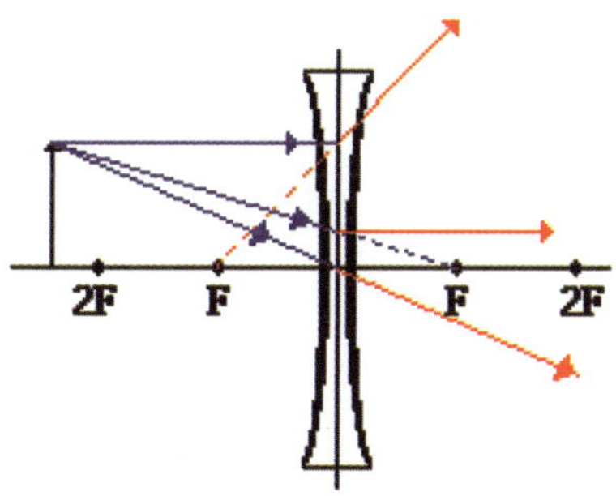

Arrowheads upon the rays indicate their direction of travel. The three rays should be *diverging* upon refraction.

3. Locate and mark the image of the top of the object.

The image point of the top of the object is the point where the three refracted rays intersect. Since the three refracted rays are diverging, they must be extended back behind the lens in order to intersect. The point of intersection is the image point of the top of the object. The three refracted rays would appear to diverge from this point. This is merely the point where all light from the top of the object would appear to diverge from after refracting through the double concave lens. The rest of the object has an image as well and it can be found by applying the same three steps to another chosen point.

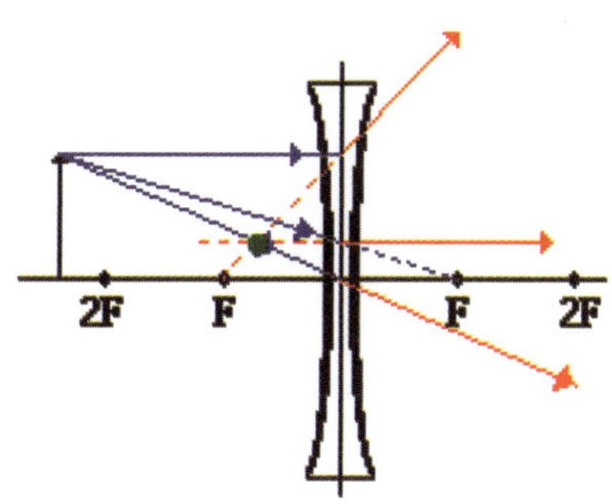

4. Repeat the process for the bottom of the object

As for the convex lens, the object of a ray diagram is to determine the location, size, orientation, and type of image that is formed by the double concave lens. Typically, this requires determining where the image of the upper and lower extreme of the object is located and then tracing the entire image. Thus, the process must be repeated for the point on the bottom of the object. If the bottom of the object lies upon the principal axis (as it does in this example), then the image of this point will also lie upon the principal axis and be the same distance from the lens as the image of the top of the object.

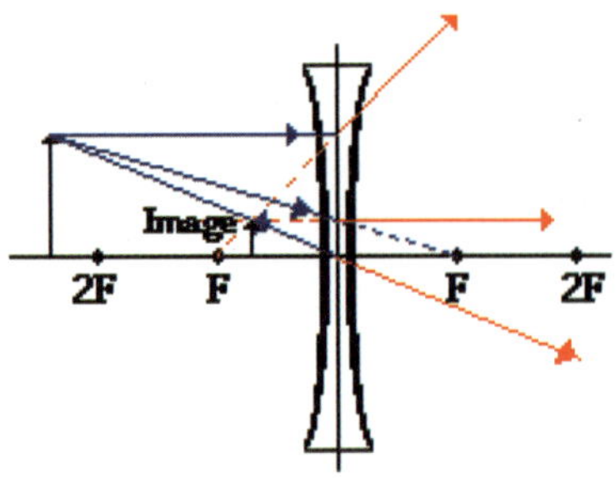

Example

A lens produces a virtual image between the object and the lens.
(i) Name the lens
(ii) Draw a ray diagram to show the formation of this image

[ICSE, 2016]

(i) Concave lens
(ii) The ray diagram is drawn as shown in the adjoining figure

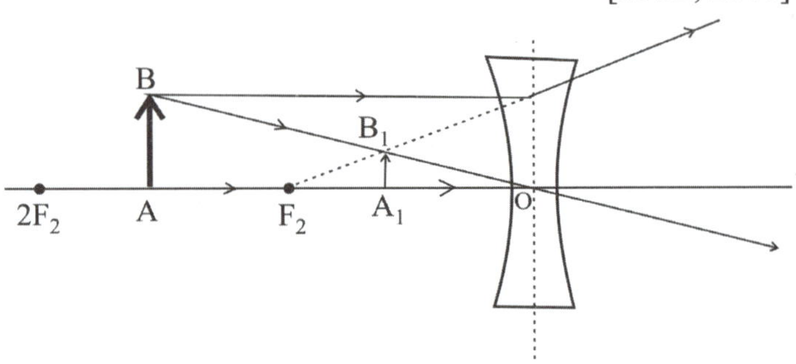

Power of a Lens

Power of a lens is the ability of the lens to converge the rays of light falling on it. It is defined as the reciprocal of its focal length in metre. The SI unit of the power of a lens is dioptre.

Whenever a ray of light passes through a lens (except when it passes through the optical centre), it bends. As discussed, the bending of light rays towards the principal axis is called convergence and bending of light rays away from the principal axis is called divergence. The degree of convergence or divergence of a lens is expressed in terms of its power. A lens of short focal length deviates the rays more while a lens of large focal length deviates the rays less. Thus,

$$\text{power of a lens} = \frac{1}{\text{focal length in metre}}$$

For example, if the focal length of a lens is, f = 1 m, its power would be given by P = 1/f = 1/1 = 1 dioptre. Or

$$1 \text{ dioptre} = \frac{1}{\text{metre}}$$

$$1\text{D} = 1\text{m}^{-1}$$

The power of a convex lens is positive and that of a concave lens is negative.

Example

A lens forms an erect, magnified and virtual image of an object.
(i) Name the lens
(ii) Draw a labelled ray diagram to show the image formation
(iii) If the focal length of the lens is 25 cm, calculate power of the lens

[ICSE, 2014]

(i) Such an image is formed by a convex lens
(ii) The ray diagram is as shown

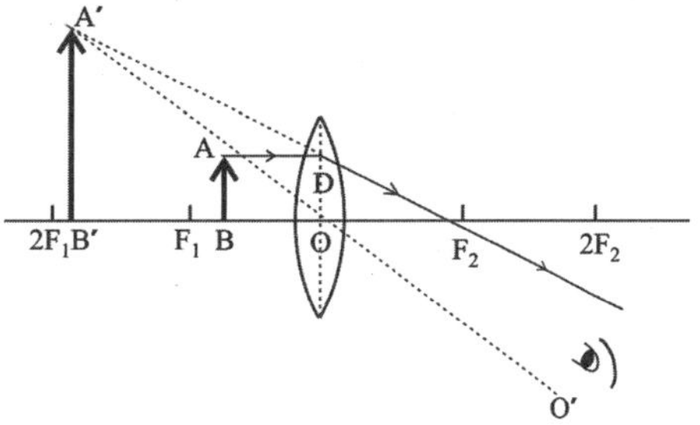

(iii) The power of a lens is defined as the reciprocal of its focal length in metre
The focal length of the above lens is 25 cm

$\therefore$ power of the lens = 1/0.25 or 4 dioptres

MAGNIFYING GLASS OR SIMPLE MICROSCOPE

The apparent size of an object depends upon its "visual angle" that is subtended on the eye; the greater the visual angle, the larger would be the apparent size of the object as observed by the eye. In order to increase the visual angle, the distance between the object and the eye must be as small as possible. However, a normal human eye cannot see an object clearly if it is closer than the 'least' distance for distinct vision (that is without a blur) which is very nearly 25 cm.

In a number of cases, one wishes to see the details of smaller objects; for example, a watch maker looking into a wristwatch, a finger-print expert studying finger prints etc. To help them in such applications, a convex lens of short focal length is most suited to see the details of an object within 25 cm. Such a lens is called a **magnifying glass** or **simple microscope.**
Thus, a magnifying glass (also called a hand lens in laboratory contexts and in general usage) is a *convex* lens that is used to produce a *magnified image* of an object. The commonly used magnifying lens is usually mounted in a frame with a handle as illustrated in Fig.3.16; also shown are two examples of magnification.

(a) a magnifying lens

(b) magnification I

(c) magnification II

Fig.3.16 : Action of a magnifying lens

A magnifying glass produces an upright, magnified *virtual* image. The virtual image produced is on the same side of the lens as the object. For a magnified image to be observed the distance between the object and the lens must be shorter than the focal length of the lens.

Magnification

In general, the magnification of a lens can be expressed as

$$\text{magnification} = \frac{\text{height of image}}{\text{height of object}}$$

As this is a ratio between heights, it has no units. A magnification of 2 means the image is twice the size of the object and a magnification of 1 indicates an image size being the same as the object size.

Magnifying Power of Magnifying Lens

This can be derived by reference to the ray diagram of a typical magnifying glass as shown in Fig.3.17.

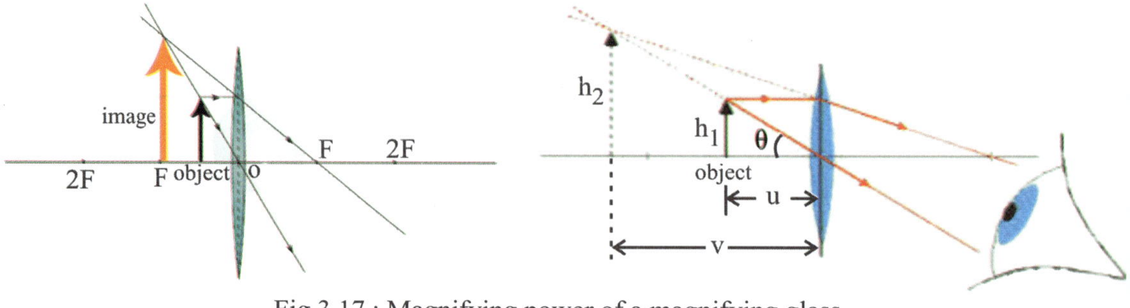

Fig.3.17 : Magnifying power of a magnifying glass

In the figure the object is seen by the eye as an image *on the same side as the object* and enlarge appropriately depending on the position of the object *between* the left-side principle focal point, F, and optical centre, O, and the focal length of the lens.

$$\text{magnification} = \frac{\text{height of image}}{\text{height of object}} = \frac{v\tan\theta}{u\tan\theta} = \frac{v}{u}$$

where

θ is the angle subtended by the object and the image at the optical centre as shown; θ being common as both the object and the image are on the same side, and
u and v denote the distance of the object and the image, respectively, from the optical centre; note that when used in the lens equation, v will be negative.
Thus, using

$$\frac{1}{f} = \frac{1}{u} - \frac{1}{v}$$

where f is the focal length of the lens.

On elimination, u is obtained as $u = \dfrac{fv}{v+f}$

whence magnification $= \dfrac{v}{u} = 1 + \dfrac{v}{f}$

implying that a lens of short focal length yields higher magnification, in general.

Applications of Lenses

Amongst very numerous and extensive applications of lenses – both convex and concave – some of the common ones are

- "glasses" or spectacles - or indeed just 'lenses', called "contact" lenses – for humans for correction of eye sight or simply for cosmetic purpose[8]
- cameras, esp. the "conventional" ones
  usual type or high speed
- magnifying glass(es)
- "eye" pieces (for watch makers)
- microscopes: simple and compound

[8]The normal eye sight of a person is referred to as "6/6". This means that the person is able to read without difficulty any combination of alphabets of 6 mm height from a distance of 6 m. A person not being able to do so is known to be "short-sighted" or "myopic"; he/she would have to be provided with spectacles (or contact lenses) of appropriate focal length to 'correct' his/her eye sight. The persons who are unable to *read*, say a book, but have no problem about distant vision are treated differently and are provided another class of spectacles or "reading glasses".

- telescopes
- assorted "projection" equipment such as cine, slide or LCD projectors
- data storage media
- electrical and electronic sensors and measurements
- various optical devices in eye hospitals; for example an ophthalmoscope
- spectroscope

IV Electromagnetic Spectrum

Using a triangular prism to produce a visible spectrum from white light; electromagnetic spectrum; scattering of light
Deviation produced by a triangular prism; dependence on colour (wavelength) of light; dispersion and spectrum; electromagnetic spectrum: broad classification (names only arranged in order of increasing wavelengths); properties common to all electromagnetic radiation; properties and uses of infrared and ultraviolet radiation; simple application of scattering of light; e.g. blue colour of the sky.

Deviation Produced by a Triangular Prism

Dispersion

Constituents of White Light

White light constitutes a band of wavelengths of different colours known as **spectrum** of white light. The constituent colours vary from red to violet in a given order. The longest wavelength is of red light, having a wavelength of about 700 nm or 0.7 μm (micron) in air, or rather vacuum. The shortest wavelength is that of violet colour being about 450 nm or 0.45 μm.

In vacuum (or practically in air), all the colours 'travel' at the same speed – the speed of light. In a medium other than air, for example glass, however, the colours travel at different speeds; red travelling the fastest and violet the slowest. Since refraction in the medium is dependent on the speed of a wave – red to violet in this case, so when a ray of white light is incident, say at the point O, on a glass prism as shown in Fig.4.1, the various colours of light are refracted in different direction such as OBR – OCS etc.[1]. The glass prism has thus 'separated' or *dispersed* the white light into its

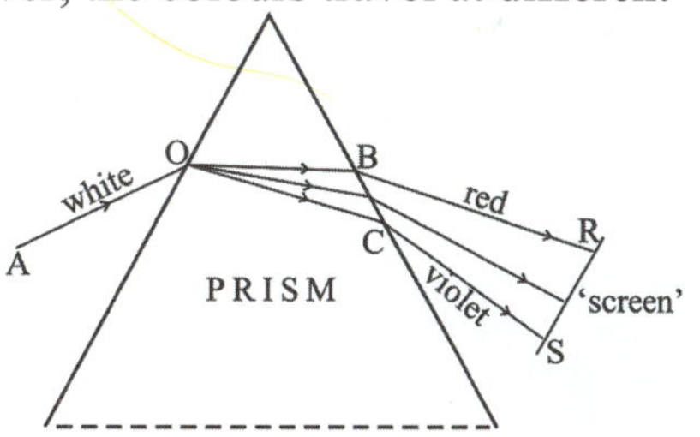

Fig.4.1 : Refraction of white light through a prism

[1]Note that this phenomenon is a consequence of tapered profile (or cross section) of a typical triangular prism, being broad at the base and reducing to a 'knife' edge at the top. Accordingly, when a ray of light is incident on a surface of a flat glass slab, the is refracted *as a whole* without splitting into different colours.

constituent colours or wavelengths. This phenomenon is called **dispersion.**[2]

Two images of dispersion of white light are illustrated in Fig.4.2(a) and (b)

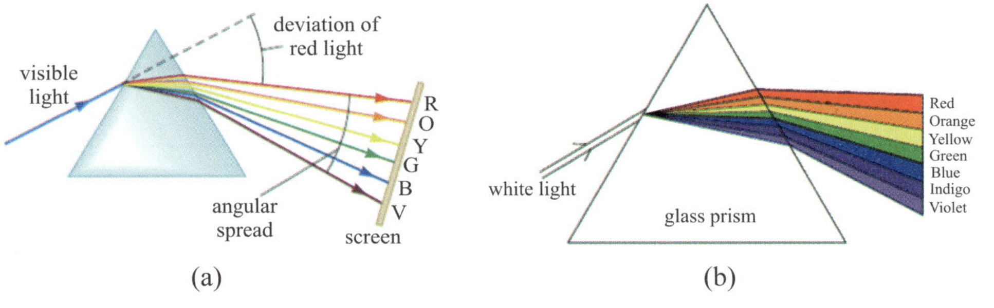

Fig.4.2 : Images of dispersion of white light

Deviation of Light

The degree of bending of light as it emerges from the prism following refraction is called the **deviation** and is measured with reference to the (extended) direction of the incident ray. This is shown for the refracted red light in the image (a) in Fig.4.2(a). Observe that the deviation, or rather the angle of deviation, is colour dependent for the dispersion of light through the prism. For a given angle of incidence, the deviation is small for red colour, increasing progressively for violet colour.

Experimental Determination of Deviation

The deviation of the refracted light from a prism can be obtained experimentally using simple means like a white paper and a few drawing pins. The procedure in detail is described in detail in Appendix A.

Electromagnetic Spectrum

The band of colours of white light as it emerges from the prism, shown in of Fig.4.2(b), is generally known as spectrum of (white) light. More specifically, the distribution of electromagnetic radiation in which the parts are arranged according to wavelengths would be called a spectrum.

Thus, white light comprises a band or spectrum that corresponds to a set of wavelengths and hence frequencies such that the product

wavelength (of a given colour), λ x its frequency, f = a constant

the constant being the speed of light in vacuum (or air), denoted by the letter c, numerically equal to 3×10^8 m/s(very approximately).

---

[2]First discovered by Newton in year 1666.

Clearly, knowing either the wavelength or frequency, the other quantity (frequency or wavelength can be derived using the above expression.

The full spectrum

Broad Classification

The full, 'classified' spectrum of electromagnetic radiation (or waves) is given in Fig.4.3.

From within a very broad range of electromagnetic waves (shown in the figure), light constitutes a narrow band or spectrum as marked in Fig.4.3. See also Fig.1.1, Chapter I.

Colours of Sunlight and Wavelengths

The six constituent colour waveforms of (white) sunlight are represented in terms of their individual waveforms and corresponding frequencies as detailed below[3].

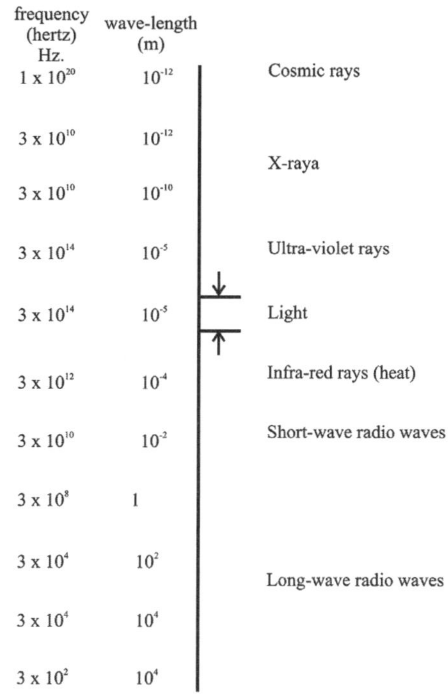

Fig.4.3 : Electromagnetic waves or radiation

Colour	Wavelength, nm (approx.)	Frequency, THz (approx.)
Red	700 - 635	430 - 480
Orange	635 – 590	480 – 510
Yellow	590 – 560	510 – 540
Green	560 – 520	540 – 580
Blue	490 – 450	610 – 670
Violet	450 – 400	670 - 750

[3]In addition to the six colours listed here, another colour, Indigo, is mostly assumed to be present in-between violet and blue, having a wavelength of 445 nm (approx.). This is particularly associated with the colours of a typical rainbow where the colours are identified by the acronym VIBGYOR, referring to first letter of the seven colours in a sequence from violet to red.

'Addition' of colours

An interesting aspect of the various colours comprising the white sunlight is the addition or mixing of two or more colours to obtain colours different from the original colours. Adding red to green yields yellow; adding all three primary colours together yields white.

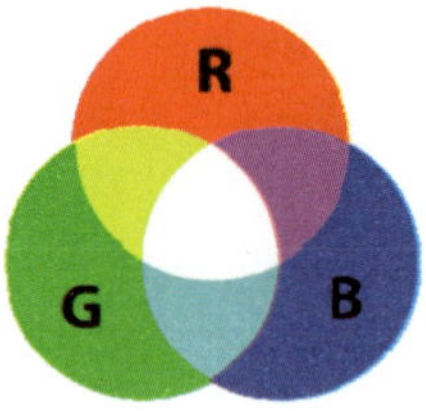

Fig.4.4 : Addition of colours

This is illustrated in Fig.4.4.

Properties Common to All Electromagnetic Radiations

The complete spectrum or extent of electromagnetic waves or radiation is given in Fig.4.5 Although specific radiations, for example radio waves, micro waves, infrared, visible light, ultraviolet rays, and x-rays etc. are characterised by special features of each, there are some common properties of *all* electromagnetic waves.

These are

- Electromagnetic waves comprise *oscillating* (that is time-varying) electric and magnetic fields, being mutually at right angles to each other[4].
- Owing to their oscillatory propagation as above, all electromagnetic waves are known as **transverse** waves.
- Electromagnetic waves need **no medium** of propagation[5].
- All waves travel with a constant velocity of **3×10^8 ms^{-1},** that is, speed of light in vacuum.
- Electromagnetic waves are **not** deflected by electric or magnetic field.
- All electromagnetic waves possess the property of exhibiting **interference** or **diffraction**.
- Electromagnetic waves may be **polarised.**
- All electromagnetic waves follow the same expression relating their wavelengths to frequencies, viz.,

$$c = f \times \lambda$$

where c is the speed of light (in m/s), f the frequency in Hz (cycles/s) and λ the wavelength (in m) of the given electromagnetic wave.

---

[4]The direction of propagation is at right angles to either the electric or magnetic field associated with the wave; the direction of *energy flow* contained in the wave coincides with the direction of propagation. Clearly, the phenomenon in entirety is three dimensional – an important physical activity.

[5]However, the age-old concept of propagation through "ether" is sometimes ascribed to electromagnetic waves.

Properties and uses of ultraviolet radiation

Properties

- **Ultraviolet** rays pertain to in the electromagnetic radiation beyond violet (colour), having a wavelength of 10 AU to about 4000 AU (or 10 nm to 400 nm), corresponding to frequencies above 7.5×10^{14} Hz (approx.), AU or Angstrom Unit being a commonly employed unit for wavelength, equal to 10^{-10} m . [See Fig.4.5].

The amount of UV radiation reaching the earth varies depending on the directness of the sun's rays, time of day, time of year, the location's latitude, cloud cover, and the amount of dust, haze and pollution in the air.

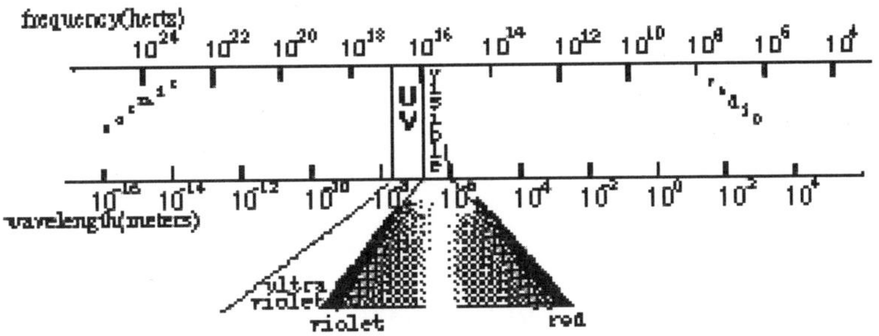

Fig.4.5 : Ultraviolet radiation

- UV radiation is absorbed by media such as glass, but pass through substance like quartz – a property utilised to obtain UV radiation from its source through a quartz screen that would not allow ordinary light to pass.
- UV rays follow the usual laws of refraction and reflection through appropriate media.
- The radiation is chemically more active compared to, say, ordinary light and hence can easily affect 'photographic' plates or film – a property that is of particular significance in special photography.
- They possess a special property of causing fluorescence – an important radiation property[6].
- The UV radiation can travel in straight line at the speed of light without hindrance through some substances (like quartz) that block ordinary light or radiation.

Harmful Effects

Ultraviolet rays can reach a high enough level on earth to be harmful to plants, animals and humans. Some of the well-known harmful effects of UV radiations are

- Overexposure of UV radiation can cause "sunburn" – a most common hazard to working in the open, exposed to sun for long hours - or even skin cancer.
- Long-term exposure to UV radiation can lead to immune suppression, skin aging and eye diseases such as cataract and macular degeneration.
- In high-altitude countries, UV exposure can cause snow blindness.

---

[6]To observe this property experimentally, hold a fluorescent tube in front of a "Tesla coil", or any source of UV radiation, at a suitable distance; the tube would immediately light up when the coil is excited.

Uses

UV radiation, though harmful in some cases finds extensive use in many applications.

- Hospitals use UV lamps to sterilise surgical equipment and the air in operation theatres.
- Food and drug companies also use UV lamps to sterilise their products before marketing.
- Suitable doses of UV rays are helpful for the body to produce vitamin D, vital for the body to absorb or metabolise calcium, and this is used by doctors to treat vitamin D deficiency and some skin disorders in human body[7].
- To provide efficient illumination, street lights are arranged to use mercury vapour lamps which work on the production of UV rays[8].
- UV lamps of special design find use in various category of photography where light is made to expose special paper or films for desired results.
- UV radiation may be of use in detecting purity of items like gems and food material such as eggs.

Properties and uses of infrared radiations

Properties

- **Infrared** radiation (IR) is the invisible radiant energy, or the electromagnetic radiation with longer wavelengths than those of visible light, extending from the nominal red edge of the visible spectrum at 700 nm (frequency 430 THz) to 1 nm (300 GHz). See Fig.4.6.

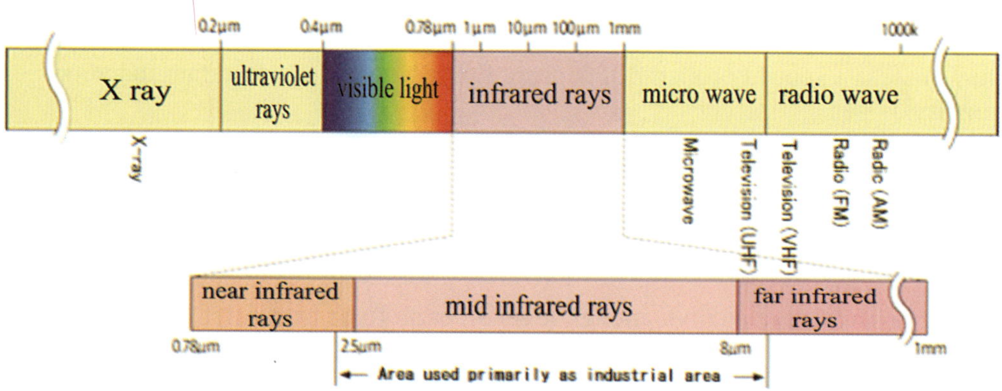

Fig.4.6 : Infrarcd radiation and visible spectrum

[7]Note that the most potent source of vitamin D is the sun and it would do well for one to expose to sunlight daily for a few hours to avoid any vitamin D deficiency. This is of particular significance in a tropical country such as India.

[8]These have been gradually replaced by slightly better sodium vapour lamps, esp. for express way or "motor"-way illumination. At present, the most modern means of street illumination employ LED lamps which are extremely efficient and much more lasting.

- Similar to UV rays, IR rays are also not visible to naked eye.
- Much of the energy from the Sun arrives on Earth in the form of infrared radiation.
- The balance between absorbed and emitted infrared radiation has a critical effect on the Earth's climate.

Hazards of Infrared Radiation

- Strong infrared radiation can cause blindness to the eye by damaging the retina or cornea depending on the intensity of radiation. To protect eyes from IR radiation which is invisible, it is essential to use IR-proof goggles whilst working with infrared rays.
- When using infrared radiation for therapy where muscle pain and other pains are treated, care has to be taken to regulate the intensity of the radiation, otherwise it can cause severe burns.
- Elderly people can be prone to low blood pressure if exposed to infrared radiation.
- If exposed to infrared radiation for a long time, the body can lose fluid or water and this can result in dehydration.

Uses

Infrared radiation has extensive uses and application in daily life and industry.

■ Medical applications

A thermal imaging technique called digital infrared thermal imaging (DITI) based on infrared radiation is used for diagnosis in the medical field. This process makes use of thermal imaging cameras that record thermograms depending on the infrared radiation emitted by the body. These find application in detecting breast cancer, inflammation in the body, thyroid conditions, screening for viruses etc. These cameras may also be used to observe changing blood flow in the skin.

Infrared radiation is used in thermotherapy which finds wide application in the treatment of high blood pressure, arthritis, heart failure, chronic fatigue, stress, toxicity, insomnia, pain relief, high cholesterol, injuries, diabetes, muscular pain etc., with the help of infrared saunas. It is also used in physiotherapy and in treatments of cancer.

Specially designed infrared lamps, sources of IR radiation, are widely used in medical field to provide therapeutic treatment to patients using heating produced by the lamps. One such device is illustrated in Fig.4.7.

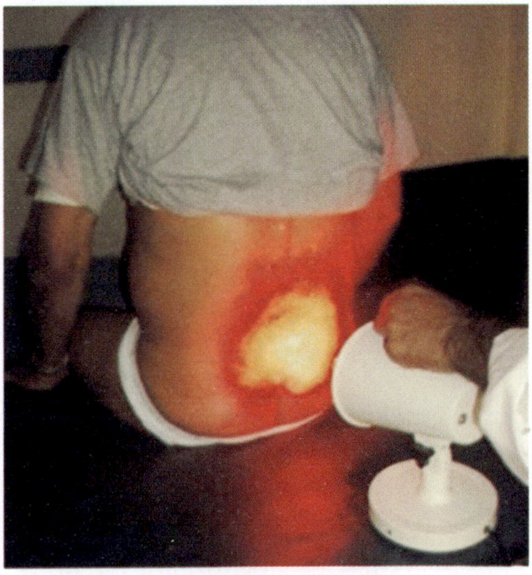

Fig.4.7 : Application of infrared lamp

■ Industrial applications

• Infrared heating: Infrared radiation can be used as a heating source in numerous applications. Some of these are

- Infrared light is often used to heat food; special lamps that emit thermal infrared waves are used in fast food restaurants to maintain food items hot at appropriate temperature.
- Using special lamps, IR heating is being employed in "infrared saunas" to 'heat' the occupants for possible treatment of chronic health problems, such as high blood pressure, congestive heart failure and rheumatoid arthritis.
- Infrared heating is successfully used for de-icing of aircraft wings, to remove ice from the wings.
- IR heating is used in curing of (paint) coatings in, for example, car manufacturing industry, forming of plastics, annealing, plastic welding, print drying etc.
- Infrared (heated) panels are used to emit infrared radiation to heat up rooms or interiors in houses/premises in cold climates.

• Communication: IR data transmission is employed in short-range communication among computer peripherals and personal digital assets. These devices use infrared light-emitting diodes (LEDs) to emit infrared radiation which is focused by a plastic lens into a narrow beam. Infrared is the most common way for remote controls to command appliances such as TVs, DVD players etc.

• Photography: In infrared photography, infrared filters are used to capture the near-infrared spectrum. Digital cameras often use infrared blockers.

• Astronomy: Astronomers observe objects in the infrared portion of the electromagnetic spectrum using optical components, including mirrors, lenses and solid state digital detectors. For this reason it is classified as part of optical astronomy.

• Infrared Thermography: This is a branch of science that deals with acquiring and analysing thermal information from remote thermal imaging devices. This is used to determine the temperature of the objects remotely and is used in military and industries. This works on the principle that infrared radiation is emitted by all bodies and these radiations are detected by thermography cameras.

■ Military

• Tracking using infrared: This is used on missile guidance system where the target is tracked using the infrared radiation it emits. For example, an infrared emitter (*transmitter*) is placed on the object that needs to be tracked and signals/radiation from this emitter are received by infrared receivers that are connected to a computer system with display unit, that will display the location of the object.

■ Miscellaneous

- Used in night vision devices to observe animals or people using the phenomenon called infrared illumination. The observant will not know that they are being observed.
- Used in infrared lasers that produce infrared beams that can be used to read information on compact discs.
- Used in infrared imaging devices that are used to obtain images of infrared radiation that are reflected or produced by objects, for example, heat insulation in buildings, sources of thermal pollution, mapping vegetation on the earth's surface.
- Used in the chemical industry to determine molecular structure and composition of substances by studying their infrared emission or absorption.

Scattering of Light

Scattering is the process by which "small particles suspended in a medium of a different index of refraction diffuse a portion of the incident radiation in all directions." With scattering, there is no energy transformation, but a change in the spatial distribution of the energy.

Scattering of light can also be explained as the deviation of light rays from its straight path. As light propagates through the atmosphere, it travels in a straight path until it is obstructed by bits of dust or gas molecules.

The process of scattering is illustrated by the images in Fig.4.8.

Note that scattering is not limited to light. Scattering can happen to many types of energy waves and is more pronounced in certain media.

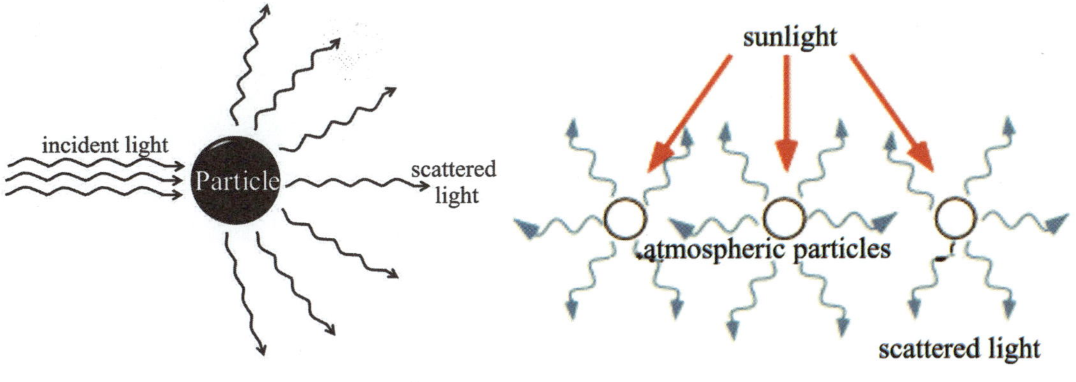

Fig.4.8 : Principle of scattering of light and scattered light through atmosphere

Simple applications

The phenomenon of scattering of light is fundamental to various application in general visual observations, science and industry. Some of these are

- The most significant application or outcome of scattering of light, for example sunlight, is to lead one to "see" various colours such as colour of sky at sunrise or sunset, or normal, blue sky.
- A process called "laser light scattering", the technique of scattering of light (not necessarily the sunlight) by the use of laser(s) finds extensive application in biotechnology; for example for characterisation of protein molecules.
- The principle of light scattering is applied in microbiology; for example for elucidating the conformation of biomolecular systems in solutions.
- In some chemical industries, the phenomenon is used for determining coating of paint and its exact (colour) matching – a complicated requirement by itself.

Blue colour of the sky

This is by far the most illustrious example of scattering of sunlight during day time; note that the sky would be all black at night.

The blue colour of the sky results from the scattering of sunlight off the molecules of the atmosphere. The scattering, called "Rayleigh scattering"[9] , is more effective at short wavelengths – the blue end of the visible spectrum, that is light. Therefore, the light scattered down towards the earth at a larger angle with respect to the direction of the sun's light is predominantly in the blue end of the spectrum.
[See Fig.4. 9].

Note that the blue of the sky is more "saturated" when looked further from the Sun. The almost white scattering near the sun is not very wavelength dependent[10].

Fig.4.9 : Scattering of sunlight and blue sky

[9]Named after the inventor, Lord Rayleigh, a British physicist who invented the phenomenon circa 1871.

[10]An important parallel to Rayleigh scattering is Raman scattering, invented by Indian physicist C.V.Raman in 1928 that deals with yet another phenomenon of scattering of light, mainly in liquids. Raman effect deals with change in the *wavelength* of light that occurs when a light beam is deflected by molecules. This would explain why sea water is generally (deep) blue – an observation that puzzled Dr Raman during his voyage to Europe.

Appendix A

Experimental determination of deviation in a triangular prism

A sheet of white paper is fixed on a drawing board using drawing pins. A glass prism is placed on it in such a way that it rests on its triangular 'base'. The outline of the prism is traced using a pencil. A straight line PE is drawn inclined to one of the refracting surfaces, AB, of the prism. Two pins are fixed at points P and Q, on the line PE as shown in the adjoining figure. The images of the pins, fixed at P and Q are looked through the other face AC of the prism. Two more pins are fixed at points R and S, such that the pins at R and S and the images of the pins at P and Q lie on the same straight line.

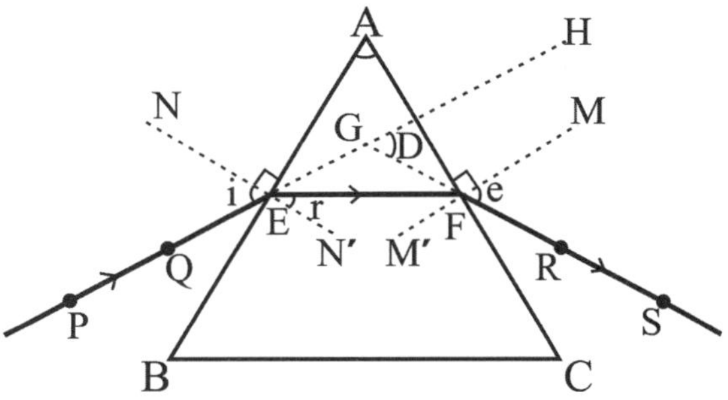

The pins and the glass prism are removed. The points P and Q are joined and the line is extended till it meets the nearest boundary of the prism. Similarly, the points R and S are joined and produced back. The lines should meet the boundary of the prism at E and F, respectively. E and F are joined. Perpendiculars are drawn to the refracting surfaces AB and AC of the prism at points E and F, respectively. The angle of incidence (angle i), the angle of refraction (angle r) and the angle of emergence (angle e) are marked as shown.

It is seen that a ray of light enters from air to glass at the first surface AB. This light ray on refraction bends towards the normal. At the second surface AC the light ray enters air from glass and bends away from normal. These rays obey the laws of refraction of light. But emergent ray is not parallel to the incident ray.

The angle between the direction of the incident ray and that of the emergent ray is called angle of deviation. As marked in the figure, angle D is the angle of deviation.

3. Sound

Unit 3: Topics covered

Sound

(i) Reflection of Sound Waves; echoes: their use; simple numerical problems on echoes.

Production of echoes, condition for formation of echoes; simple numerical problems; use of echoes by bats, dolphins, fishermen, medical field. SONAR.

(ii) Natural vibrations, Damped vibrations, Forced vibrations and Resonance – a special case of forced vibrations.

Meaning and simple applications of natural, damped, forced vibrations and resonance.

(iii) Loudness, pitch and quality of sound:

Characteristics of sound: loudness and intensity; subjective and objective nature of these properties; sound level in db (as unit pitch and frequency; quality and waveforms (with examples).

I Reflection of Sound Waves

Reflection sound waves; echoes: their use; simple numerical problems on echoes. *Production of echoes, condition for formation of echoes; simple numerical problems; use of echoes by bats, dolphins, fishermen, medical field; SONAR.*

SOUND

Sound: what is it?

Sound can be identified as a form of energy, similar to light; although with many characteristic differences.

Production of Sound

Sound is generally produced by 'things' or bodies when they *vibrate*, that is when they move back and forth at a given pace – slow or fast. For example, banging a drum makes its tight skin vibrate, at very high speed (so fast that it cannot be easily seen or perceived), forcing the air all around it to vibrate in turn. As the latter moves, it carries energy out of the drum in all directions. Eventually, as the vibrating air enters human ear and strikes the ear drum, one begins to perceive the sound of the vibrating drum.

The other simple examples are a person talking (involving movement of lips and tongue), a dog barking or the cone of a loudspeaker vibrating to reproduce the sound input to its electromagnet coil. These are illustrated in images of Fig.1.1 where Fig.1.1(a) shows the oscillogram of human voice recorded on a cathode ray oscilloscope, CRO.

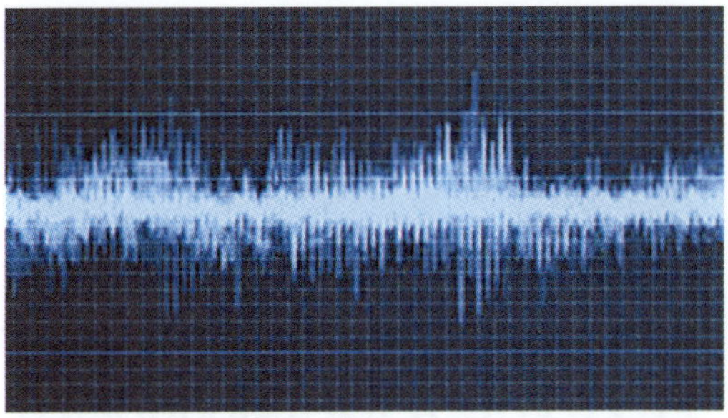

(a) voice of a person recorded on a CRO

The images of barking sound of a dog and sound emanating from a loudspeaker are shown in Fig.1.1(b) and (c).

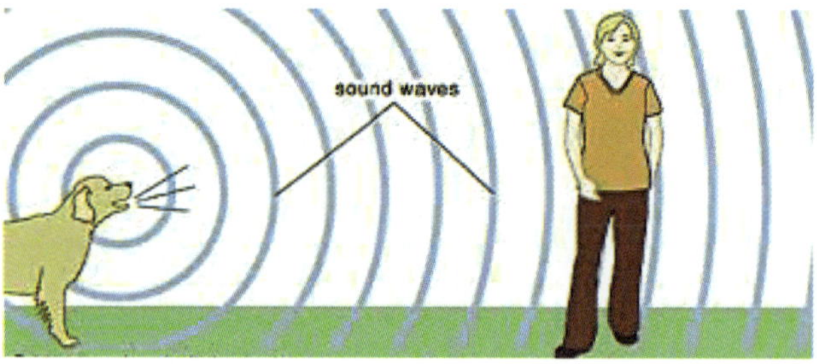

(b) a dog barking

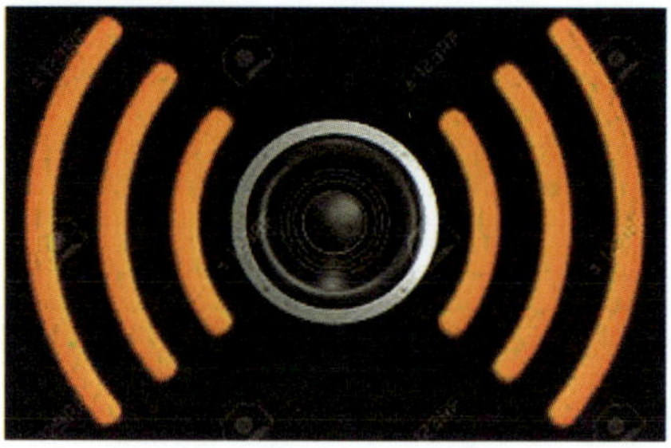

(c) sound reproduced by a loudspeaker

Fig.1.1 : Production of sound from different sources

In short, there are two different aspects to sound: there's a *physical process* that produces sound energy and transmits it to air (or other media) and there is a separate '*psychological*' process that happens inside the ears and brains that leads to interpretation of sound as noise, speech or music etc.

Sound Waves

Sound waves are characterised by

- longitudinal

or

- transverse

waves that move through air or other materials or media.

Sound does not travel in a vacuum.

Longitudinal Waves

In longitudinal waves the displacement of the medium is *parallel* to the propagation of the wave as depicted in Fig.1.2.

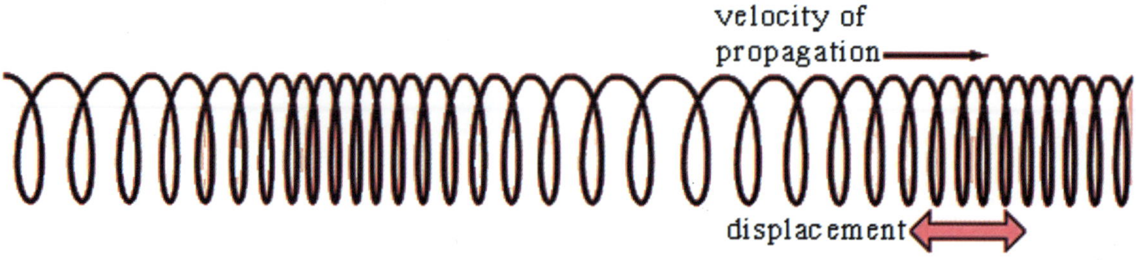

Fig.1.2 : A longitudinal wave

Transverse Waves

The propagation of sound in solids and liquids usually takes place by *transverse* waves. A transverse wave is characterised by the displacement of the medium perpendicular (or at right angles) to the direction of propagation of the wave. A ripple in a pond or a wave on a (vibrating) string are typical examples of transverse waves. A graphic representation of such a wave is shown in Fig.1.3.

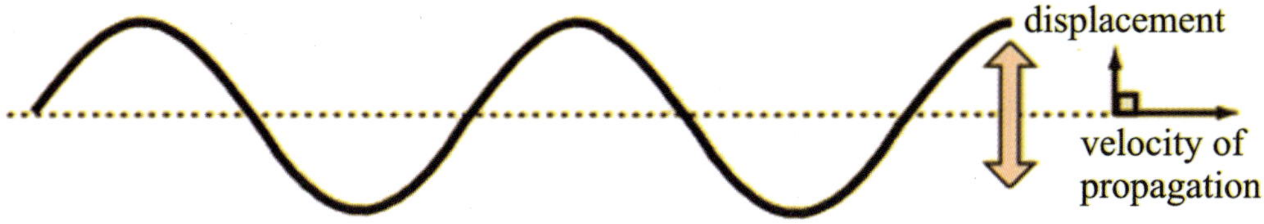

Fig.1.3 : A transverse wave

Sine Wave

A sine wave can be reckoned as the 'pure' or 'fundamental' form of wave or variation as shown in Fig.1.4. Any wave, or rather waveform, can be represented in terms of a fundamental sine wave PLUS more components called "harmonics", individually being sine waves themselves. All (sound) waves are characterised by an amplitude and wavelength or time period as shown for a pure sine wave in Fig.1.4[1].

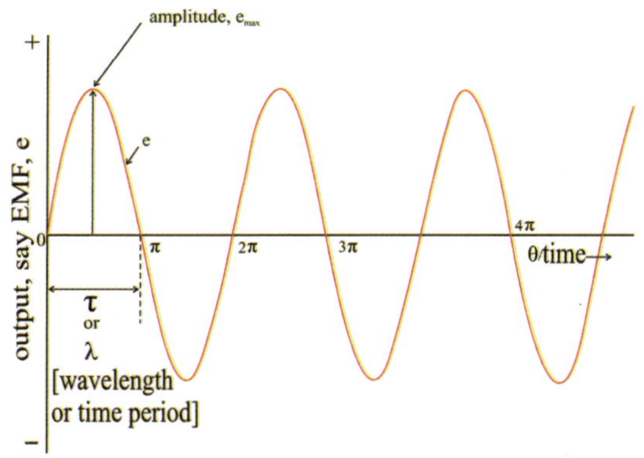

Fig.1.4 : A sine wave, amplitude and wavelength (or time period)

Frequency of sound waves

The vibrations representing sound (waves) usually comprises a band or spectrum of frequencies, NOT just one frequency, superposed on each other to produce a given sound; for example, voice of an individual [see Fig.1.1(a)], chirping of a particular bird or barking of a dog. Each individual frequency of the band may itself be a sine waveform of a given amplitude *and frequency*, expressed by 1/λ where λ is the wavelength or time period as illustrated in Fig.1.4.

---

[1]A sine wave is one that follows a sine relationship such that the amplitude of the waveform is given by the sine of the angle along the time axis (the abscissa); for example at angle equal to 90°, the amplitude is maximum whilst at angle equal to 30°, the amplitude is half, and so on.

Human Ear and Sound

The ears of a human (and other animals) are sensitive detectors capable of detecting the fluctuations in air pressure caused in wound waves or energy that impinge upon the eardrum. [See also Chapter III]. A *normal* human ear is capable of detecting sound waves with a wide range of frequencies, ranging *theoretically* between approximately 20 Hz to 20 000 Hz; in practice, threshold for most individuals may be 50-100 Hz in the lower range and about 15 kHz on the higher side.

Infrasound and Ultrasound

Any sound with a frequency, or a composite band of frequency, below the audible range of hearing to humans, that is less than 20 Hz is known as **infrasound** and any sound with a frequency above the audible range of hearing, that is, above 20 kHz is known as **ultrasound**. Many animals have the ability to detect sounds bordering on ultrasound; for example

- dogs can detect sounds of frequencies 50 Hz – 45 kHz
- cats can be sensitive to sounds of 45 Hz – 85 kHz
- bats can detect sounds of frequencies as high as 10 kHz, and are nocturnal creature
- dolphins can detect sounds of frequencies as high as 200 kHz.

In contrast, elephants have the distinction of sensitive to sounds of infrasound frequency of as low as 5 Hz and up to 10 kHz.

Speed of Sound

Sound travels at different speeds in different media. In air at 20 °C, sound travels at 344 m/s. This figure varies with temperature and pressure. In water at 20 °C, sound travels at 1498 m/s. In general, the more incompressible a medium is, the *faster sound will travel in that medium*. If an object travels faster than the speed of sound in a particular location and medium, say air, that object is said to travel at **supersonic** speed.

Speed of Sound in Different Media

The speed of sound in some common media are listed in the table below.

Material/medium	Speed of sound at 20 °C, m/s
Air	344
Hydrogen (at 0°C)	1284
Water ("pure")	1284
Aluminium	5000
Iron (or steel)	5120
Glass	5170

To realise that sound can travel through a solid medium such as steel, sound of running wheels of a train can be heard by placing ears on the rail track away from the wheels. Similarly, the travel of sound through liquid can be verified by immersing a ringing bell under water when the sound of the bell can be heard outside the water.

Difference between sound and light waves

Since both light and sound represent forms of energy and associated with unique waves, it is important to observe how the two types of wave differ from each other.

	Sound waves	Light waves
1.	sound cannot travel in a vacuum	light waves can travel through vacuum
2.	sound is the transfer of kinetic energy	light is the transfer of electromagnetic energy
3.	sound waves are in general longitudinal waves	light waves are essentially transverse waves
4.	sound waves travel at 'slow' speed(s) [only about 350 m/s in air]	light waves travel at tremendous speed of 3×10^8 m/s
5.	sound waves cannot travel far as their enormous energy is dissipated easily	light waves can travel through distances without apparent loss of energy [recall that sunlight reaches earth having travelled millions of kilometres through space]
6.	the speed of sound increases as the waves move from a lighter to a denser medium [for example from air to steel]	the speed of light decreases as the waves move from a lighter to a denser medium [for example from air to glass]

ECHOES

Acoustic or sound waves would travel in air when produced from any source by virtue of vibrations of a given amplitude and band of frequencies. However, if the waves 'suddenly' reach or strike an obstruction, for example a wall, a mountain cliff or any hard surface, the waves experience a reflection from the surface and arrive at a place, distant from the surface, where a listener may be standing. The reflected wave(s) of the original sound as heard by the listener is called **echo**[2]. Clearly, there would be a time delay in the instant of origin of the sound and the instant it is heard by the listener, the delay being dependent on, or a function of, the speed of sound and the distance of the reflecting surface from the source (of the sound) and the listener. The reason for the reflection may be explained as the discontinuity in the propagation medium. A schematic illustration of the formation of echo is given in Fig.1.5.

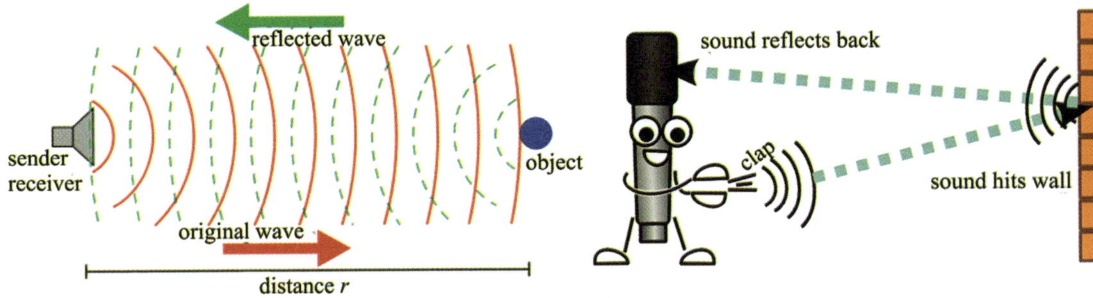

Fig.1.5 : Schematic of echo formation

Typical examples of the phenomenon are the echo produced by the bottom of a well (when the sound originates at the top of the well), by a building or by the walls of an enclosed room.

True Echo

A *true* echo is one that returns with a *single* reflection from the distant surface *without attenuation*; in practice, the reflected wave may undergo some attenuation through travel in the medium – esp. when large distance is involved – or there may be a little absorption of original sound on reaching the obstruction.

Reverberations

Many a times, the sound of origin may undergo *multiple* reflections owing to the situation or configuration of obstructions, for example the location of the original source of sound and listener being surrounded by cliffs. This would result in echoes that are known as **reverberations**.

---

[2]Observe that the word "echo", too, is frequently used in colloquial English to mean "imitate the opinion etc. of a person in authority", sometimes with "blind" faith!

Production of echo

The echo of a sound wave is produced when

- a source results in a sound wave somewhere with a 'reasonable' intensity
- the sound wave travels in a 'free' medium as a longitudinal wave
- the wave reaches and strikes an obstruction where it may be partly absorbed, mostly reflected back
- the listener located anywhere in the medium, usually near to the source of origin of the sound, would hear the echo of the sound wave

Conditions for Echo Formation

The formation of echo and it being heard by the listener is dependent on a number of conditions.

- The distance between the sound source and the reflecting surface must not be less than 17 metres where the time period between hearing the original sound and its echo should not be less than 0.1 of a second

- The human ear cannot distinguish between two successive sounds if the period between them is less than 0.1 second

Thus, the critical distance of 17 m can be derived from the knowledge of speed of sound in air, being about 340 m/s, and the to-and-fro distance between the location of source of sound and the obstruction causing reflection of sound.
Let this distance be d m. Then, to-and-fro distance would be 2 x d m and the time taken by the sound wave to travel this distance would be

$$(2 \times d)/340 \text{ s}$$

Considering the critical time of 0.1 s for a person to listen an echo, the distance d works out to be 17 m.

■ It follows that the echo cannot be heard if the distance between the source of sound and the reflecting surface is less than the critical distance of 17 m because the time between the commencement of the origin sound and its echo would be less than the threshold time of 0.1 s for human ear.

■ The reflecting surface of the obstruction, for example a cliff, must be 'big' and wide so as to reflect the sound wave completely.

■ When the distance between location of the origin of sound and the reflecting surface is multiples of 17 m (that is, twice or three times), the echo would be heard in the form of the *last two or three phases* of the sound produced as a whole.

Some "Famous" Echoes

- Hamilton Mausoleum, Hamilton, South Lanarkshire, Scotland: Its high stone holds the record for the longest echo in the world, taking 15 seconds for the sound of a slammed door to decay.
- Gol Gumbaz of Bijapur, India: Any whisper, clap or sound gets echoed repeatedly.
- The Golkonda Fort of Hyderabad, India
- The Echo Wall at the Temple of Heaven, Beijing, China
- The Whispering Gallery of St Paul's Cathedral, London, England
- Echo Point, the Three Sisters, Katoomba, Australia
- The Temple of Kukulcan, Chichen Itza, Mexico

Applications and uses of echoes

Determination of Speed of Sound (in air)

The method consists of

- producing a sound of 'good' intensity (preferably of a single note of, say, 1 to 5 kHz), directed towards an obstruction, for example a wide and 'high' wall to reflect the sound wave completely

- measuring the distance between the source of sound and the wall accurately; say, equal to D m

- using an accurate stop watch, with a resolution of 0.1 s or better, noting the time from the instant of production of sound to the instant of listening its echo from the wall, being T s

  The speed of sound is then simply given by

$$v = 2D/T \quad \text{m/s}$$

Check that it turns out to be 340 m/s, approximately[3].

Example

1. A person standing between two vertical cliffs and 480 m from the nearest cliff shouts. He hears the first echo after 3 s and the second echo 2 s later.
Calculate
(i) The speed of sound
(ii) The distance of the other cliff from the person.

[ICSE, 2015]

---

[3]The methodology of computing distance of an object based on travel of sound and its echo is called **ranging.**

Assume that the person is standing between two cliffs such that one of the cliffs nearer to him is to the left and the farther one is to the right.

(i) After his shout, the first echo is heard from the nearest cliff on his left.

Let d_1 be the distance of the person from the left cliff. Then distance travelled by sound (or its echo) in going to and coming back from the left cliff is

$$S_1 = 2 \times d_1 = 2 \times 480 \quad \text{or} \quad 960 \text{ m}$$

Time taken by sound is 3 s

$\therefore$ speed of sound $\quad v = $ total distance/time taken

$$= 960/3 \quad \text{or} \quad 320 \text{ m/s}$$

(ii) The second echo is heard from the farther cliff on the right

Let distance of this cliff from the person be d_2. Then distance travelled by the sound (or its echo) in this case is $2 \times d_2$

The total time taken is $t_2 = 3 + 2 = 5$ s

Here

$$v = \frac{2d_2}{5} = 320$$

whence $\quad d_2 = 800$ m

2. A man standing in front of a vertical cliff fires a gun. He hears the echo after 3 seconds. On moving close to the cliff by 82.5 m, he fires again.

This time, he hears the echo after 2.5 seconds. Calculate:

(i) the distance of the cliff from the initial position of the man

(ii) the velocity of sound

[ICSE, 2007]

Let the distance of the cliff from the man from his initial position be D m

Let the velocity of sound (in air) be v m/s

Then, in the first case

$$3 = (2 \times D)/v$$

or

$$v = (2 \times D)/3 \qquad (1)$$

In the second case when the man has moved towards the cliff by 82.5 m, the distance is

$$D' = D - 82.5$$

Then, time being distance/velocity

$$2.5 = 2 \times (D - 82.5)/v \qquad (2)$$

Solving eqns. (1) and (2)

$$D = 495 \text{ m} \quad \text{and} \quad v = 330 \text{ m/s}$$

Other Applications

These include

- To measure the depth of the sea
  This is achieved by sending a concentrated sound wave down a ship and measuring the time taken by the echo as illustrated in Fig.1.6.

- To locate a sunken ship wreck or cargo
- To locate submarines from surface ships
- Using fishing ships or boats, to detect large shoals of fish
- In medical field use for medical diagnosis and therapy and also as a surgical tool to detect defective foetus, as a tool in the treatment of muscular pain or to locate the exact position of an eye tumour

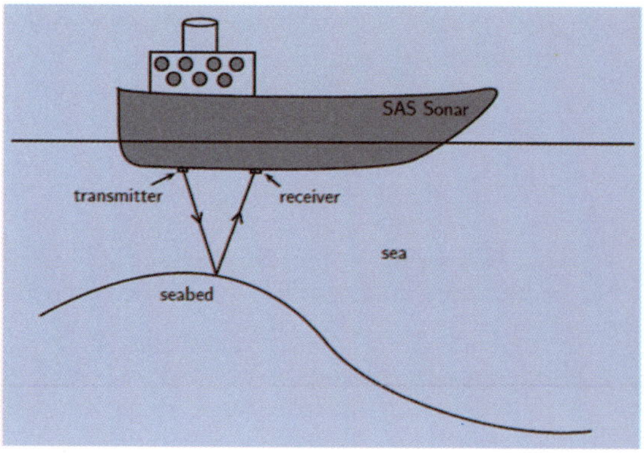

Fig.1.6 : Measuring depth of sea

Example

An ultrasonic wave is sent from a ship towards the bottom of the sea. It is found that the time interval between the sending and receiving of the wave is 1.5 second. Calculate the depth of the sea if the velocity of sound in sea water is 1,400 m/s.

[ICSE, 2009]

The distance covered by the wave in travelling to the bottom of the sea and returning to the ship is

$$v \times t \text{ or } 1{,}400 \times 1.5 \text{ m}$$

This must equal twice the depth of the ship

$\therefore$ depth of the ship = (1,400 x 1.5)/2 or 1,050 m

Use of echoes by bats, dolphins and fishermen: echolocation

■ Bats: Although bats have good eye sight, they depend on echolocation to navigate as well as capture their food. Bats who invariably dwell in the dark for reason, of safety etc. use the phenomenon of echo to find prey like insects in the dark. Most bats make calls higher in pitch up to 100 kHz that results in an echo. The sound is made by bats' noses or sometimes through their mouth. When the sound bounces off the object and echo of the sound returns to the bat, the time delay helps it process how far away the object is. Even more amazing, however, is the way bats can determine which *direction* the object is moving and how big it is.

The bat can even tell the direction the object is moving about by comparing whether the sound reaches its right or left ear first. In addition to the object's horizontal position, the bat can determine if the object is above or below it by using special folds on its outer ear. The bat can also determine the size of the object by the *intensity* of the echo since a smaller object will reflect the sound wave less and the returning echo will be softer. If the echo is louder, the object is obviously larger.

■ Dolphins: Dolphins produce sounds by squeezing air through their nasal passages. They then focus on sounds they make with their melon. The melon is a fat-filled area on the dolphin's forehead that acts like a lens to focus on sound waves ahead of the dolphins. The sound waves given are then reflected off the objects and received through the dolphin's lower jaw, mainly to give away the location of the object.

■ Fishermen: Fish finders, in a ship, operate at high frequencies of sound, approximately 20-200 kHz. They transmit sound pulses, measure the time it takes for echoes to return, and calculate the distance to the objects. Fish finding transducers located at the bottom of the ship send and receive signals many times per second. They concentrate sound into a beam that may appear as a 'cone'. These units include visual displays that print the echoes. The bottom appears as a continuous line drawn across the display. In addition, any objects that are in the water between the surface and the bottom may also be displayed.

The presence of fish is detected primarily by detecting the air in the fish swim bladders. The air conserved in the swim bladder changes the sound path and reflects energy back. The fish finder detects this reflected energy and converts it into fish images on the screen.

A typical arrangement is illustrated in Fig.1.7.

Specially designed transducers send sound waves or signals into the water that rebound when they strike an object. The fish reflects some of the signal back to the boat, the remainder of the signal continues to the seafloor then it bounces back to the boat.

Fig.1.7 : A ship and location of fish

Ultrasound

Ultrasound is sound with a frequency higher than 20 kHz, the frequency range going up to several hundred kilohertz. This is above the upper threshold of human range of hearing.

Applications of Ultrasound

Ultrasound frequency waves find numerous applications owing to some unique properties such as

- they can be produced as a narrow beam using specially designed transducers or otherwise
- when of high frequency, they can travel without deviation in a straight path: both outward and reflected, thus resulting in 'sharp' echo
- if of sufficient intensity, they can travel long distance without much attenuation
- they may not undergo much absorption at the reflecting surface or in the medium
- they would still travel at the same speed through the medium as the sound waves of audible frequency, thus being amenable to computations; the process known as *sound ranging*

The most common use of ultrasound, creating images, has industrial and medical applications. The use of ultrasound to create images is based on the reflection and transmission of a wave at a boundary. When an ultrasound wave travels inside an object that is made up of different materials (such as the human body), each time it encounters a boundary (e.g., between bone and muscle, or muscle and fat), part of the wave is reflected and part of it is transmitted. The reflected rays are detected and used to construct an image of the object.

SONAR

SONAR is an acronym for **S**Ound **N**avigation **A**nd **R**anging and relates to travel of sound and its echo in a given medium. The 'term' is now invariably is used in the context of numerous techniques and process(es) of measurement utilising the production of ultra-high frequency sound waves, or ultrasound, and their reflection, that is echo; for example deducing the depth of objects deep in the sea, from the knowledge of speed of sound in (sea) water.

Medical Ultrasound

A direct application of the concept of SONAR is in "medical ultrasound", also known as **diagnostic sonography** or **ultrasonography.** This is a diagnostic imaging technique based on the application of ultrasound, now widely in use in diagnostic labs. to see internal body structures such as tendons, muscles, joints, vessels and internal organs. Its aim is often to find a source of a disease or to exclude any pathology. The practice of examining pregnant women using ultrasound is called obstetric ultrasound and is commonly used to check the development and health of the foetus.

Ultrasonic images also known as sonograms are made by sending pulses of ultrasound into tissue using a special hand-held probe. The sound echoes off the tissue; with different tissues reflecting varying degrees of sound. These echoes are recorded and displayed as an image to the operator.

A sonogram of a foetus is depicted in Fig.1.8.

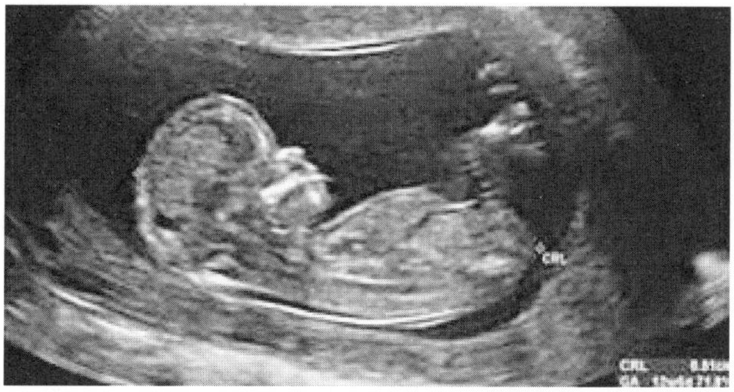

Fig.1.8 : The sonogram of a foetus obtained by sonography

Anechoic Chamber

An **anechoic chamber** ("an-echoic" meaning non-reflective, non-echoing or echo-free) is a room designed to completely absorb reflections of either sound or electromagnetic waves. They are also insulated from exterior sources of noise. The combination of both aspects means they simulate a quiet open-space of *infinite* dimension, which is useful when exterior influences would otherwise give false results.

Anechoic chambers were originally used in the context of acoustics (sound waves) to minimize the reflections of a room, now being used to test antennas, radars, or electromagnetic interference.

Anechoic chambers range from small compartments the size of household microwave ovens to ones as large as aircraft hangars. The size of the chamber depends on the size of the objects to be tested and the frequency range of the signals used[4].

[4]Special measures are taken in cinema halls and auditoria to eliminate echo sounds from the loudspeakers or public address systems [PAS] used in the premises.

EXERCISES

1. A radar sends a signal to an aircraft at a distance of 30 km away and receives it back after 2×10^{-4} s. What is the speed of the signal?

[ICSE, 2014]
[3×10^{8} m/s

2. An observer stands at a certain distance away from a cliff and produces a loud sound. He hears the echo of the sound after 1.8 s. Calculate the distance between the cliff and the observer if the velocity of sound in air is 340 m/s.

[ICSE, 2013]
[360 m

3. A man standing between two cliffs produces a sound and hears two successive echoes at intervals of 3 s and 4 s, respectively. Calculate the distance between the two cliffs. The speed of sound in air is 330 m/s.

[ICSE, 2012]
[1,155 m

4. A man standing 25 m away from a wall produces a sound and receives the reflected sound.
(i) Calculate the time after which he receives the reflected sound if the speed of sound in air is 350 m/s
(ii) Will the man be able to hear a distinct echo?

[ICSE, 2011]
[0.14 s; yes since the distance is more than 17 m

5. A man stands at a distance of 68 m from a cliff and fires a gun. After what time interval will he hear the echo, if the speed of sound in air is 340 m/s?

[ICSE, 2010]
[0.4 s

6. Calculate the minimum distance at which a person should stand in front of a reflecting surface so that he can hear a distinct echo. (Take speed of sound in air = 350 m/s).

[ICSE, 2009]
[17.5 m

7. A radar sends a signal to an aeroplane at a distance 45 km away with a speed of 3×10^{8} m/s. After how long is the signal received back from the aeroplane?

[ICSE, 2008]
[3×10^{-4} s

II Vibrations

Natural vibrations, damped vibrations, forced vibrations and resonance – a special case of forced vibrations. *Meaning and simple applications of natural, damped, forced vibrations and resonance,*

Vibrations: what are these?

Vibration is a mechanical phenomenon whereby oscillations occur about an *equilibrium* 'state'[1]. The oscillations may be periodic, such as the motion of a pendulum, or random, such as the movement of a tyre on a gravel road. A common example in physics class rooms is the vibrations produced by a tuning fork when once struck. The prong(s) of the fork vibrate for considerable time, to and fro about its position at 'rest', that is, when the prong is at rest. This is depicted in Fig.2.1.

The phenomenon of sound and vibration is closely related. Sound, or pressure waves, are generated by vibrating structures (for example, vocal cords); these pressure waves can also induce the vibration of structures (for example, ear drum).

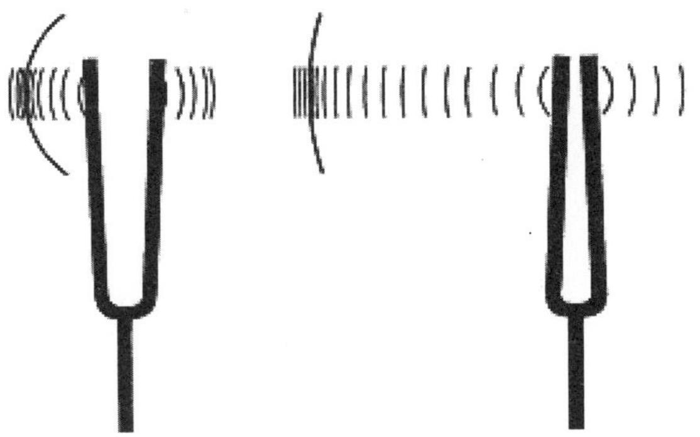

Fig.2.1 : A vibrating tuning fork

In general, the word "vibration" may be referred to

- an instance of vibrating; for example, "powerful vibrations from an earthquake"; synonyms: quiver, quivering, shake, shaking, tremble, tremor, shudder, shuddering
- in informal context: "a person's emotional state"; for example, 'I picked up no unusual vibrations as to the envelope's contents'.

[1]The word comes from Latin *vibrationem* (meaning "shaking, brandishing").

Desirable and Undesirable

Vibrations can be desirable: for example, movement of keys of a piano, the reed in woodwind instrument like a flute or the cone of a loudspeaker to reproduce music.

In many cases, however, vibrations may be undesirable, to be a nuisance, creating unwanted sound, generally known as "noise", and wasting 'energy'. For example, the vibrational motions of engines, electric motors or transformers or any mechanical device in operation. Such vibrations may usually be caused by imbalances in the rotating parts, friction in the meshing of gear teeth. Careful designs usually minimise unwanted vibrations.

Natural or Free Vibrations

In **free vibrations** the body at first is given an initial displacement and the force is withdrawn. The body starts vibrating and continues the motion of its own accord. No external force acts on the body further to keep it in motion. The frequency of free vibrations is known as *free or natural frequency*. Typical examples of free vibrations are a tuning fork, once struck; a simple pendulum given an initial push and left to oscillate or pulling a child back on a swing and letting go. Free vibrations are also termed "natural vibrations". See Fig.2.2 for images of free vibrations[2].

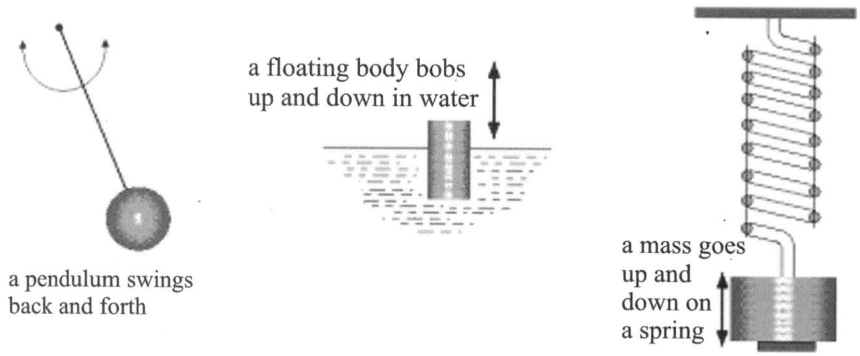

Fig.2.2 : Images of free vibrations

The free vibrations are known to follow *simple harmonic motion*, or SHM, the important feature of these being, when un-attenuated, the pattern of movement when plotted on a paper relating the 'position' of the vibrating body (for example that of the bob of a pendulum) at any instant *with time.* This would trace as a purely sinusoidal waveform of a given amplitude and frequency as shown in Fig.2.3.

---

[2]Note that when free to oscillate only in one direction about the mean position, the vibrations are characterised by *single degree of freedom.*

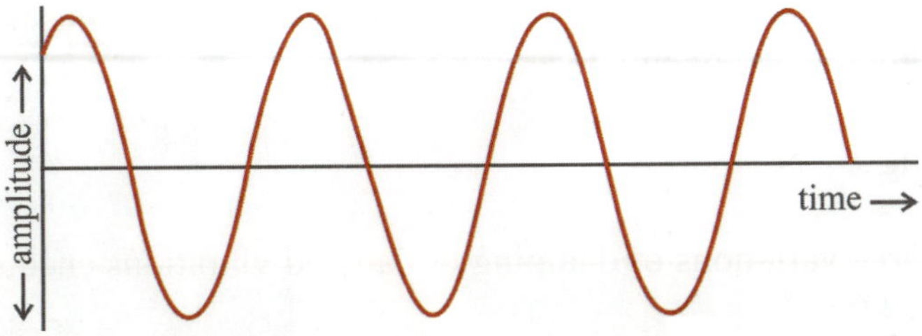

Fig.2.3. : A sinusoidal waveform representing simple harmonic motion

The free vibrations of an *elastic* body can further be of three types:

(a) Longitudinal vibrations: when the particles of the body move *parallel to the axis of the body.*
(b) Transverse vibrations: when the particles of the body move nearly *perpendicular to the axis of the body.*
(c) Torsional vibrations: when the particles of the body move in a *circle about the axis of the body*

See Figs.1.2 and 1.3, Chapter I for waves corresponding to vibrations of type (a) and (b).

An example of torsional vibrations is shown in Fig.2.4 in the form of a vibrating pendulum oscillating about its (vertical) axis; to an extent of θ_{max} on either side about point P.

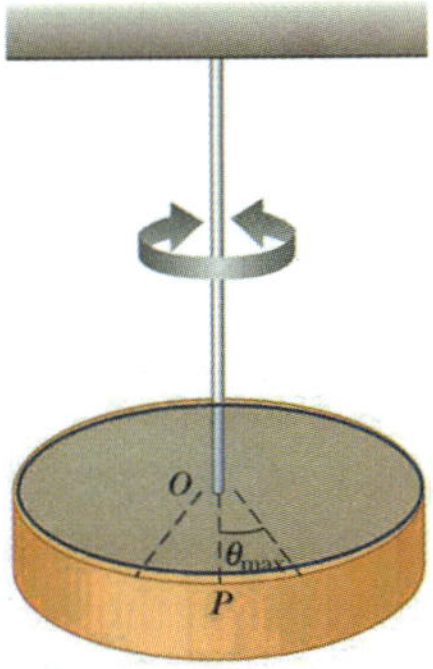

Fig.2.4 : A pendulum exhibiting torsional vibrations

Damped Vibrations

A universal feature of free vibrations is damping.
All systems are subject to frictional forces, and these steadily sap the energy of the vibrations, causing the amplitude to diminish with time, usually exponentially. The motion is therefore never precisely sinusoidal for long. Thus, a swinging pendulum, left un-driven, will eventually return to rest at the equilibrium or pre-swinging position. A graphic representation of a typical (sinusoidal) damped vibration is shown in Fig.2.5.

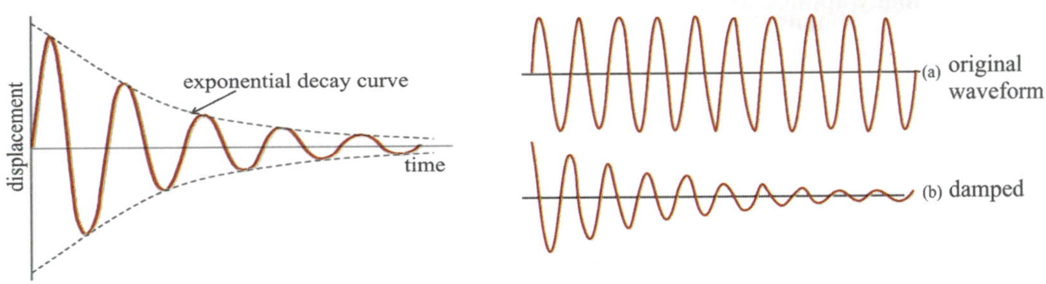

Fig.2.5 : Typical damped sinusoids

Observe the envelope of the decreasing amplitudes of successive sinusoids. Almost invariably, this follows an exponential variation.

Types of Damping

There are mainly *three* variations of damping or damped vibrations encountered in practice as illustrated in Fig.2.6.

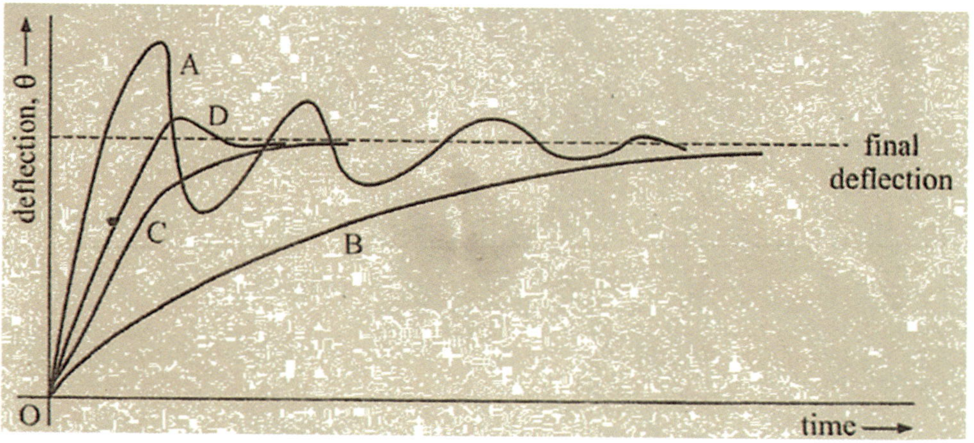

Fig.2.6 : Types of damped oscillation of a body

These are

- under-damped vibrations when the decrease of amplitudes is relatively slow, taking much longer time to decay to a given amplitude, curve A in Fig.2.6.
- over-damped vibrations when the amplitudes of oscillations decrease very rapidly, curve B[3] in the figure
- "critically"-damped vibrations when the decrement is 'just' right or optimal, curve C

In each case, observe the variation of deflection (or the "dependent variable") with time which relates to the analytical expression for obtaining the magnitude of the oscillation of the body at any instant. Curve D depicts a special case of damping, called "dead beat" employed in many applications such as a deflection-type measuring instrument – a voltmeter or an ammeter –where the pointer of the instrument deflects 'quickly' by a little more than the final deflection before coming to rest.[4]

---

[3]A simple example or demonstration of over-damping is suddenly holding the shell of a bicycle bell that has been ringing for a while when the bell is immediately silenced, as the vibrations die out quickly.

[4]See, for example,
S.C.Bhargava: Electrical Measuring Instruments and Measurements (book), B S Publishers, India, 2013.

Forced Vibrations

Forced vibrations occur when an *external* time-varying disturbance (load, displacement or velocity) is applied to a mechanical system that is free to vibrate. The disturbance can be a periodic and steady-state input, a transient input or a random input. The periodic input can be a harmonic or a non-harmonic disturbance. Examples of these types of vibrations include a washing machine shaking due to an imbalance, transportation vibrations caused by an engine on an uneven road, or the vibrations of a building during an earthquake.

Thus, the tendency of one object to force another *adjoining* or *interconnected* object into vibrational motion is referred to as a **forced vibration.**

Applications

- A Guitar

If a string of a guitar is stretched to a given length and tightness and plucked, a noise or sound would be heard, but the sound would be feeble, being barely audible. However, the same string when attached to a proper guitar having a "sound box" (the wooden, 'hollow' body of the instrument), a healthy sound is heard. This happens because the vibrating string forces sound vibrations in the sound box that, in turn, forces air particles inside the box into vibrational motion at the same natural frequency as that of the string. The entire 'system' comprising the string, the wooden box of the guitar and the enclosed air therein, begins vibrating, forcing the surrounding air particles into vibrational motion that reaches one's ears as the desired music. An image of a commonly used guitar is given in Fig.2.7.

Fig.2.7 : A guitar commonly used by musicians

The fact that the surface area of the sound box is greater than the surface area of the string means that more surrounding air particles will be forced into vibration. This causes an increase in the loudness of the sound.

- Experiment Using a Tuning Fork

A mounted tuning fork force more air particles vibrating, creating a louder sound.

This same principle of forced vibrations is often demonstrated in a physics classroom using a tuning fork. If the tuning fork is held in one's hand and hit with a rubber mallet, a sound is produced as the prong(s) of the tuning fork set surrounding air particles into vibrational motion. However, the sound produced by the tuning fork would be barely audible to students in the back rows of the room. If now the tuning fork is set upon the white board panel or simply on the wooden table top, the panel begins vibrating at the same natural frequency as that of the tuning fork, resulting in the tuning fork forcing the white board (or the table top) into vibrational motion. The vibrating white board in turn forces surrounding air particles into vibrational motion and the result is an increase in the amplitude and thus loudness of the sound. This principle of forced vibration explains why demonstration tuning forks are usually mounted on a sound box.

Resonance

In physics, resonance describes the phenomenon when a vibrating system or external force drives another system to oscillate with *greater* amplitude at a specific preferential frequency, usually the natural frequency.

Frequencies at which the response amplitude is a relative maximum are known as the system's **resonance** or **resonant frequencies** . At resonant frequencies, small periodic driving forces have the ability to produce large amplitude oscillations. This is because the system stores *vibrational* energy.

A Familiar Example

A familiar example of resonance in practice is a playground swing which acts as a pendulum. Pushing a person in a swing in time with the natural interval of the swing (its resonant frequency) makes the swing go higher and higher (maximum amplitude), while attempts to push the swing at a faster or slower tempo produce smaller arcs. This is because the energy the swing absorbs is maximised when the pushes are "in phase" with the swing's natural oscillations. See Fig.2.8.

Fig.2.8 : Swinging in a playground as example of resonance

Resonance in sound or acoustic resonance

When a sound of a given amplitude and (natural) frequency produces forced vibrations in another body, of the *same* natural frequency and usually of increased amplitude, the phenomenon is known as **resonance.** The frequency is identified as resonant frequency and forced vibrations as resonant vibrations.

A Demonstration

Consider, again, a tuning fork mounted on a sound box, set upon a table. Suppose a second tuning fork on a sound box, *having the same natural frequency*, is placed by the side of the first one. If the first tuning fork is struck with a rubber mallet, its prongs begin to vibrate at the given frequency, the sound being clearly audible owing to the sound box. The surrounding air particles are set into motion *at the same natural frequency* as that of the tuning fork. These vibrations are 'picked' by the second tuning fork and this, too, begins to vibrate at the same natural frequency and would continue to vibrate even if the first tuning fork stops vibrating, say by grabbing it by hand. This is an example of **resonance** when one object vibrating at the same natural frequency of a second object forces that second object into vibrational motion.

Resonating Tuning Forks

More Examples/Applications of Resonance

- "Inverted" pendulum

The end result of resonance is always a large vibration in the object or body affected by forced vibrations, regardless of the vibrating system. This can be demonstrated by means of an odd-looking mechanical system resembling an "inverted pendulum" as shown in the adjoining figure. The apparatus consists of THREE sets of two identical plastic bobs mounted on a very elastic metal pole, which are in turn mounted at the lower ends to a rigid support. Each metal pole and attached bob has a different length, thus imparting it a different natural frequency of vibration. The bobs are often coloured red blue and green to distinguish one from the other. The red bobs are mounted on the longer poles and they have the lowest natural frequency of vibration. The blue bobs are mounted on the shorter poles and have the highest natural frequency of vibration. When the red bob is 'disturbed', it begins vibrating at its natural frequency. This in turn results in the attached bar to vibrate at the same frequency and forcing the other attached red bob into vibrating at the same natural frequency. This is **resonance:** one bob vibrating at a given frequency forcing a second object with the same natural frequency into vibrational motion.

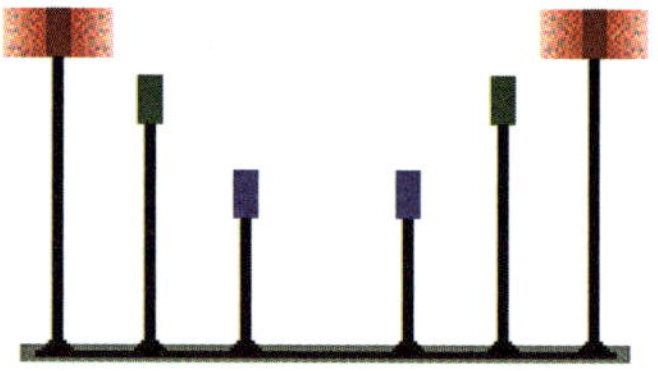

The vibrations of the red bob forces the other red bob into resonance - large vibrations at the same frequency.

Observe that the green and the blue bobs are unaffected by the vibrations of the red metal bar. This is because the frequency of vibration of the first red bob is tuned to the frequency of the second red bob; they share the same natural frequency. The result is that the second red bob begins vibrating with a huge amplitude.

- Resonance through air column

This demonstration of resonance involves a plastic tube containing an air column as shown in the adjoining figure. The length of the air column can be adjusted by raising and lowering a reservoir of water (dyed red). The raising and lowering of the reservoir controls the height of water in the open-air tube, and thus adjusts the length of the air column inside the tube. As the length of the air column is decreased, the natural frequency of the air column is increased and *vice-versa.*

While adjusting the height of the liquid in the tube, a vibrating tuning fork is held above the air column of the tube. When the natural frequency of the air column is *tuned* to the frequency of the vibrating tuning fork, resonance occurs and a loud sound results. In the resonance situation, the tuning fork and the air column share the same vibrational frequency.

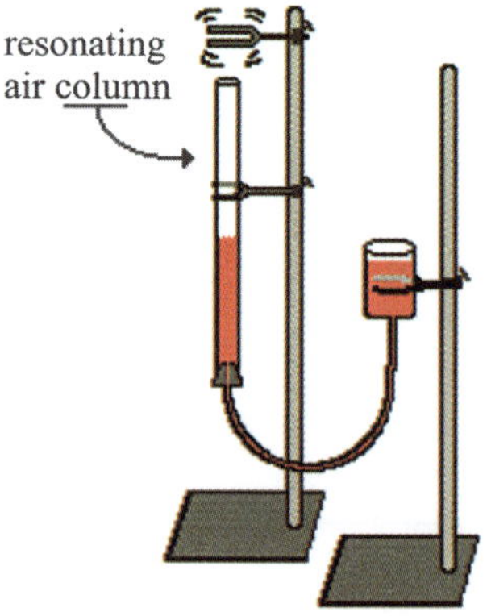

tuning fork forcing air coumn into resonance

Other types of resonance

Mechanical Resonance

Mechanical resonance is the tendency of a mechanical system to absorb more energy when the frequency of its oscillations matches the system's natural frequency of vibrations than it does at other frequencies. It may cause violent swaying motions and even catastrophic failure in improperly designed and constructed structures including bridges [see Fig.2.9], buildings, trains and aircraft.

Fig.2.9 : Mechanical resonance and a swinging bridge
[This bridge is moving because the wind velocity is matching natural frequency of the bridge]

Clocks keep time by mechanical resonance in a balance wheel, pendulum, or quartz crystal.

Electrical Resonance

Electrical resonance occurs in an electric circuit at a particular *resonant frequency* when the impedance[5] of the circuit is at a minimum in a series circuit or at maximum in a parallel circuit with corresponding circuit current.

An important application of electrical resonance is in the radio-frequency stage of a radio receiver where tuning of a desired radio station (broadcasting at a given "carrier" frequency), and therefore a quality reception of the audio, is effected by adjusting inductance of a coil or capacitance of a variable capacitor in a *parallel* circuit comprising an inductor and capacitor. See Fig.2.10(a).

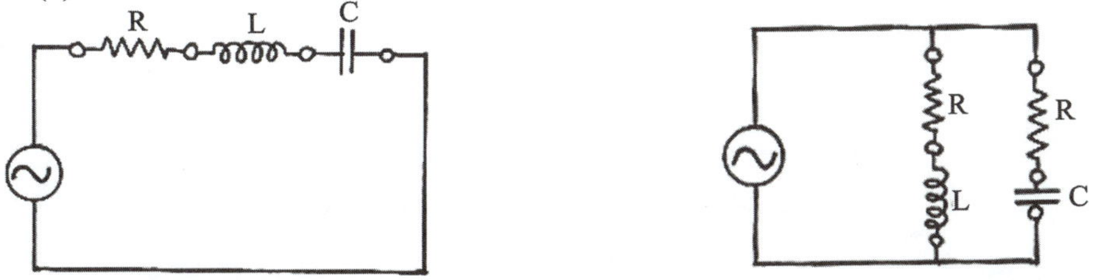

(i) series resonance [minimum impedance at resonance frequency]

(ii) parallel resonance [maximum impedance at resonance frequency]

(a) series, parallel electrical resonance circuits

[5]Akin to resistance in a DC circuit, impedance controls current in an AC circuit such that I = V/Z where V is the applied voltage, I the current and Z denotes the impedance of the circuit.

(b) resonant circuits can generate very high voltages as, for example, in a " Tesla" coil - a consequence of extreme resonance

Fig.2.10 : Electrical resonance

Nuclear Magnetic Resonance

Nuclear magnetic resonance (NMR) is the name given to a physical resonance phenomenon involving the observation of specific *magnetic* properties of an atomic nucleus in the presence of an applied, external magnetic field of *extremely* high strength. An important application of NMR in medical field for special, advanced diagnostic procedures is the "magnetic resonance imaging", or MRI, now routinely used in medical imaging.
The images produced in MRI imaging are far superior and clearer than those obtained using routine X-rays or even CT scans. The procedure is also much safer and 'non-invasive' as no X-rays are used in the process. A characteristic feature of the imaging machine is the use of very high magnetic field, of the order of several tesla, which makes it necessary to observe special precautions on the part of the patient during the procedure.

A typical MRI 'machine' is shown in Fig.2.11 whilst samples of images of a brain are depicted in Fig.2.12.

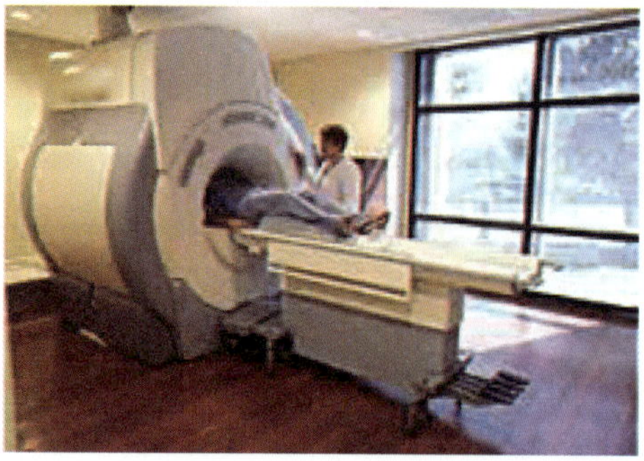

Fig.2.11 : A MRI machine

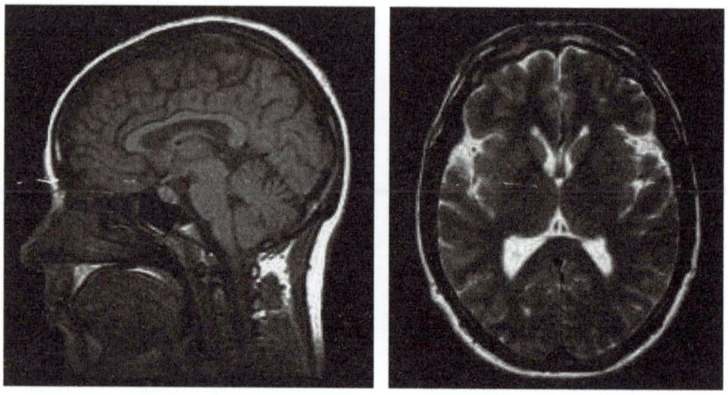

Fig.2.12 : Images of brain produced by MRI

III Loudness, Pitch and Quality of Sound

Characteristics of sound; loudness and intensity; subjective and objective nature of these properties; sound level in db (as unit only); noise pollution; interdependence of pitch and frequency; quality and waveforms (with examples).

Characteristics of Sound

A sound *wave* is characterised by

- Wavelength
- Amplitude
- Time period
- Frequency
- Velocity (Speed)

However, *sound* as perceived by living beings, *esp. humans*, has the following main characteristics

- Loudness and Intensity
- Pitch
- Timbre (or "quality" of sound)

Loudness of sound

The characteristic feature or concept of loudness of sound can be comprehended in many ways and is in general related to the degree of sensation of sound produced *in the ears of a particular person*. This is because no two persons will have the same experience of hearing a sound ensuing from a 'common' source. There may be persons who may be able to hear a whisper from a distance of a metre; whilst others may not hear the same even in close vicinity.

More 'technically', **loudness** as defined by ANSI, the American National Standards Institute, is "the attribute of auditory sensation in terms of which sounds can be ordered on a scale extending from 'quiet' to 'loud'".

See images of loud vs. soft sound in Fig.3.1.

(a) loud sound from a megaphone

(b) soft sound – no sound or a whisper!

(c) and a lion's roar is loud – and fearsome!

Fig.3.1 : Images of loud and soft sounds

Unit of Loudness

The unit "sone" is used to measure loudness[1].

---

[1]The unit commonly used in practice is decibel, or dB, that can be related to sone, discussed later in detail.

Loudness or Volume

In audio technology or usage in audio/audio-video appliances such as (as audio part of) radio receivers, DVD players or TVs (audio part) etc., the loudness of sound from such appliances is often expressed as "volume", meaning that loudness of sound is "high volume" whilst softness of sound is "low volume".
A simple device invariably employed in various appliances to control the volume (or loudness) to suit the comfort of the listener is commonly known as a "volume control" that comprises a potentiometer, enclosed in a case with a stem protruding from it to be fitted with a knob externally. A schematic of the device is depicted in Fig.3.2.

Typically, the IN or input to the potentiometer would originate from sound producing device like a microphone and OUT or output of the potentiometer in terms of voltage (μV to mV), being variable from zero to maximum, is fed to a (power) audio amplifier that is connected to a loudspeaker.

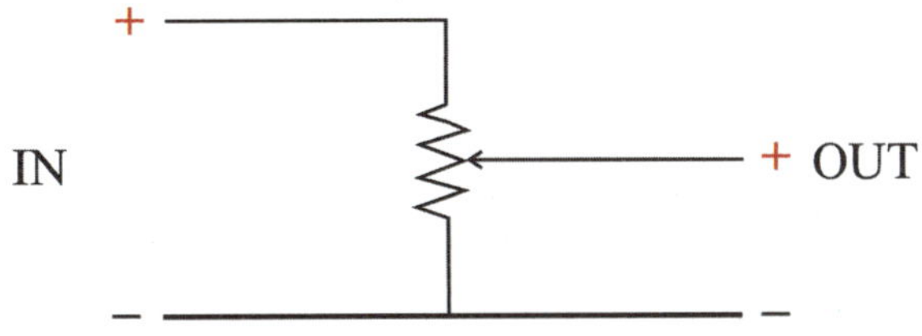

Fig.3.2 : A potentiometer as volume control

A block diagram of such a scheme is shown in Fig.3.3 whilst an *actual* potentiometer or volume control is shown in Fig.3.4. Note that the volume control is implicit to the audio amplifier.

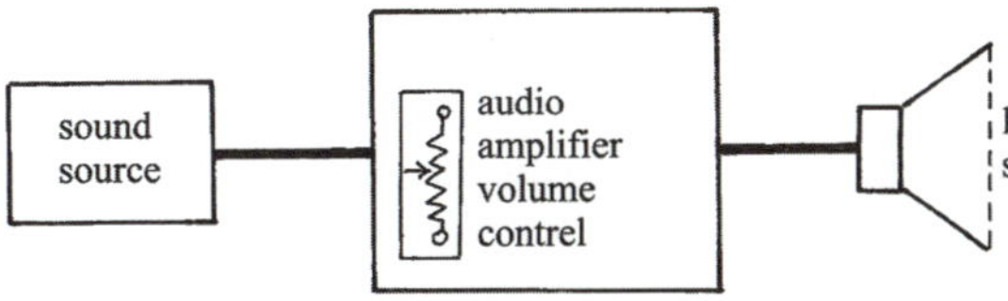

Fig.3.3 : Block diagram of connections using a volume control

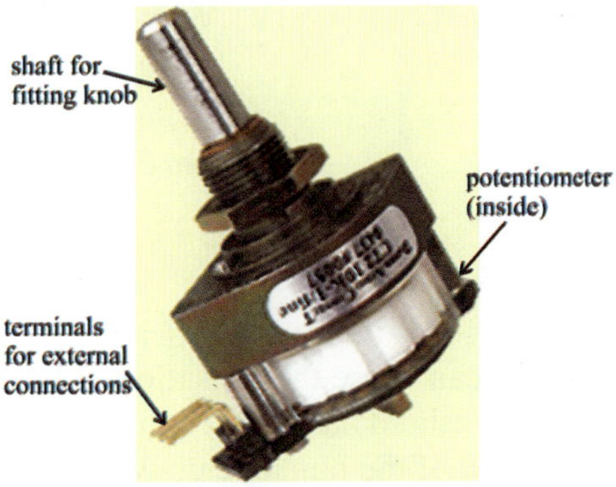

Fig.3.4 : Actual volume control used in audio appliances

Intensity of sound

The amount of sound *energy passing each second through a unit area* is called the **intensity** of sound.

Or, in other words, by definition, the intensity (*I*) of a sound wave is the *time-averaged* power (*P*) it transfers per unit area through a given region of space. That is,

$$I = P/A \quad W/m^2$$

where P is power contained in the sound wave in watt and A represents the area involved in sq. metre.

Factors Affecting Intensity of Sound

▪ Amplitude

The intensity of sound waves emanating from a source is directly proportional to the square of the amplitudes of the waves or vibrations and constitutes the most important factor affecting intensity and, in turn, loudness as sensed by ears. Note that amplitude of the waveform (fundamental and harmonics) is a measure of energy, or time-averaged power, contained in the sound wave. Two images of sinusoidal waveforms showing amplitudes are given in Fig.3.5. The loudness of a sound increases with the amplitude of the sound wave.

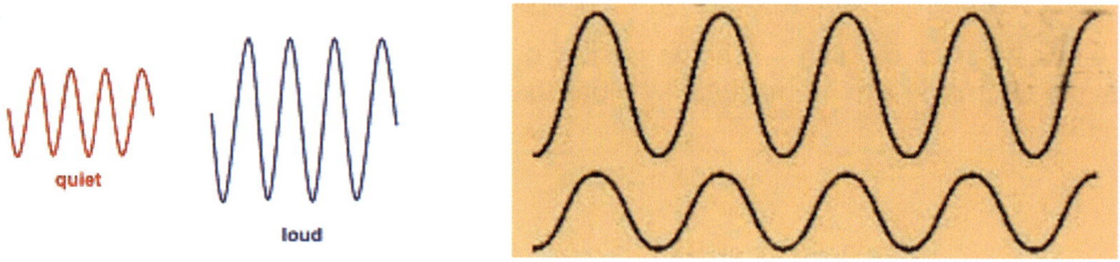

small amplitude: quiet; large amplitude: loud

Fig.3.5 : Images depicting amplitude and loudness

For a person with 'normal' hearing, the minimum intensity of sound (at a frequency of 1000 Hz) that can cause the sensation of hearing is 10^{-12} Wm^{-2}. The amplitude of the sound waves at this threshold has been estimated to be about 1×10^{-10} m, reflecting on the enormous sensitivity of the human ear. When sound becomes so intense that it is painful to the ear, the amplitude is of the order of 1-2 mm.

▪ Surface area

In terms of production of sound from a source, the intensity would depend on the area of the vibrating body: larger the area, greater would be the intensity[2].

---

[2]This should not be confused with the definition of intensity expressed above which is more from the listener's point of view.

- Distance between the source and the receiver

Intensity of sound varies *inversely* as the *square* of the distance between the source of sound and the receiver or the listener. When the distance is less, the intensity, and hence loudness of the sound would be relatively more.

- Density of the medium

The intensity depends on the *density* of the medium through which the sound propagates; denser the medium, higher would be the intensity[3].

Factors Affecting Loudness

Loudness of sound is 'intimately' related to sound intensity, although it can by no means be considered identical to intensity. Thus, whilst various factors that affect intensity of sound can also affect loudness, directly or indirectly, the following are some additional factors affecting loudness.

- Direction of wind

If the wind is blowing in the direction of propagation of sound, loudness would appear to increase.

- Presence of resonating bodies

If a body or device is introduced or positioned suitably such that its natural frequency matches that of the vibrating source of sound, resulting in resonance, the loudness of sound may increase appreciably.

Loudness vs. Intensity of Sound

- Sound intensity is a property of the sound source but loudness depends on the source of sound, the medium as well as the receiver
- Sound intensity holds a small significance in problems involving human hearing system, but loudness is an important property to be considered in such cases
- Sound intensity is measured in watt per square metre whereas loudness is measured in sones

---

[3]This can be qualitatively checked by listening to the vibrations of sound source through air than interposing a dense medium such as a steel rod between the source and the ears.

The above comparison is also listed in the following table.

Loudness	Intensity
1. It is 'subjective' in nature	It is 'objective' in nature
2. It is not truly a physical quantity	It is a physical quantity
3. It largely depends on the sensitivity of ears	It does not depend on the sensitivity of ears
4. Its unit of measurement is sones or decibel (dB)	It is measured in W/m^2

Subjective and objective nature of loudness and intensity

Sound loudness is a subjective term describing the strength of the ear's perception of a sound. The property is related to relative sensitivity of hearing of different persons. Thus, whilst some people may have "sharp" ears and may discern even a faint, or a less 'loud', sound from a distance, others may not, even for the *sounds of same intensity*[4]. The ability of an individual in terms of sensitivity to loudness of sound may also be frequency-dependent and may differ from person to person. In general, human ears are most sensitive to a 'pure' note of 1 kHz. Thus, loudness being a perception confined to hearing ability of different persons is purely subjective phenomenon, having no physical unit, whereas intensity which is a measurable quantity with a proper unit is 'objective' – a scientific quantity.

Sound Level in Decibel (or dB)

Relation Between Intensity and Loudness

The loudness of sound is defined as the *degree of sensation* of sound produced in the ear. The loudness of sound depends on its intensity. However, the relationship is not linear, but is based on a *logarithmic* scale[5]. Thus, according to this law,

$$L \propto \log I$$

or

$$L = K \log I$$

[4]There is also a practical aspect of sensitivity of ears of different persons related the range of audio frequencies which is 20 Hz to 20 kHz; but a particular person may not be able to hear a sound less than, say, 5o Hz and more than, say, 15 kHz or higher notes of sound. Note also that sounds encountered in practice are seldom of single frequency; rather a mix of notes of a number of frequencies, superposed to form a spectrum. See, for example, Fig.1.1(a).

[5]It obeys an approximate law of psychology which states that the magnitude of any sensation is proportional to the logarithm of the physical stimulus which produces it. This is called the Weber-Fechner law. It does not exactly represent the relationship between loudness and intensity but is a fair approximation for pure tones at most frequencies.

where L is the loudness, I is the intensity at a given frequency and K is a constant of proportionality.

As already mentioned, the minimum intensity of sound required to produce aural sensation is 10^{-12} Wm^{-2}. Let this be denoted by I_0 and let the loudness corresponding to this threshold of audibility be L_0. Then the above expression can be written as

$$L_o = K \log I_o$$

Subtracting the second expression from the first

$$L - L_o = K \log I - K \log I_o$$
$$= K (\log I - \log I_o)$$

Or

$$L - L_o = K \log (I/I_o)$$

The quantity $(L - L_o)$ indicates how much the loudness of a given sound is above the minimum value for hearing. Let this be called the intensity level L_i.

$$\therefore \quad L_i = K \log (I/I_o)$$

If the value of the constant K is chosen to be 1, then L_i is measured in a unit called Bel[6]. Thus, if

$$I = 10\, I_o, \text{ then } \quad L_i = \log(10\, I_o/I_o) = 1 \text{ bel}$$

Hence, the intensity level or loudness of a sound is said to be 1 bel, if its intensity is *ten times* the threshold intensity.

If the value of the constant K is chosen to be 10, then L is obtained in *decibel*, given by

$$\boxed{L = 10 \log (I/I_0) \text{ decibel or dB}}$$

where decibel or dB represents the unit of loudness.

It follows that loudness is related to the *ratio* of intensities of sound, involving an spectrum of frequencies, but based on a logarithmic scale. For example, if $I = I_o$, $L = 0$ dB, that is, the loudness is "zero" or completely inaudible to ears at the lower threshold intensity at 1 kHz (= 10^{-12} W/m^2)[7]. If the intensity were to increase to $I = 100\, I_o$, the loudness would increase to 10 log 100 or 20 dB; if I to be 1000 I_o, loudness would be 40 dB.

---

[6]Named to honour the British scientist Alexander Graham Bell who is credited to have invented the telephone.
[7]When referred to the intensity at 1 kHz, the loudness is also known as "sound level".

Now if L be 1 dB, then using the expression $L = 10 \log (I/I_o)$,

$$10 \log (I/I_o) = 1$$

or

$$\log (I/I_o) = 1/10$$

whence

$$I/I_o = \text{antilog}\ (0.1) = 1.26$$

Thus, 1 decibel or dB can also be defined as the change in levels of loudness when the intensity of sound changes by 26%.

Loudness or Sound Levels from Different Sources

The following table provides *approximate* sound levels or loudness perceived from different sources of sound.

Source of sound	loudness or sound level, dB*
Threshold of hearing ("pin-drop" silence!)	0
Rustling leaves in a 'distance'	10
Background in a radio or TV studio	20
Soft whisper	30
'Quiet' library	40
Average home/normal conversation	50
Conversational speech at 1m	60
Vacuum cleaner at 1m	70
Running car	80
Diesel truck at 10m	90
Disco, 1m from speaker	100
Chainsaw at 1m	110
Police car siren	120
Threshold of discomfort to ears	130
Jet aircraft, 50m away	140

*Indicative values.

Characteristics of Musical Sounds

As pointed out in the beginning, the three characteristics that *specifically* differentiate musical sounds from one another are

- Loudness[8]
- Pitch

[8]Recall the previous discussion related to loudness.

- Timbre or quality

Loudness

It is that characteristic of a *musical* sound that determines the *degree* of sensation that the sound can produce in the ear. It depends on the intensity of the sound which is objective in nature as compared to the sensitivity of the ear which is subjective in nature as brought out previously. A factor that plays an important role in imparting sensitivity to the ear is its anatomy as illustrated in Fig.3.6. A human ear typically comprises three parts: external, middle and inner. The external ear is responsible for collecting the sound waves and direct these to the ear drum which leads the sensation of sound to the middle ear. It is the middle ear that 'amplifies' the received sound using an amazing 'set' of tiny bones, called stapes, and determine the level of loudness for an individual[9]. The amplified sound from the middle ear is carried to the inner ear, the nerves from where connect to the brain for ultimate perception and interpretation of a particular sound, connecting it to the source of the sound[10].

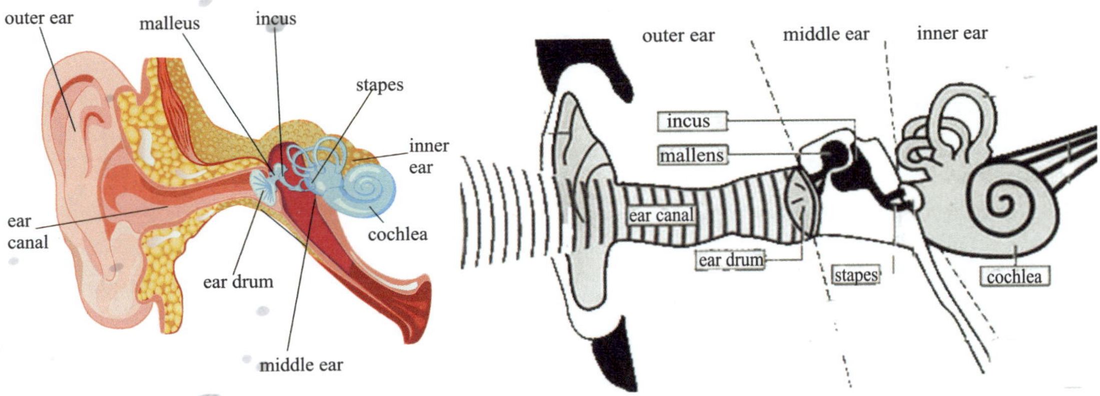

Fig.3.6 : Anatomy of human ear and perception of sound

Pitch

It is that characteristic of a musical sound by which a "shrill" sound can be distinguished from a "grave" or "dull" one, *even though the two sounds may be of the same intensity*. It is also defined as that characteristic of sound by which the ear assigns it a place on a musical scale and corresponding sensation to the brain. In general, the shrillness or otherwise of a sound is essentially dependent on the frequency of the sound wave or the high-frequency content of the entire spectrum comprising a given sound.

---

[9]For the persons developing hardness of hearing, the trouble involves steppes in the middle ear which turn inflexible to vibrate. The problem is usually resolved by performing a micro surgery, called stapedectomy.

[10]A crucial role played by the ears is to stabilise walking of humans (or bi-pads) on their feet. This is accomplished by equalisation of pressure, external vs. internal, by the ears; in the absence of this function, proper walking would be extremely difficult.

When a stretched string is plucked, a sound of a certain pitch sensation is produced. If the tension in the string is increased, the pitch (the shrillness) becomes higher. Increasing the tension, also increases the frequency of vibration. Therefore, pitch is intimately related to frequency. As other examples, the voice of a woman or a child is shrill, that is high pitch, but that of a man is almost invariably 'hoarse' (i.e. flat or grave). The sound emitted by a cuckoo is shrill while that by a crow or a lion is hoarse. Similarly, the buzzing of a mosquito, though low in intensity has a high pitch.

Pitch (or "shrillness") and Frequency

It follows that pitch of a sound would depend on frequency or frequencies as a whole of the sound. In simple terms, this can be explained by reference to the waveforms shown in Fig.3.7.

Musical instruments such as a piano or a violin comprise a number of wires or strings of different thickness and hence tensions and therefore result in sounds of varying frequencies or pitch; (thin) strings under higher tension giving rise to higher pitch notes and *vice versa*. In contrast, a flute would always result in producing low-pitch notes.

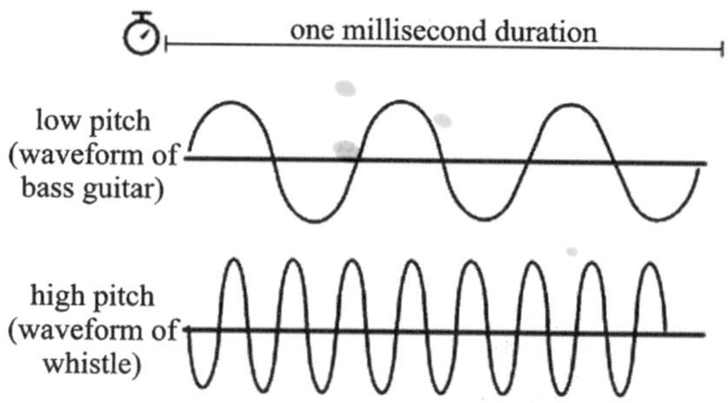

Fig.3.7 : Pitch of a sound vs. frequency

Timbre or quality

Timbre is that characteristic of a musical sound which enables us to distinguish between the sounds produced by two different musical instruments or two different persons although their pitch and loudness may be same. It is because of this characteristic that we are able to recognise the voice of a known person over the telephone or to distinguish between sounds produced by different musical instruments in an orchestra.

The feature depends primarily on the waveform of the sound. The waveform of the sound emitted by a tuning fork is a simple sine curve. But, in practice, most bodies vibrate in a very complex manner, the emitted sound being composed of the fundamental and overtones or harmonics that are multiples of the fundamental.

Quality of Sound

Quality of sound, like in many fields, may be somewhat subjective in nature. For example, one can produce the same tone with the same frequency and intensity as one's favourite singer, but the singer's voice has a better *quality* because of a different combination of the overtones in it. Similarly, the sound from a well-tuned violin or guitar would be of much superior quality compared to when the instrument is out of tune; note that tuning of either instrument may result in change of the loudness and/or pitch of the emanating sound to some extent.

Quality and Waveform

Characteristically, quality or timbre of a sound depends on waveform(s) of the sound that is emanated from a source; whilst comparing different sources, these may be of same loudness (or amplitude) and pitch (or frequency), but different *waveforms*. This is illustrated by simple waveforms of four fictitious sounds in Fig.3.8.

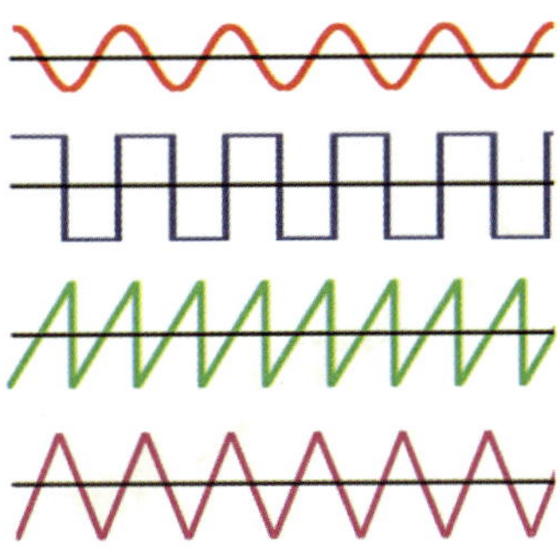

Fig.3.8 : Quality of sound vs. waveform

Note that the sine wave of a given amplitude and frequency would result in a "pure", continuous note of loudness and pitch[11], similar to the note produced by a vibrating tuning fork; the sound of other waveform would produce sound of different, not-easily-distinguishable quality.

Each of the non-sine waveform can be resolved in a fundamental component and other harmonic components of a different frequency each, also called primary and subsidiary components, that might produce notes of different loudness and pitch. The sound from different musical instruments may similarly produce a number of fundamental and subsidiary vibrations or notes, their combined or superposed effect leading to characteristic sound of a given instrument.

Hearing and Listening

The ears carry the sensation of sound to the brain which produces the perception. However, considering the anatomy of ears, whilst the mechanism of sensing, amplification and onward transmission of the sound is somewhat mechanical, the human trait of *registering* the sound and what it really means is termed as "listening", and this may vary from person to person. Many people may easily hear or even differentiate a sound, for example a speech vs. musical notes, yet they may fail to listen[12] as it may require some kind of concentration or attention.

[11]A feature used in the testing of loudspeakers in industry, manufacturing, say, audio amplifiers with integrated loudspeakers, where notes of different frequencies are produced from an audio-frequency signal generator and response of the loudspeaker under test monitored on a calibrated output meter.

[12]Clearly, the difference between "hearing" and "listening" may be termed to be subtle or rather abstract; nevertheless, it is common in practice. Also, the trait is clearly subjective.

Means to Improve Sound

There are numerous gadgets or devices now available in audio technology and usage, both analogue and digital, that may remarkably modify or improve the musical sound, say from an orchestra, or singers performing on a stage. These devices, commonly in the form of appropriate transducers, may increase or decrease loudness, alter the pitch or improve quality of sound emanating from the source.

Loudspeakers

These constitute by far the most frequently used device to increase the loudness or sound level generally where the sound has to reach large gathering or audience. A loudspeaker forms the output stage of an amplifier, interposed to amplify the sound, the input to the amplifier being obtained from another (sound) transducer known as a microphone that would be positioned in front of the source of sound – the musical instrument or the singer/speaker. The schematic action of an amplifier is shown in Fig.3.9.

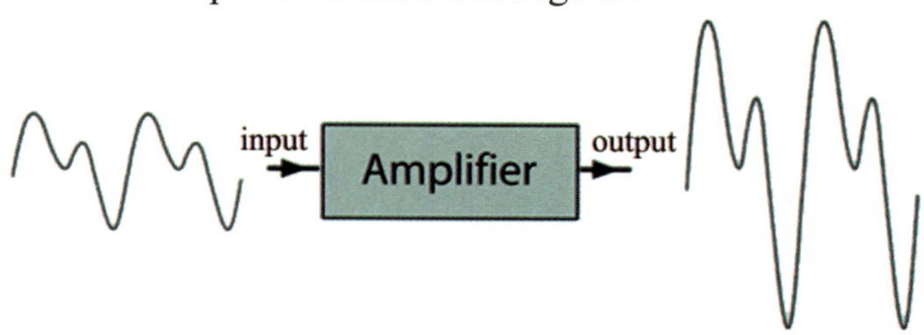

Fig.3.9 : Action of an audio amplifier

It may be noted that a 'good' or 'ideal' amplifier should NOT alter or modify the form or spectrum of the original sound, unless intentionally intended as discussed later, but simply amplify the sound level, the control being effected by changing the position of the volume-control potentiometer [see Fig.3.3]. In order that every frequency contained in the original sound spectrum is amplified to the same degree, an audio amplifier should be designed to have a "gain vs. frequency" characteristic as shown in Fig.3.10.

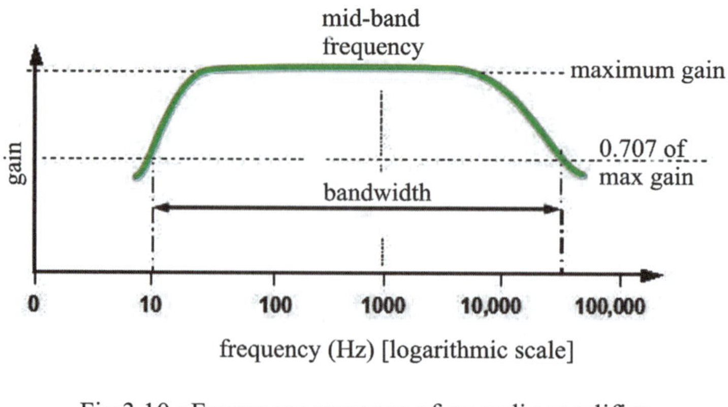

Fig.3.10 : Frequency response of an audio amplifier

Control of Lower or Higher Frequency Bands

All modern amplifiers incorporate appropriate controls to modify the lower-frequency band(s), called "bass" (pronounced 'baes'), typically in the range 20 Hz-500 Hz, or higher-frequency band(s), called "treble", typically in the range 5 kHz-20 kHz, to increase or decrease the gain or loudness in the given range; the term "buck" or "boost" being employed to decrease or increase the gain, respectively[13]. Similar to a volume control, the buck or boost controls are also effected using suitable potentiometers[14].

In terms of the actions as above, the "basic" frequency response of the amplifier shown in Fig.3.10 would appear modified as illustrated in Fig.3.11.

An alternative set of characteristics is shown in Fig.3.12. Note that the sound level corresponding to any frequency is expressed in dB, the unit of loudness. Also, buck/boost curves may differ from design to design.

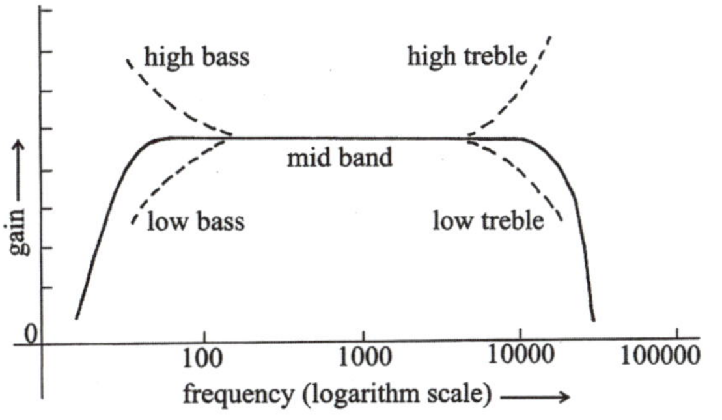

Fig.3.11 : Frequency response characteristic of an audio amplifier as modified for bass and treble

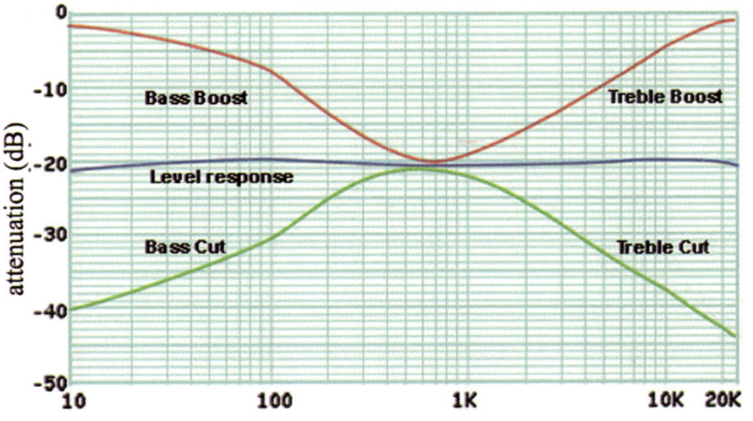

Fig.3.12 : Alternative buck/boost control characteristics

[13]Accordingly, if a listener were interested in only bass frequency sounds, such as from drums (or *tabla*), he may boost the low-frequency band (or cut the treble part). Similarly, if the interest would lie more to listen relatively high-frequency sounds from, for example, clarinet or violin, the treble part would be boosted (or bass part be cut). In practice, however, a 'judicial' balance of overall sound may be strived for.

[14]In modern practice, the term "equaliser" is often used to equalise or adjust overall sound effect, by using a set of (sliding) controls in place of potentiometers as discussed.

Stereophonic Sound

Stereophonic sound or, more commonly called, **stereo**, is a method of sound reproduction that creates an illusion of multi-directional audible perspective. This is usually achieved by using two or more independent audio channels through a configuration of two or more loudspeakers (or stereo headphones) in such a way as to create the impression of sound heard from various directions, as in hearing natural setting. The concept began to be applicable to musical sounds from various instruments, with the bass- and treble-type instruments being positioned to the left or right side on the dais, facing the audience. Correspondingly, the microphones positioned to the left and right side carried sounds through independent *two* channels, via amplifiers, to two loudspeakers located similarly in the auditorium. It is contrasted with monophonic, or "mono" sound, where audio is heard as coming from one position, often centred in the sound field (analogous to a "visual field").

Progressively, the term "stereophonic" developed into being applies to so-called "quadraphonic" and "surround-sound" systems as well as the more common, 'original' two-channel, two-speaker systems. In the years 2000s, stereo sound is common in entertainment systems such as broadcast radio and TV, recorded music on CDs etc. and the cinema halls.

The schematic of a typical 2-channel stereo sound system is shown in Fig.3.13.

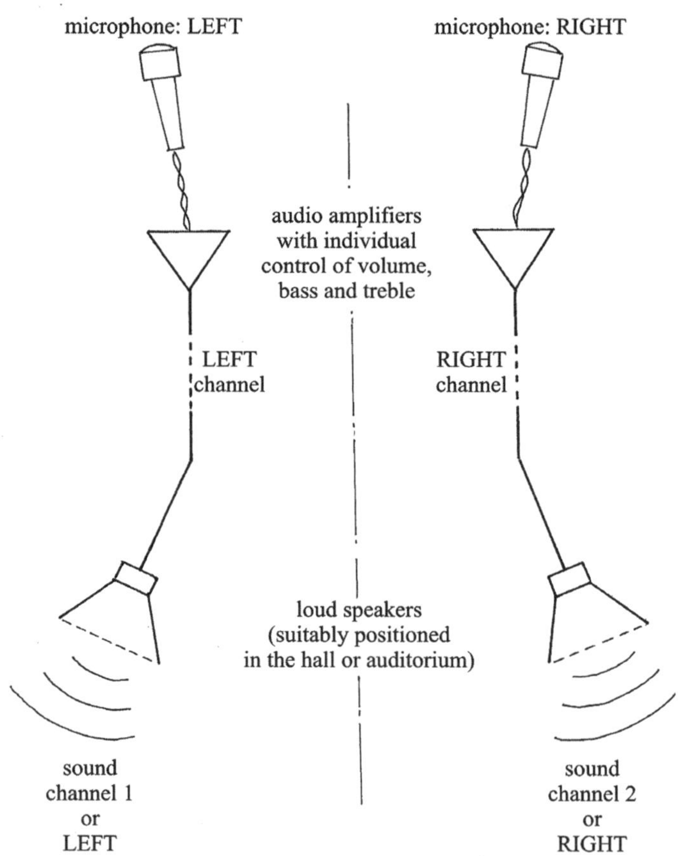

Fig.3.13 : Schematic of a 2-channel stereo system

NOISE AND NOISE POLLUTION

Noise: what constitutes it?

Noise is a sound of a complex character with an irregular period and amplitude, originating from a source in non-periodic motion. Although it can be identified and related to its source, it is not of much consequence and can be very unpleasant and irritable, producing jarring effect on the ears, bordering on being a nuisance. Thus, noise is an undesired sound whose effect is to disturb the normal work, sleep or peace of mind of human beings. A typical waveform of noise in sound is shown in Fig.3.14.

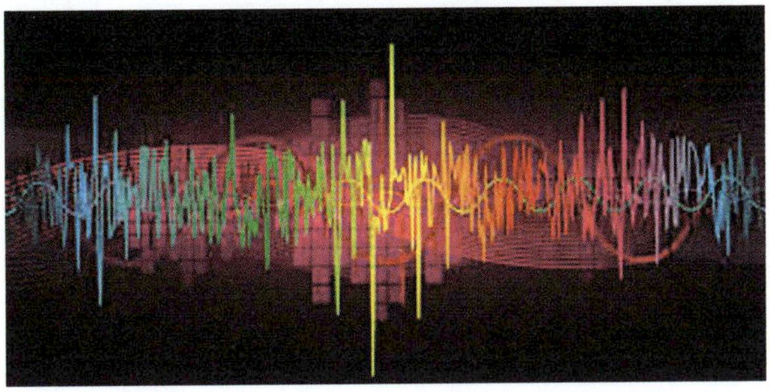

Fig.3.14 : A waveform depicting noise

Common Examples

Some of the common sources of noise are[15]

- the honking of a motor car or truck horn
- the sound of a hammer striking on an anvil
- the report of a gun
- an aeroplane flying above
- bursting of crackers
- barking of dogs in the middle of night
- loudspeakers blaring speech in a public meeting

A few images of sources of noise are illustrated in Fig.3.15.

(a) noise from loudspeakers

(b) noise produced by cars and other vehicles in a traffic jam

[15]The list is certainly not exhaustive; there may be innumerable sources of noise in daily life - continual or short-lived - that may be simply unbearable.

(c) an aeroplane flying over a residential area

(d) an industrial worker wearing a pair of muffins to escape high level of industrial noise

Fig.3.15 : Various sources of noise

Airborne noise is mainly due to the variations in the air pressure with respect to the mean atmospheric pressure. Structural borne noise is due to mechanical vibrations in elastic bodies. In most cases noises produce mental, or even bodily, strain, irritation and headache. A high noise level (industrial or from over amplified music) can cause permanent hearing impairment. It affects the concentration of mind of workers, decreases their efficiency and may lead to errors being committed.
Noise (level) is usually expressed in the unit of loudness, dB.

A typical reflexive reaction to noise is depicted in Fig.3.16.

Fig.3.16 : Reflexive reaction to noise

High noise levels can contribute to cardiovascular effects in humans and an increased incidence of coronary artery disease. In animals, noise can increase the risk of death by altering predator or prey detection and avoidance, interfere with reproduction and navigation, and contribute to permanent hearing loss

Noise Pollution

Noise pollution can be interpreted as 'combined', undesirable and irritable effect of noises produced from various possible sources, creating a disturbing environment that may be harmful to people living in that environment, leading to health hazards[16].
See Fig.3.17 for a 'scenario' of noise pollution.

Common causes of noise pollution

In addition to some of the common sources of noise listed previously, more examples of noise pollution can be

■ Industrial Machinery – There are various examples of industrial machinery, such as large compressors, cranes, furnaces and exhaust fans that create severe disturbing and hazardous sounds to the industrial workers.

■ Construction Sites – These involve use of huge cranes, drilling machines, cement mixers and bulldozers etc. that emit very loud high-pitched noises all around.

Fig.3.17 : A scenario of noise pollution

■ 'Big' Events – Celebrating big events such as birthdays, anniversaries, house parties, festivals can create much noise and unpleasant, very disturbing sounds in external surroundings or neighbourhood, disturbing sleep of peace-loving people. People play songs on full volume and dance till midnight which makes the condition of people living nearby unbearable.

■ Poor Urban Planning – In most of the developing countries, poor urban planning also plays a vital role in giving rise to noise pollution. Congested houses, large families sharing small space, fight over parking, frequent fights over basic amenities lead to intolerable environmental noise.

■ Household Chores – In modern living, most household chores are characterised by extensive use of various such as mixer grinder, pressure cooker, vacuum cleaner, washing machines, coolers, air conditioners etc. that contribute to unpleasant noise pollution for the residents and neighbours.

In general, 'thick' density of population, engaged in a variety of 'noisy' activities lead to noise pollution.

---

[16]With ever increasing industrialisation and evolvement of near-exploding technologies over the past two centuries, noise pollution can be reckoned to be the unavoidable 'gift' of modern times, leading to chaos to ALL living beings, including plants!

Effects of noise pollution

Some of the harmful effects of noise pollution of which many may be subjective and extreme in nature, are

- **Hearing problems:** Constant exposure to loud levels of noise can easily result in the damage of ear drums and loss of hearing. It may also reducer sensitivity of ears to sounds that the ears pick up unconsciously to regulate our body's rhythm.

- **Health issues:** Excessive noise pollution in working areas such as offices, construction sites, shopping malls and even in homes can influence psychological health of people exposed to noisy environs.

- **Sleeping disorders:** Loud noises can in most cases hamper sleeping pattern and may lead to irritation, fatigue and uncomfortable situations, affecting one's performance in offices or other working places.

- **Cardiovascular issues:** Studies suggest that high-intensity noise causes high blood pressure and increases heart beat rate as it disrupts the normal blood flow, leading to cardio-vascular diseases.

- **Trouble communicating:** High decibel surrounding noise can make it troublesome for two people to communicate freely. This may lead to misunderstanding and irritable behaviour and response.

- **Effect on wildlife:** Wildlife faces far more problems than humans because noise pollution since they are more dependent on sound.

Control of noise pollution

Ever increasing noise pollution is having deteriorating effect on the life of people, esp. in urban areas for reasons that may be easily identified. It is therefore essential to exercise all-out control to curb the menace of noise pollution. Some of the steps that may be adopted are

- Careful Planning of Cities: Residential areas should be far removed from industrial or commercial areas or *vice versa*. The houses, or cluster of these such as high-rise buildings, should have minimum "people density" and must be separated from the main streets by means of wide green belts. Enough greenery or plantation should surround the residential areas since plants and trees have noticeable effect on alleviating noise pollution.

- Control of Vehicles: Vehicular traffic constitutes a major cause of noise pollution, mostly owing to haphazard and undisciplined movement of a variety of vehicles both on the main roads as well as streets serving residential premises. To reduce vehicular noise (pollution), a control of "vehicular density", or number of vehicles on the roads at any time, must be strictly controlled. The traffic must move in an orderly manner, following the 'lane system' and etiquette. Honking of horns, esp. in areas close to hospitals, schools and residential areas must be strictly prohibited.

- Legislation: Legal control of noise pollution must be exercised in the larger interest of public, by evolving prohibitive measures. Many countries have come out with strict legislation for controlling noise. At the same time, workers in industries with particularly noisy environment must be given right to claim compensation if they suffer loss of hearing ability or meet an injury as a result of damaging noisy conditions[17].

- Education: It is extremely important that public at large is educated, *using all possible means*, to avoid or at least minimise the production of noise from various sources, esp. when driving, using noisy gadgets judicially and partying etc. The educational institutes must organise regular discourses and campaigns to bring awareness about noise pollution to the minds of youngsters right from an early stage.

ACOUSTICS

A term often closely associated with production, transmission and reception of sound is known as **acoustics**[18].

What is acoustics?

Acoustics is defined as the science that deals with the production, control, transmission, reception and effects of sound. Many people mistakenly think that acoustics is strictly musical or architectural in nature. While acoustics *does* include the study of musical instruments and architectural spaces, it also covers a vast range of topics, including: noise control,

In simple words, acoustics (pronounced əˈ kuː stɪ k) is understood as[19]

- the properties or qualities of a room or building that determine how sound is transmitted in it; for example saying "the Symphony Hall has perfect acoustics".

[17]In India, noise pollution being a major problem, the Government of India has evolved rules and regulations against firecrackers and loudspeakers, but enforcement is extremely lax. Awaaz (sound) Foundation is an Indian NGO working to control noise pollution from various sources through advocacy, public interest litigation, awareness, and educational campaigns since 2003.

[18]The word "acoustic" is derived from the Greek word ἀκουστικός (*akoustikos*), meaning "of or for hearing, ready to hear".

[19]See. for example,
Oxford Dictionary, Oxford University Press.

Since hearing is one of the most crucial means of survival in the animal world, and speech is one of the most distinctive characteristics of human development and culture, the requirement of 'good' acoustics acquires special significance in auditoria, usually ill designed based on criteria other than the ones necessary for the purpose, where the sound from the speakers or performers from the stage or dais may barely be audible to the audience at the back or on sides, or fairly distorted. This may happen on several counts:

- positioning of microphones and loudspeakers or sound reproducing equipment
- quality of amplifiers or signal conditioning equipment
- civil construction of the auditorium such as its length vs. width proportion, low ceiling and 'flat' vertical walls with poor air circulation
- reflection of sound and, at times, echo of produced sound

The acoustics can be improved by adopting several measures

- elimination and/or rectification of some the above deficiencies at appropriate stage
- employing sound-absorbing device(s) on walls, esp. the rear wall, from floor to the ceiling; this may comprise mounting specially constructed thick 'perforated' boards or simply thickly-woven drapes[20] or furnishing.

The principles of acoustics have been applied since ancient times. In about 20 BC, the Roman architect and engineer Vitruvius wrote a treatise on the acoustic properties of theatres including discussion of interference, echoes, and reverberation the beginnings of architectural acoustics. An open-air auditorium from Roman times in the city of Amman is depicted in Fig.3.18.

Fig.3.18 : An open-air theatre from Roman times

[20]It is on this count that "open-air" auditoria or 'theatres' qualify for much improved acoustic.

4. Electricity and Magnetism

Unit 4: Topics covered

Electricity and Magnetism

(i) Ohm's Law; concepts of emf, potential difference, resistances in series and parallel, internal resistance.
Concepts of pd (V), current (I), resistance (R) and charge (Q) Ohm's law: statement, V=IR; SI units: experimental verification; graph of V vs I and resistance from slope; affecting resistance (including specific resistance) and internal resistance; super conductors, electromotive force (emf); combination of derivation of expressions for equivalent resistance. Simple numerical problems using the above relations. [Simple network of resistors].

(ii) Electrical power and energy.
Electrical energy; examples of heater, motor, lamp, loudspeaker, etc. Electrical power; measurement of electrical energy, W=QV = V It from the definition of pd. Combining with ohm's low $W = V It = I^2 Rt = (V^2 / R)t$ and electrical power $P = (W/t) = VI = I^2R = V^2/R$. Units: SI and commercial; Power rating of common appliances, household consumption of electric energy; calculation of total energy consumed by electrical appliances; W = Pt (kilowatt x hour = kW h), [simple numerical problems].

(iii) Household circuits – main circuit; switches; fuses; earthing; safety precautions; three-pin plugs; colour coding of wires.
House wiring (ring system0, power distribution; main circuit (3 wires-live, neutral, earth) with fuse / MCB, main switch and its advantages – circuit diagram; two-way switch, staircase wiring. Need for earthing, fuse, 3-pin plug and socket; Conventional location of live, neutral and earth points in 3 pin plugs and sockets. Safety precautions, colour coding of wires.

(iv) Magnetic effect of a current (principles only, laws not required); electromagnetic induction (elementary); transformer.
Oersted's experiment on the magnetic effect of electric current; magnetic field (B) and field lines due to current in a straight wire (qualitative only), right hand thumb rule – magnetic field due to a current in a loop; Electromagnets: their uses; comparisons with a permanent magnet; Fleming's Left Hand Rule, the DC electric motor-simple sketch of main parts (coil, magnet, split ring commentators and brushes); brief description and type of energy transfer (working not required): Simple introduction to electromagnetic induction; frequency of AC in house hold supplies, Fleming's Right Hand Rule, AC Generator – Simple sketch of main parts, brief description and type of energy transfer(working not required). Advantage of AC over DC. Transformer-its types, characteristics of primary and secondary coils in each type (simple labelled diagram and its uses).

I Electric Current

Ohm's law; concepts of emf, potential difference, resistance; resistances in series and parallel; internal resistance.
Concepts of pd (V), current (I), resistance (R) and charge (Q). Ohm's law: statement, V = IR; SI units; experimental verification; graph of V vs. I and resistance from slope; ohmic and non-ohmic resistors, factors affecting resistance (including specific resistance) and internal resistance; super conductors, electromotive force (emf); combination of resistances in series and parallel and derivation of expressions for equivalent resistance. Simple numerical problems using the above relations.[Simple networks of resistors].

ELECTRIC POTENTIAL

The term "**electric potential**" comprises the words "electric" and "potential". It is essential to first comprehend the word potential.

POTENTIAL

The Oxford dictionary defines potential as
"adj., capable of coming into being or action.n**1** Capacity for use or development.**2** Usable resources.**3** *Physics* quantity determining the energy of mass in a gravitational field or of charge in an electric field."
making an implicit link of the word with the gravitational or electric *field.*

Gravitational Potential

A gravitational field exists about the Earth that exerts gravitational influences upon all masses located in the space surrounding it. Moving an object upward against the gravitational field increases its *gravitational potential energy*. This energy may be defined as "the energy stored in an object by virtue of its vertical position above the earth" and it depends on the mass of the body and its height. Thus. a body of mass one kg when raised to a height of one m would acquire a gravitational potential energy of 9.8 J; a body of two kg when raised by two m would acquire an energy of 4 x 9.8 J and so on where 9.8 denotes acceleration due to gravity in m/s/s.

As discussed previously, *work has to be done on the body* against the gravitational force or field to impart the said energy to the body; an object moving downward within the gravitational field would lose gravitational potential energy.

Clearly, potential *energy* depends on more than just location, being dependent on both mass *and* height of a body. If now a concept is considered that relates the phenomenon to only height (alone) for a body of given mass, a quantity **gravitational potential** can be introduced as potential energy per kilogram. Gravitational potential is, therefore, a *location-dependent* quantity that is independent of the mass of the object experiencing the field. Gravitational potential describes the effects of a gravitational field upon objects that are placed at various locations within it with the *earth's surface as the datum* having zero gravitational potential. It follows that a body in the gravitational field has the *potential* to do work[1].

Electric Potential

If gravitational potential is a means of rating various locations within a gravitational field in terms of the amount of potential energy per unit of mass, then a similar concept may be applicable to electric potential.

Consider the electric field created by a positively charged Van de Graaff generator[2] as shown in Fig.1.1. The direction of the electric field is such that a positive test charge would be pushed; in this case, the direction is outward away from the Van de Graaff sphere. Work would be required to move a positive test charge towards the sphere against the electric field. The amount of force involved in doing the work is dependent upon the amount of charge being moved (according to Coulomb's law of electric force). The greater the charge on the test charge, the greater the repulsive force and the more work that would have to be done on it to move it the *same* distance as brought out in the figure. This work would change the potential energy by an amount that is equal to the amount of work done. Thus, the electric potential energy is dependent upon the amount of charge on the object experiencing the field *and* upon the location within the field. Similar to gravitational potential energy, electric potential energy has a dependency upon the charge of the object experiencing the *electric field.* Thus, electric *potential* is purely location dependent being expressed as the potential energy per unit charge.

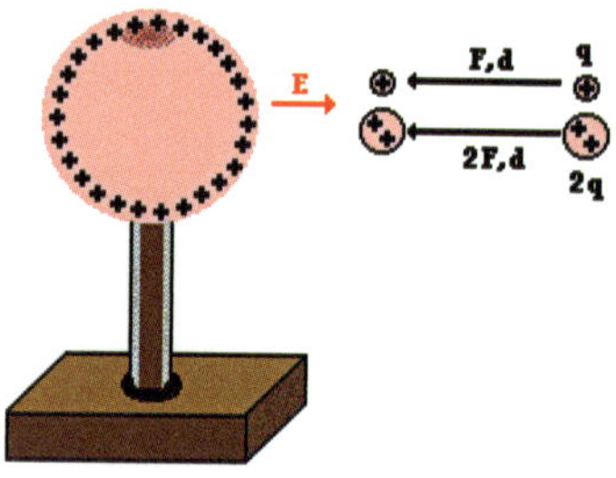

Fig.1.1 : A positively charged body

---

[1]Note that the property named as potential should not be misinterpreted with the word "level" which has a different meaning and usage, being explained in the dictionary as horizontal line or plane, height or value reached, position on a real or imaginary scale etc.

[2]A device specially designed to generate and store electric charge on a sphere by using mechanical means.

Or

$$\text{electric potential} = \frac{\text{potential energy, PE}}{\text{charge, Q}}$$

Concept of electric potential

The concept of electric potential is used to express the effect of an electric field of a source in terms of the *location within the electric field.* A test charge with twice the quantity of charge would possess twice the potential energy at a given location; yet its electric potential at that location would be the same as any other test charge. A positive test charge would be at a high electric potential when held close to a positive source charge and at a lower electric potential when held further away. In this sense, electric potential is simply a property of the 'location' within an electric field.

Suppose that the electric potential at a given location is 12 joules per coulomb, then that is the electric potential of an object with a charge of 1 coulomb or a 2 coulomb. Stating that the electric potential at a given location is 12 joules per coulomb, would mean that an object with 2 coulomb charge would possess 24 joules of potential energy at that location and an object having a charge of 0.5 coulomb would experience 6 joules of potential energy at the location.

Electric potential: definition

Electric potential is defined as "the amount of work needed to move a unit charge from a reference point to a specific point against an *electric* field". Typically, the reference point is the Earth itself, although any point beyond the influence of the electric field charge can also be used as the reference.

Or

The "work that would have to be done in bringing a unit positive charge from an infinite distance to that point or from a place of zero potential".

Unit of electric potential

The unit of electric potential is directly related to the expression

potential = work done/charge

Thus, in SI units the unit of work done being joule and that of charge being coulomb, the unit of electric potential would be joule/coulomb, that is **volt**[3].

Or

$$\text{one volt} = \frac{\text{one joule}}{\text{one coulomb}}$$

Electric potential in circuits

Consider an electric circuit[4] powered by a battery shown along side. The circuit has locations of high and low potential. Charge moving through the wires of the circuit will encounter changes in electric potential as it traverses the circuit. Within the electrochemical cells of the battery, there is an electric field established between the two terminals, directed from the positive terminal towards the negative terminal. As such, the movement of a positive test charge through the cells from the negative terminal to the positive terminal would require work, thus increasing the potential energy of every coulomb of charge that moves along this path. This corresponds to a movement of positive charge against the electric field. *It is for this reason that the positive terminal is described as the high-potential terminal.*

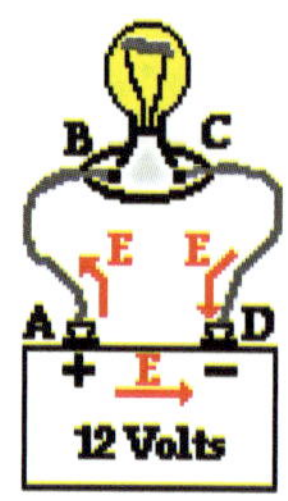

A Simple Electric Circuit

Similar reasoning would lead to conclusion that the movement of positive charge through the wires from the positive terminal to the negative terminal would occur naturally. Such a movement of a positive test charge would be in the direction of the electric field and would not require work. The charge would lose potential energy as it moves through the *external circuit* from the positive terminal to the negative terminal. The negative terminal is described as the low-potential terminal. This assignment of high and low potential to the terminals of an electrochemical cell presumes the traditional convention that electric fields are based on the direction of movement of positive test charges.

Electric Potential Difference [PD]

Consider the task of moving a positive test charge within a *uniform* electric field from location A to location B as shown in the diagram to the right. In moving the charge *against the electric field* from location A to location B, work will have to be done on the charge by an external force. The work

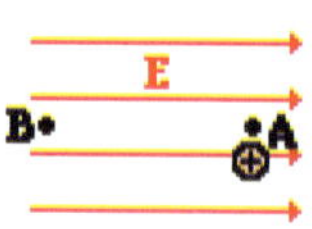

[3]Named in honour of the scientist Alessandro Volta.

[4]An electric circuit would generally be either *active* or *passive*, comprising conductors connecting a source of EMF, elements such as resistors (discussed later), current and voltage measuring devices and keys etc.

A circuit involving a source of EMF such as a battery is called an active circuit.

done on the charge changes its potential energy to a higher value; and the amount of work that is done is equal to the change in the potential energy. As a result of this change in potential energy, there is also a difference in electric potential between locations A and B. This difference in electric potential is referred to as the **electric potential difference**. Therefore, by definition,

"the electric potential difference is the difference in electric potential (V) between the final and the initial location when work is done upon a charge to change its potential energy".

Or

$$V_B - V_A = \Delta V = \frac{\Delta PE}{\text{charge}}$$

where V_A and V_B denote the electric potentials at points A and B and ΔPE is the change in potential energy. See also Fig.1.2.

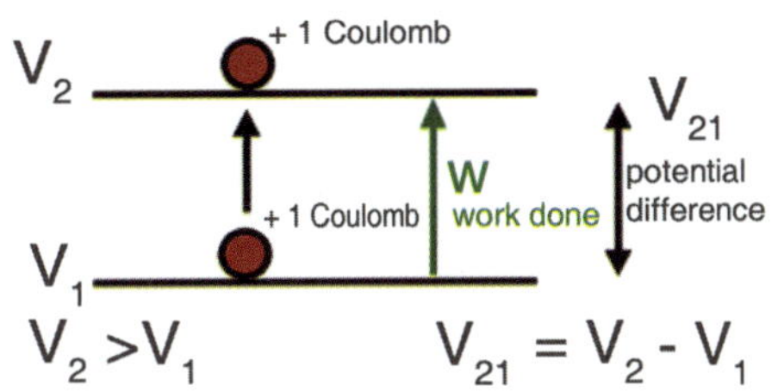

Fig.1.2 : Concept of electric potential difference

Unit of Potential Difference

The SI unit of electric potential difference is also volt.

Concept of Electric Current

Current as flow of charge

The basic concept of current is interpreted in the *flow of electric charge with time*, or it is the rate of flow of charge. Thus, if a charge of dQ flows, say in a circuit, in time dt, the current flowing in the said time would be given by

$$\text{current} = \frac{dQ}{dt}$$

Unit of Electric Current

In SI units, the charge is expressed in coulomb and time in second. Hence, the SI unit of current is obtained as

coulomb/second

and is named as **ampere**[5], denoted by the letter A.

It follow that **one ampere of current will equal to flow of one coulomb of charge in one second.**

[5]Named in honour of the French scientist Andre-Marie Ampere. [1775-1836].

Multiple and sub-multiple units

Huge currents in industries and power systems are often expressed in kilo-ampere where one kilo-ampere, or 1 kA = 1000 A [10^3].
Similarly, small currents, typically in electronic circuits and devices, are specified as a fraction of ampere. For example
one milli-ampere, or 1 mA = 1/1000 A [10^{-3}]
and one micro-ampere, or 1 μA = 1/1,000,000 A [10^{-6}]

Charge of Electron and Flow of Current

In physics, the flow of current *in a circuit* is best explained in terms of movement or 'flow' of (free) electrons that are attracted by positive charge(s) anywhere in the circuit.
In a conductor, electrons are able to move freely, each carrying a negative charge given by 1.6×10^{-19} coulomb. Assuming n electrons passing through a conductor in time t s, the total *charge* that would pass through the conductor will be

$$Q = n \times 1.6 \times 10^{-19} \text{ coulomb}$$

and the current flowing through the conductor will be given by

$$I = Q/t = n \times 1.6 \times 10^{-19}/t$$

Since the charge is in coulomb and time in second, the current in the above expression is obtained in ampere, implying that, as an alternative interpretation, one ampere current is equivalent to there being a flow of $1/1.6 \times 10^{-19}$ or 6.25×10^{18} electrons per second across or through the conductor.

Classic definition of ampere
The ampère is that constant current which, if maintained in two straight parallel conductors of infinite length, of negligible circular cross-section6, and placed one metre apart in vacuum, would produce between these conductors a force equal to 2 × 10−7 newton per metre of length

Electric current in a circuit

According to the universally adopted convention, the flow of the current in the circuit is defined in a direction opposite to that of the electrons such that the positive direction of current flow is assumed to be from the positive terminal or 'point' in the circuit to the negative terminal. This is illustrated schematically in Fig.1.3.

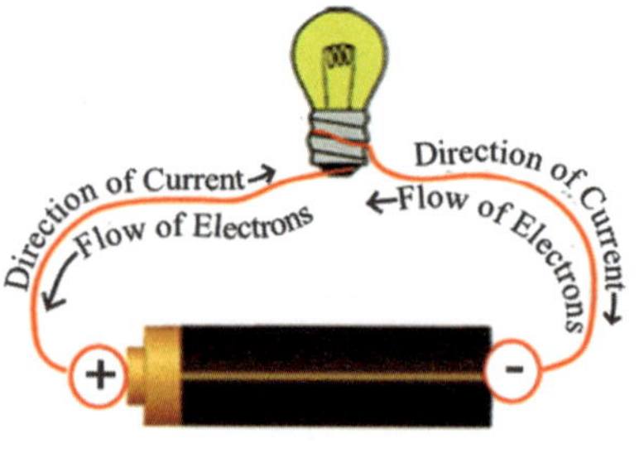

Fig. 1.3 : Current in an electric circuit

[6]Thus implying a "line" current.

OHM'S LAW

Statement of the Law

Ohm's law states that the current through a conductor *at a constant temperature* between two points is directly proportional to the voltage across the two points as depicted in the adjoining figure .

Resistance

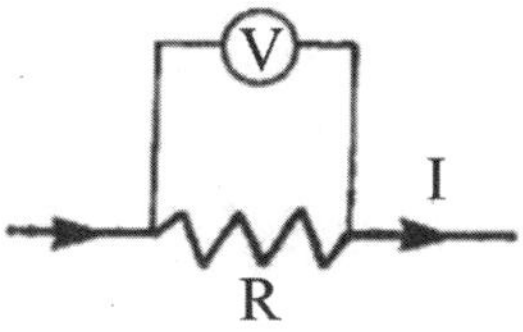

The constant of proportionality as defined above is called the **resistance** of the conductor, the mathematical equation that describes this relationship is

$$I \propto V$$

Or

$$1 = \frac{V}{R}$$

Or, simply $$V = I \times R$$

where I is the current through the conductor, V is the voltage applied or measured *across* the conductor and R is the resistance of the conductor.

It follows that resistance is the property of a conductor whereby it would *resist* the flow of current through it when a voltage is applied across the conductor. The larger the resistance, the smaller the current for a given applied voltage and *vice versa.* This is also indicated above where it is shown that current through a conductor is *inversely proportional to the resistance.*

Unit of Resistance

When the applied voltage is expressed in volt and current that ensues is (measured) in ampere, the resistance is expressed in **ohm** which is also the SI unit of resistance[7]. The accepted symbol of ohm is the Greek letter Ω (omega).

[7] Named in honour of German physicist Georg Simon Ohm who, in a treatise published in 1827, described measurements of applied voltage and current through simple electrical circuits containing various lengths of wire.

More specifically, worded differently, Ohm's law states that *"the resistance, R, in this relation is constant, independent of the current"*.

Example

1. A music system draws a current of 400 mA when connected to a 12 V supply.
(i) What is the resistance of the music system?
(ii) The music system is left playing for several hours and finally the battery voltage drops and the music system stops playing when the current drops to 320 mA. At what battery voltage does the music system stop playing?

[ICSE, 2016]

(i) The resistance of the music system is

$$R = V/I = 12/(400\text{x}10^{-3}) \text{ or } 30\ \Omega$$

(ii) The system stops plying when the current drops to 320 mA. Assuming the system resistance to be the same, the battery voltage would be

$$V = I \text{ x } R$$
$$= (320\text{x}10^{-3}) \text{ x } 30 \quad \text{or} \quad 9.6 \text{ V}$$

2. Calculate the quantity of heat produced in a 20Ω resistor carrying 2.5 A current in 5 minutes.

[ICSE, 2016]

The heat produced in a resistor, R Ω, carrying a current, I A, for time t s is given by

$$H = I^2 \text{ x } R \text{ x } t \text{ J}$$

∴ substituting

$$H = 2.5^2 \text{ x } 20 \text{ x } (5\text{x}60) \text{ or } 37{,}500 \text{ J}$$

3. The adjoining figure shows a circuit

When the circuit is switched on, the ammeter reads 0.5 A.
(i) Calculate the value of the unknown resistor R
(ii) Calculate the charge passing through the 3Ω resistor in 120 s
(iii) Calculate the power dissipated in the 3 Ω resistor

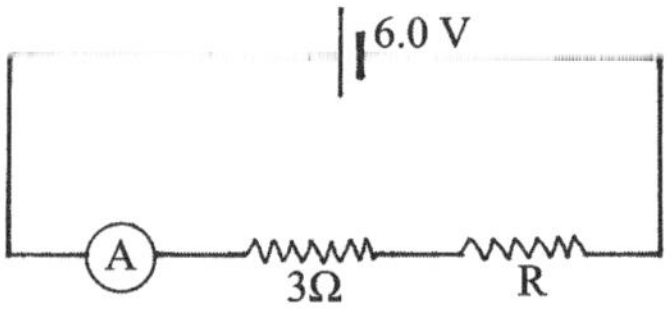

[ICSE, 2013]

The current drawn by the circuit = the reading of the ammeter = 0.5 A

∴ total resistance in the circuit = 6.0/0.5 or 12 Ω

(i) the value of resistance R = 12 – 3 or 9 Ω
(ii) charge through the 3 Ω resistor (or the circuit) = current x time
= 0.5 x 120 or 60 C
(iii) power dissipated in the 3 Ω resistor

$$= I^2 R = (0.5)^2 \times 3 \quad \text{or} \quad 0.75 \text{ W}$$

Multiple and Sub-multiple Units of Resistance

Multiple units

Higher values of resistance are expressed as

one thousand ohm = 1 kilo-ohm (1 kΩ) = 10^3 ohm

one million ohm = 1 meg-ohm (1 M Ω) = 10^6 ohm

Sub-multiple units

Values of resistance lower than one ohm may be expressed as

one thousandth of ohm = 1/ 1,000 Ω = 1 milli-ohm (1 m Ω) = 10^{-3} ohm
one millionth of ohm = 1/1,000,000 Ω = 1 micro-ohm (1 μ Ω) = 10^{-6} ohm

Very small resistances, expressed in micro ohm, are typically encountered in electronic circuits and devices. Resistances in milli-ohm represent resistances of electric equipment, machinery and circuits; for example, winding resistance of a motor or resistance of a conductor of a transmission or distribution line[8].

Conductance of a conductor

The property of a conductor reflecting on how well it can conduct an electric current is known as **conductance** and may be interpreted in terms of how 'free' are electrons within the conductor to move under the influence of an electric potential. Clearly, the conductance of a conductor would be reciprocal of its resistance, expressed as

conductance = 1/resistance

A high value of conductance of a conductors is a measure of its quality of carrying current – and paradoxically – its price per unit length. In terms of price, typical examples are: gold, silver, copper, aluminium and steel etc., in that order.

Unit of Conductance

Expressed as reciprocal of resistance, the unit of conductance is obtained as

[8]Typical small resistances of, say, 50 to 100 mΩ of *4-terminal type* find important application in electrical measurements.

unit of conductance = 1/ohm, called siemens having the symbol S[9,10].

Electric *potential difference* and resistance

In a *closed* circuit there would be several sections or parts of the circuit comprising various conductors or other resistance elements such that, taken individually, the two terminals of each section may be at potentials V_1 and V_2 having a **potential difference** of $V_2 - V_1$ V ($V_2 > V_1$). Then the resistance of that section of the circuit would be given by

$$\boxed{R = \frac{V_2 - V_1}{1}}$$

where I is the current through the section.

Experimental verification of Ohm's law

The connection diagram for verification of Ohm's law is shown in Fig.1.4.

A "conductor"[11] of resistance R ohm is connected in series with an ammeter A, a rheostat (variable resistance) and a battery B through key K. A voltmeter V is connected across the ends of the resistance to measure the potential difference (or 'voltage') across it.

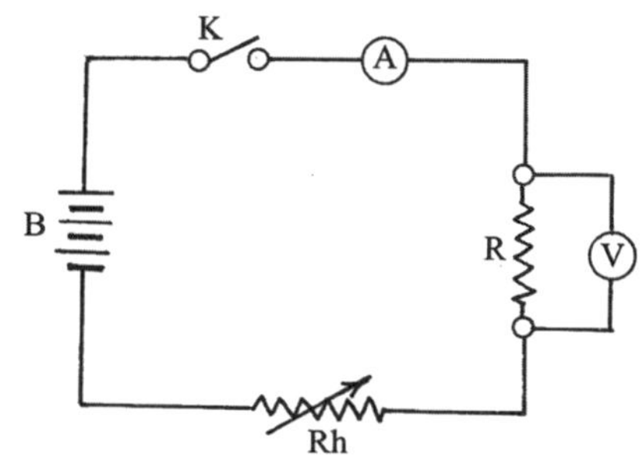

Fig.1.4 : Connection diagram for Ohm's law verification

To begin the experiment, the key K is pressed and readings of voltmeter and ammeter are noted. Next the setting of the rheostat is changed and voltmeter and ammeter readings are noted for new value of the current. Thus for various settings of rheostat the readings of voltmeter and ammeter are recorded. Form the observations it will be found that the ratio of voltmeter and ammeter readings, $V_1, \ldots V_4$ and $I_1, \ldots I_4$, respectively, remains the same, that is

$$\frac{V_1}{I_1} = \frac{V_2}{I_2} = \frac{V_3}{I_3} = \frac{V_4}{I_4} \ldots\ldots = \text{constant} = R, \text{the resistance of the conductor}$$

[9]The earlier name for the unit was "mho", being 1/Ω or ℧ (inverse of Greek letter Ω).
[10]The unit siemens is named in honour of the German scientist Ernst Werner von Siemens.
[11]In a physics lab., a resistance may be obtainable in the form of a "decade resistance box" affording a selectable resistance of a value anywhere from one ohm to tens or hundreds of ohm.

Accordingly, a graph of voltmeter vs. ammeter readings would be seen to be straight line as shown in Fig.1.5, verifying the Ohm's law.

Clearly, the slope of the straight line passing through the origin provides the value of resistance of the conductor (or resistance R) in ohm.
Note that the concept of *constant* resistance as the ratio of "voltage"/ "current" in the circuit is more based on the *incremental* values of voltage (or pd) and current. That is

$$R = \frac{\Delta V}{\Delta I}$$

where ΔV and ΔI represent the incremental values of voltage and current, respectively, about a point P on the graph as shown in Fig.1.5.

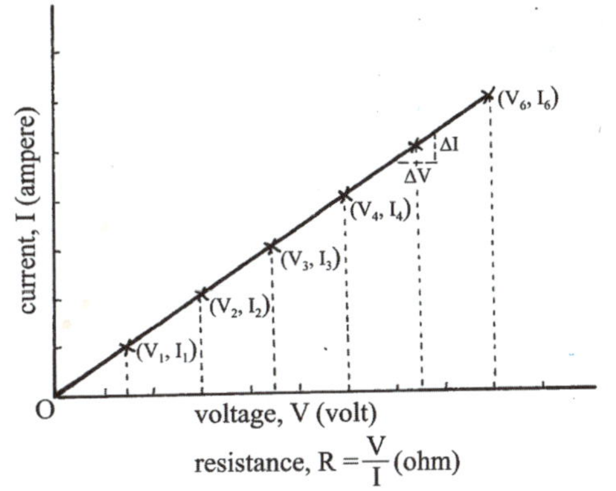

Fig.1.5 : Graphical relationship of voltage vs. current and conductor resistance

Ohmic and Non-ohmic Resistances

Electrical appliances such as an incandescent bulb to an electric iron in practice represent cases of **ohmic** and **non-ohmic** resistances.
The main difference between an ohmic and a non-ohmic resistance or conductor is that an ohmic resistance follows the ohm's law and will have a *linear* relationship between the applied voltage (or PD) and the current that flows through the resistance at any time such as the straight line shown in Fig.1.5. Such resistances are also known as *linear* resistances.

Examples of ohmic *conductors* (or resistances) are wires made from conducting metals such as copper and aluminium. These are resistive devices made from wires meant for the purpose; an example is a rheostat used in a physics lab as a 'variable' resistor.

Note, however, that these resistances behave as ohmic or linear circuit elements only in a restricted range of operating temperatures as specified in the ohm's law. At 'higher' temperatures, the linear relationship between the voltage and current may not hold and the resistance may no longer be a constant value[12].

In contrast, non-ohmic conductors, or resistors and devices, do not follow ohm's law and have their own characteristics. There are a number of examples of non-ohmic resistances.

[12]There are special alloys such as manganin or constantan, the resistance of which does not change much with temperature and is commonly used to wind "resistor" coils and rheostats.

Some of these are illustrated below, along with their respective voltage vs. current characteristics.
See Fig.1.6 (a) through (d).

Examples of Non-ohmic Resistances and Devices

A filament lamp

In a common incandescent bulb, the graph of voltage vs. current becomes 'curved' or non-linear as the filament heats up and its resistance goes up[13].
For example, a 15 W bulb having a "cold" or un-lit resistance of about 200 ohm rises to a value of nearly 3500 ohm when powered at the normal supply voltage of 230 V (consuming a current of about 66 mA); a 100 W bulb having a "cold" resistance of about 30 ohm jumps to a resistance of approximately 500 ohm when similarly powered (consuming a current of about 0.5 A).

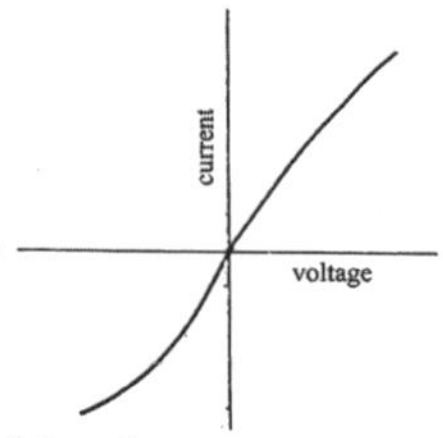

(a) voltage vs. current characteristic of a filament lamp

A diode

A diode is a semiconductor device that allows current to flow *only in one direction* through it (the diode being known to be "forward biased"). When the polarity of the applied is reversed (or the diode is "reverse biased") no current can flow through the diode[14].

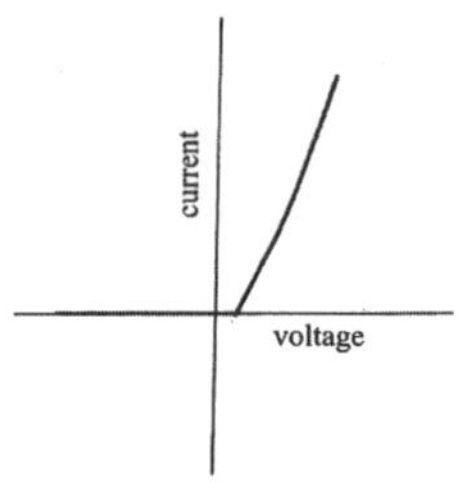

(b) voltage/current characteristic of a diode

Thermistor

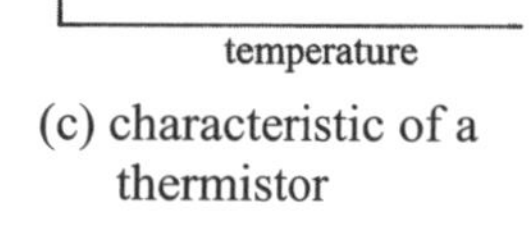

(c) characteristic of a thermistor

A thermistor – or a THERMal resISTOR – is a special non-linear resistor the resistance of which *decreases* as its operating temperature increases[15].
The resistance vs. temperature characteristic, that can also be related to voltage vs. current for its resistance *at a given temperature*, is shown in the adjoining figure.

---

[13]The filaments of incandescent lamps are invariably constructed from tungsten that has a melting point of 3695 K and so can withstand extreme heat and temperature rise of the filament when powered; also, the glass globe of the lamp is highly evacuated to prevent 'burning' of the filament.
[14]Due to the property of a diode to allow current to flow in one direction only, with practically "zero" resistance in that direction (forward biased), these are universally employed as rectifiers; that is, converting alternating current into direct current..
[15]This property of a thermistor is frequently utilised as a "temperature transducer" in circuits to monitor and control of temperature in rooms and refrigerators etc.

Light Dependant Resistor, LDR

This device has a (current) resistive property such that its resistance decreases as the intensity of light falling on it increases as depicted in the characteristic shown. Accordingly, LDRs are used in circuits which can control condition of lighting; for example to automatically switch on lights when it gets dark, such as street lighting and *vice versa*.

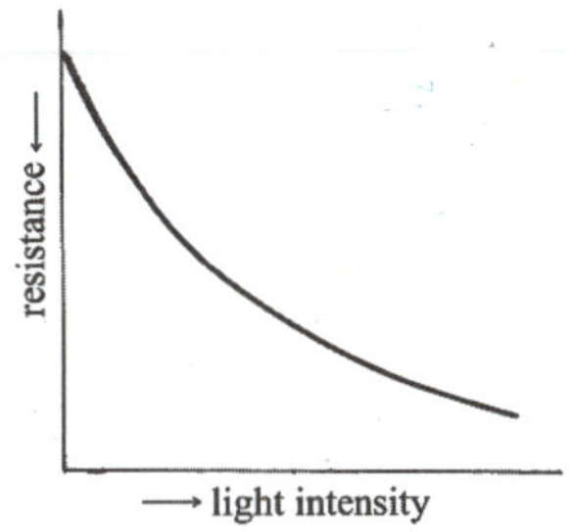

(d) resistance vs. light intensity characteristic of a LDR

Fig.1.6 : Examples of non-ohmic resistances

Light Emitting Diode , LED

An LED, or light emitting diode, is another example of a non-ohmic resistor or device. The LED is a special semiconducting device like an ordinary diode that would emit light when 'excited' by a DC voltage across its positive and negative terminals, the common operating voltages being 3 V or 12 V. The current that flows through an LED varies typically from about 15/20 mA to as high as 200 mA, depending on the design and rating of the LED. The resistance of the LED would change from a value in the passive state to when it is emitting light.

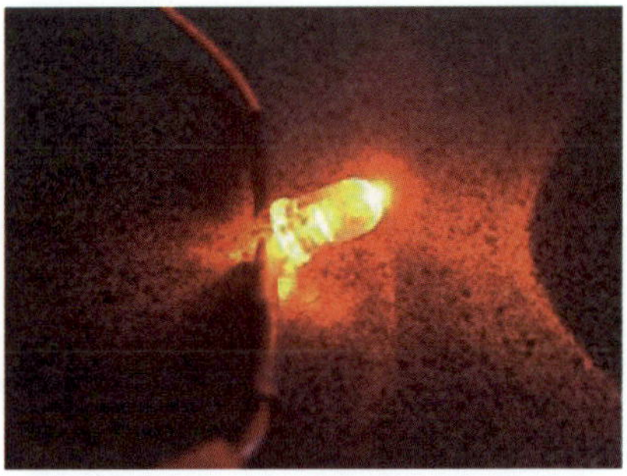

Fig.1.7 : A light emitting diode or LED

A typical, "old"-type LED in lit condition is shown in Fig.1.7 [16].

Comparison of Ohmic and Non-ohmic Resistances

- Conductors/resistors that obey ohm's law, that is the ratio V/I for all values of V and I is constant, are called ohmic conductors/resistors while those that do not obey ohm's law are called non-ohmic conductors/resistors

- Magnitude of current remains unchanged when current or voltage is reversed in ohmic conductors/resistors; the magnitude changes in the case of non-ohmic conductors/resistors

[16]Owing to their low operating current and yet emitting 'good' light, LEDs are being increasingly used for general illumination (providing white light) and decoration (using LEDs in various colours) consuming relatively small electric power. For example, a 12 W LED bulb, having an illuminating capacity of 1200 lumen, is now commercially available that can replace a 100 W incandescent bulb of old type, and with a life of nearly 50,000 hours of operation, compared to less than 1000 hours of life of a typical incandescent bulb. 'High power' LEDs are being increasingly used for street lighting.

- In ohmic conductors/resistors, current, I, is proportional to voltage, V, that is, the graph of V vs. I is a straight line, whereas this is not the case with non-ohmic conductors/resistors
- In ohmic conductors/resistors, temperature affects current and resistance whereas in non-ohmic conductors/resistor, factors like operational design affect current and resistance

Factors Affecting Resistance

There are many factors or parameters that affect the resistance of a *conductor*.

■ Length

The length of a conductor or wire directly affects its resistance; the longer the wire the more will be its resistance. Thus, the resistance

$$R \propto \text{length, } L$$

If it is assumed that resistance occurs as the result of collisions between charge carriers and the atoms of the wire, then there is likely to be more collisions in a longer wire. More collisions mean more resistance.

■ Cross-sectional area

Just as a pipe of bigger diameter or area of cross section would allow easy flow of water, that is offering less resistance to the flow, likewise a conductor of bigger cross section would have a lower resistance to the flow of current and *vice versa*.
It follows that resistance

$$R \propto 1/\text{ area of cross section,}$$

meaning that when all other variables are the same, charge will flow at higher rates through wider wires with greater cross-sectional areas than through thinner wires, offering lesser resistance.

■ Material

Not all materials are identical in terms of their conductive property to conduct current. Some materials are better conductors than others and offer less resistance to the flow of charge. Silver is one of the best conductors, but it is not practicable to use it for wires/conductors due to its cost. Copper and aluminum are among the less expensive materials with suitable conducting ability and are invariably used for drawing conductors. Gold is by far the best conductor of current and is frequently used in "high quality" miniature/micro electronic circuits like ICs, notwithstanding its high cost.

The conducting ability of a material is often expressed in terms of its **resistivity**[17];a property that is dependent upon the material's electronic structure and its *temperature*. For most (but not all) materials, resistivity increases with increasing temperature.

Clearly, resistance of a conductor would be directly proportional to resistivity of its material

---

[17]Also called specific resistance.

or

$$\text{resistance, } R \;\propto\; \text{resistivity, } \rho$$

where the Greek letter ρ (rho) is used to denote resistivity in an appropriate unit.

■ Temperature

Resistance of almost all conductors, made of any material, is adversely affected by its operating temperature; even the ambient temperature may affect the 'base' value of the resistance. In general, an increase of temperature results in an increase in the resistance of the conductor. A common example is the filament of an incandescent bulb that increases very significantly as the tungsten filament heats up to emit light.
If resistance of a conductor is known at some (initial) temperature T_o, its resistance at a (higher) temperature, T, can be obtained using the expression

$$R_t = R_o \times [(1 + \alpha (T - T_o)], \text{ very approximately}$$

where
R_o : resistance at temperature T_o
R_t : resistance at temperature T
and α : a constant or "temperature coefficient" that is dependent on the resistor material

Temperature coefficients for various materials

The values of the temperature coefficients for some of the metals are given in the table below

Metal	Temperature coefficient (x 10^{-3}/°C)
Gold	3.4
Silver	3.8
Copper	3.9
Aluminium	3.9
Platinum	3.92
Tungsten	4.5
'Iron'	5.0
Nichrome	0.4

Resistance of a conductor in terms of variables

Considering the first three parameters listed above, the resistance of a conductor *at a given temperature* will be obtained as

$$\text{resistance, R} = \text{resistivity, } \rho \times \frac{\text{length, L}}{\text{area of cross section, a}}$$

In SI units, the length and area of cross section of the conductor is expressed in metre (m) and square metre (m^2), respectively, whilst resistance is specified in ohm (Ω). Then the unit of resistivity can be derived using the above expression by substituting

$$R = 1 \text{ ohm}; \quad L = 1 \text{ m and } a = 1 \text{ m}^2$$

whence ρ is obtained to be ohm-metre (or Ω-m).

The resistivity of a few representative materials is listed in the table below.

Metal/material	Resistivity at 20 °C (ohm-metre)x10^{-8}
Gold	2.2
Silver	1.6
Copper	1.7
Aluminium	2.6
Lead	2.8
Tungsten	5.6
Brass	6.6
'Iron'	10
Platinum	11
Steel	20
Nichrome	100
Manganin	44
Constantan	49
Glass	10^{10}-10^{14}
Carbon	3.5x10^{-5}

Superconductors

Superconductivity

Superconductivity is a phenomenon related to the condition of "zero electrical resistance" occurring in certain metals/materials, such as a variety of ceramics, when cooled below a characteristic, "critical temperature"[18]. Thus, a superconductor is one

[18] The phenomenon was discovered by Dutch physicist Heike Kamerlingh Onnes in 1911.

that would exhibit 'zero' electrical resistance when cooled down to its critical temperature, usually expressed in (degree) kelvin (K), theoretically resulting in "infinite" current in the circuit when even a 'small' voltage is applied across its terminals. For example, if mercury is cooled below 4.1 K, it loses all its electrical resistance[19].

A graphical illustration of the phenomenon is shown in Fig.1.8. Observe the sudden drop of resistance to "zero" at the critical temperature.

One of the important properties of a superconductor is that it will exclude magnetic fields, a phenomenon called the *Meissner effect.* This is depicted schematically in Fig.1.9.

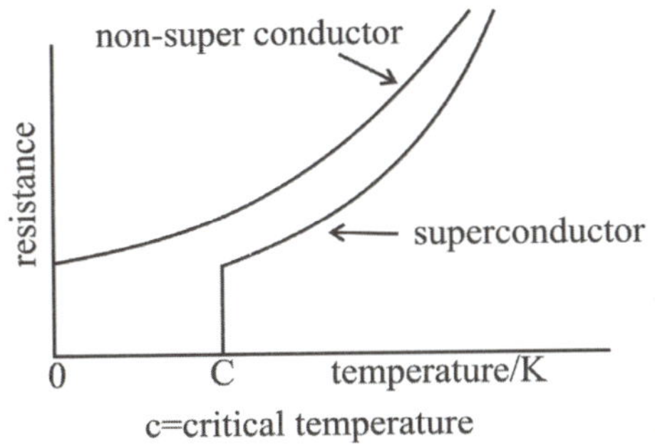

Fig.1.8 : Superconductor and critical resistance

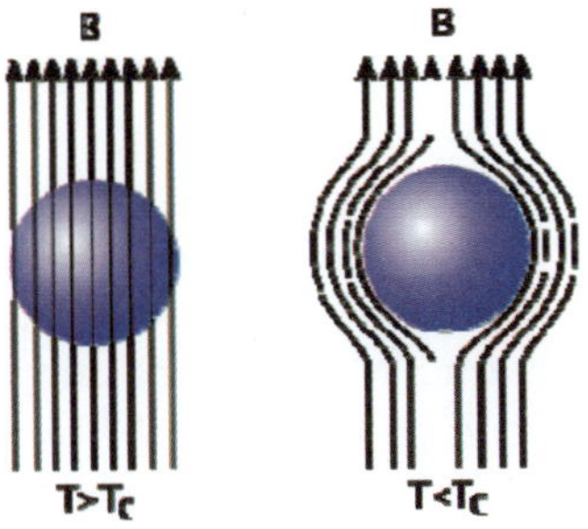

Fig.1.9 : The Meissner effect

The critical temperatures of some of the commonly used materials as superconductors are listed below.

Material	gallium	aluminium	indium	tin	mercury	lead	niobium
Critical temperature, K*	1.1	1.2	3.4	3.7	4.2	7.2	9.3

*0 K= - 273 °C

ELECTROMOTIVE FORCE (EMF)

Concept of EMF

Electromotive force[20], commonly abbreviated as EMF, is the voltage developed by any source of electrical energy such as a battery or "dynamo''. It is generally defined as "the

[19]In 1986, it was discovered that some 'ceramic' materials have a critical temperature above 90 K (−183 °C). Such materials were termed to be *high-temperature superconductors*.
[20]The word "force" in this case does not mean mechanical force, measured in newton, but a potential, or *energy per unit of charge*, measured in volts.

electric potential for a source in an electric circuit" and a device that can supply electrical energy is called a seat of electromotive force. Such a device converts chemical, mechanical, or any other form of energy into electrical energy and the 'product' of such a device is identified as EMF..

Voltaic pile and EMF

A classic source of EMF is historically related to what was known as "voltaic pile" that was invented by the Italian physicist and chemist[21] Alessandro Volta around the year 1800. The voltaic pile can be considered as *the* first "electrical battery" that could *continuously* provide an *electric current* to an external circuit. The current was essentially a direct current (DC)[22]. In 1794 Volta demonstrated that when two metals and brine-soaked cloth or cardboard are arranged in a 'circuit' they produce an electric current.

In the "pile" that Volta described in 1800, a disk of copper is placed at the bottom, followed by a disk of cloth soaked in brine or acidulated water, followed by a disk of zinc. More sets of copper-cloth-zinc disks were placed on top, until the pile reached a height of about 30 cm. The positive end of the pile was the bottom copper disk, and the negative end was the top zinc disk. A schematic of a voltaic pile is shown in Fig.1.10.

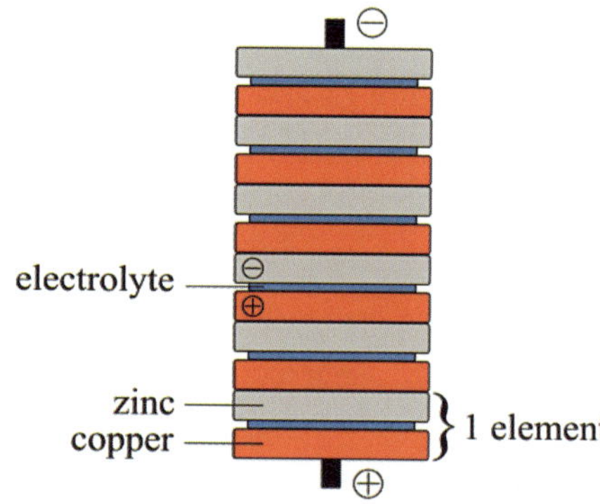

Fig.1.10 : Schematic of a voltaic pile

An *actual* pile constructed by Volta is illustrated in Fig.1.11, displaced in a museum near Volta's home in Italy.

Fig.1.11: 'Original' pile constructed by Volta

EMF Output

When no current is drawn from the pile, each 'cell' consisting of zinc/electrolyte/copper generates about 0.76 V. The voltage from the cells in the pile would thus add up so that six cells shown in Fig.1.10 would generate 4.56 V of EMF.

Modern cells and battery

In the modern usage, commonly used sources of EMF comprise dry cells and batteries.

---

[21]And a pioneer of "electricity".

[22]Prior to the invent of a source of direct current, that is the pile and subsequent cells and batteries, the concept of electricity was limited to the "static electricity", involving electronic charge(s) and charged bodies.

A Primary Cell

A primary cell lacks a 'free-flowing' electrolyte and instead utilises electrolytes that are contained within an absorbent material or a 'separator' (that is, no free liquid or electrolyte) and are thus termed as **dry cells;** also known as Leclanche dry cell. These cells cannot be easily recharged and are discarded following discharge. The cell has a positive and a negative electrode or terminal for appropriate external connection.

A dry cell is usually constructed with a zinc shell that serves as the cathode or negative terminal; a graphite rod which serves as the anode or positive terminal; and a moist mixture of ammonium chloride {$\mathbf{NH_4Cl}$}, zinc chloride {$\mathbf{ZnCl_2}$}, and manganese dioxide {$\mathbf{MnO_2}$}. A schematic representation of a dry cell is shown on the right.

A typical dry cell and its internal construction are shown in Fig.1.12.

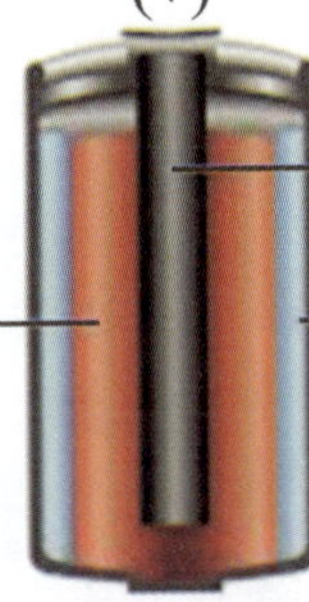

Fig.1.12 : A dry or primary cell

Examples of primary cells include the types of batteries used for children's toys, clocks, radios and similar consumer electronics products. By contrast, car batteries and standby power sources are usually secondary cells.

Secondary Cell

A secondary cell is one that can be electrically recharged after use to their original pre-discharge condition, by passing current through the circuit *in the opposite direction to the current during discharge.* Because they require a liquid electrolyte, secondary cells are usually larger than primary cells and more expensive.

Battery

An electric **battery** is a device consisting of one or more electrochemical cells, usually *connected in series* or at times other series-parallel combinations to achieve different desired terminal voltage or total current, with external connections provided to power devices such as electric cars[23]

Lead acid battery[24]

The **lead-acid battery** is one of the oldest type of rechargeable batteries. Despite having a very low energy-to-weight ratio and a low energy-to-volume ratio, its ability to supply high surge currents means that the cells have a relatively large power-to-weight ratio. These features, along with their relatively low cost, makes the lead-acid battery quite attractive for use in most industrial and other applications.

A very frequently used battery comprises multiple cells, arranged in series, each cell made up from two electrodes, of lead and lead-oxide, and dilute sulphuric acid as electrolyte. The EMF output of each cell would be typically slightly over 2 V when fully charged.
A sectional view of a typical lead-acid battery is given in Fig.1.13.

Fig.1.13 : Sectional view of a lead-acid rechargeable battery

A detailed account of their constructional features is given in Appendix A.

[23]Mainly for (self) starting of the car.
[24]Invented in 1859 by French physicist Gaston Planté.

Representation of a Cell and Battery in a Circuit

As a universal convention, a cell and a battery is represented in a circuit as shown in Fig.1.14.

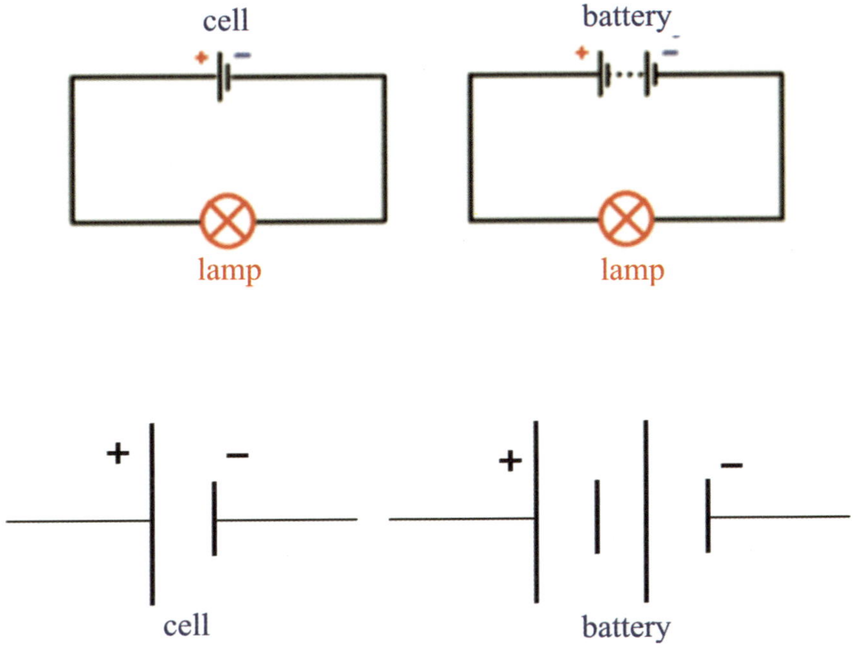

Fig.1.14 : Representation of a cell and a battery in a circuit

EMF, internal resistance and terminal voltage of a cell/battery

A cell or battery generates an EMF between its two terminals to supply electric current to the external circuit. However, along with the developed EMF, a cell or battery also has some **internal resistance** that would affect magnitude of current flow in the circuit and "terminal voltage" across the cell's (or battery's) terminals. Conventionally, the (generated) voltage appearing across a cell *when no current is drawn from the cell* is called the "open-circuit" EMF or simply the EMF. When the cell (or battery) is "on load", that is supplying current, the open-circuit EMF falls and is known as the "terminal voltage" which is less than its open circuit EMF and depends on the total external resistance as well as on the internal voltage drop in the cell (or battery) on account of its internal resistance.

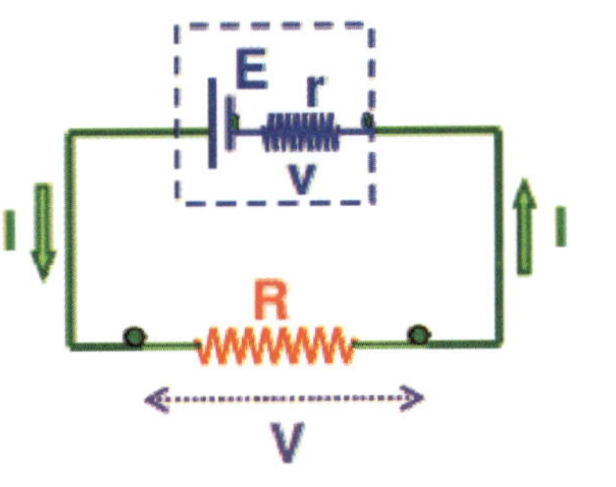

Fig.1.15 : EMF, internal resistance and terminal voltage of a cell

This is depicted schematically in Fig.1.15 in which
E is the open-circuit EMF of the cell
r its internal resistance

R the total external resistance
I the circuit current, and
V the terminal voltage of the cell[25]

In practice, the internal resistance of a cell or battery is dependent on its size, chemical properties, age, temperature and the discharge current. It has an electronic component due to the resistivity of the component materials and an ionic component due to electrochemical factors such as electrolyte conductivity, ion mobility and electrode surface area. Referring to the circuit in Fig.1.15, it follows that

$$\boxed{E = V + I \times (R+r) \quad V}$$

for a given circuit current I A and total or equivalent circuit resistance of R Ω. The internal cell or battery *EMF* is then I x r V.

Example

1. A cell of EMF 1.5 V records a PD of 1.35 V when connected to an external resistance R, such that the current flowing in the circuit is 0.75 A. Calculate the value of R and internal resistance of the cell.

The simple series circuit is shown in the adjoining figure.

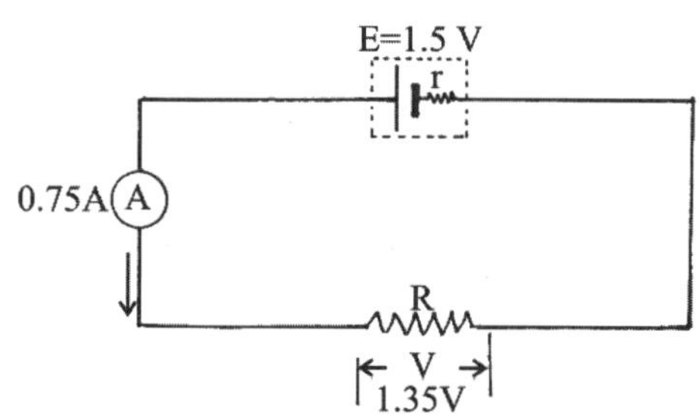

The voltage across the resistance R is the terminal voltage across the cell, that is, 1.35 V.

$$R = \frac{1.35}{0.75} = 1.8\Omega$$

∴

Now the expression for current in a circuit with a cell having internal resistance r is

$$1 = \frac{E}{R+r}$$

∴ substituting

$$0.75 = \frac{E}{1.8+r}$$

whence $$r = 0.2 \ \Omega$$

[25]The same terminology would apply to a battery.

2. A cell of EMF 2 V and internal resistance 1.2 Ω is connected with an ammeter of resistance 0.8 Ω and two resistances of 4.5 Ω and 9 Ω in parallel as shown in the figure.

(i) What would be the reading on the ammeter?
(ii) What is the potential difference across the terminals of the cell?

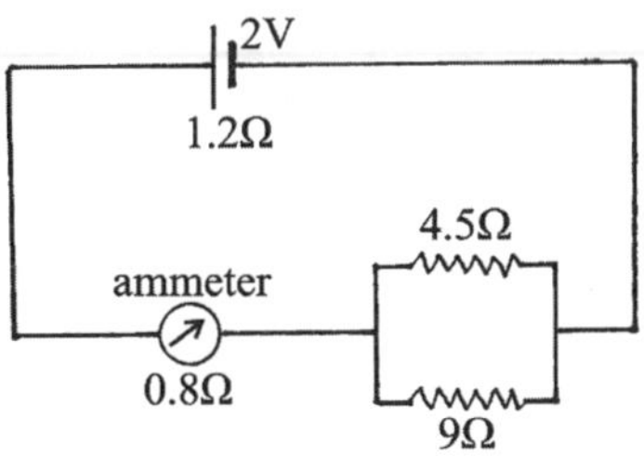

[ICSE, 2015]

(i) Here, the equivalent resistance of 4.5 Ω and 9 Ω resistors being in parallel is

$$\frac{1}{R} = \frac{1}{4.5} + \frac{1}{9}$$

whence

$$R = 3\ \Omega$$

Total circuit resistance of the circuit $= 3 + 0.8 + 1.2$ or $5\ \Omega$

∴ current in the circuit or ammeter reading $= 2/5$ or 0.4 A

(ii) potential difference across the cell = open circuit EMF – internal voltage drop
= $2 - 0.4 \times 1.2$ or 1.52 V

Combination of Resistances

Resistances can be combined in a circuit in THREEE ways

- in **series**
- in **parallel**
- in **series** *and* **parallel**, in a variety of forms

A. Resistances in series

When n number of resistances, R_1, R_2, R_3, . . . , R_n are connected in series, the total or their equivalent resistance is given by

$$R_{eq} = R_1 + R_2 + R_3 + \ . \ . \ . \ + R_n$$

in the given unit, that is all the components resistances being in the same unit (Ω , m Ω etc.) as shown in Fig.1.16. In SI, the unit of total resistance will be ohm if individual resistances are also specified in ohm.

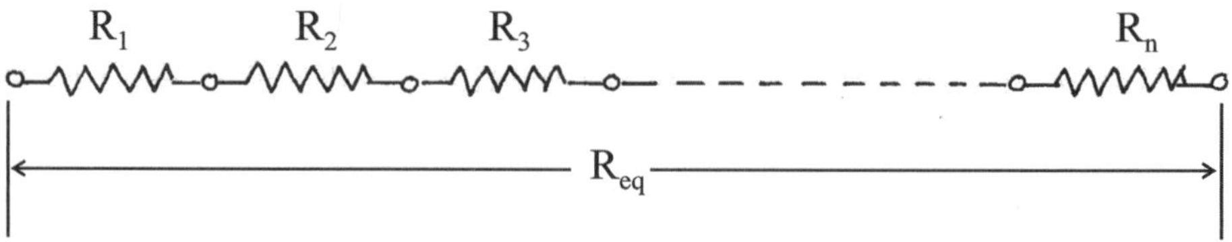

Fig..1.16 : n resistances connected in series and equivalent resistance

The above expression can be derived from a simple "series" electric circuit given in Fig.1.17.

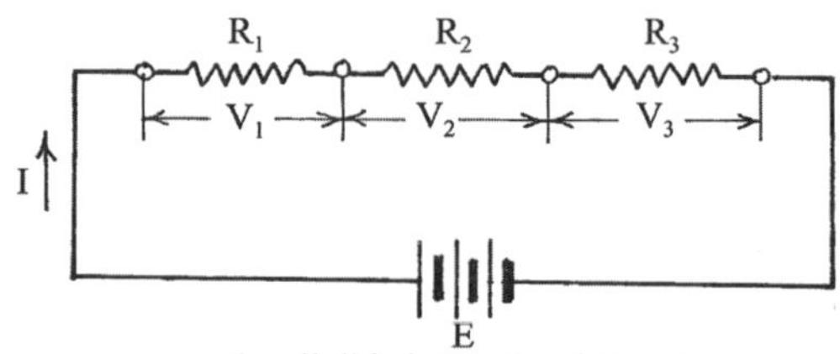

(negligible internal resistance)

Fig.1.17: A series electric circuit

The circuit consists of three resistances R_1, R_2, and R_3 connected in series across a battery of EMF E and negligible internal resistance[26]. The battery supplies a current I that flows through all the resistances or *being common to all.* Then the potential drop across individual resistances will be

$$V_1 = IxR_1$$
$$V_2 = IxR_2$$
and $$V_3 = IxR_3$$
the sum of all the PDs being
$$V_1 + V_2 + V_3$$
And this must equal the EMF of the battery
Or $$E = V_1 + V_2 + V_3$$
Substituting for V_1, V_2, V_3
$$E = IxR_1 + IxR_2 + IxR_3$$
$$= I\,x\,(R_1 + R_2 + R_3)$$

If it be assumed that the battery supplies to a total or equivalent resistance of R, then
$$E = I\,x\,R$$
so that $$I\,x\,R = I\,x\;(R_1 + R_2 + R_3)$$
whence $$R = (R_1 + R_2 + R_3)$$
and so on for n resistances.

Note that if all the resistances were equal, each being R, the equivalent resistance will simply be 3R; and for n resistances equal to nR.

Example

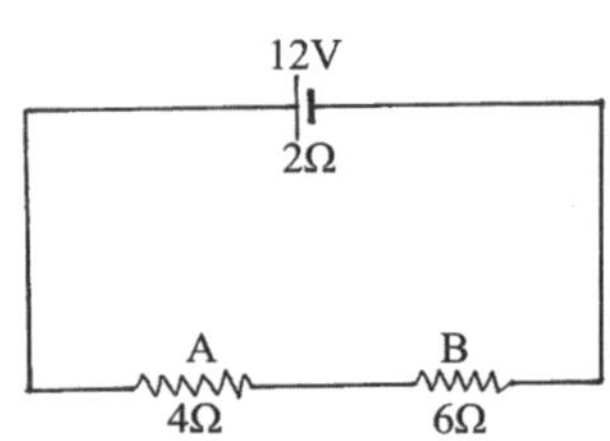

A battery of EMF 12 V and internal resistance 2Ω is connected to two resistors A and B of resistance 4 Ω and 6 Ω, respectively, joined in series as shown.

---

[26]This assumption is meant to simplify the 'analysis'; it does not in any way affects the final result as the internal resistance, and hence the EMF, can be accounted for in the PDs in the circuit.

Find

(i) Current in the circuit

(ii) The terminal voltage of the battery

(iii) The potential difference across 6Ω resistor

(iv) Electrical energy spent per minute in 4Ω resistor

[ICSE, 2016]

Total resistance in the circuit, R = A + B + internal resistance of the battery
= 4 + 6 + 2 or 12 Ω

(i) Current in the circuit = V/R or 12/12 = 1 A

(ii) The terminal voltage across the battery = open-circuit EMF – internal drop
= 12 – 1 x 2 or 10 V

(iii) Potential difference across the $6\ \Omega$ resistance = 1 x 6 or 6 V

(iv) Electrical energy spent in the $4\ \Omega$ resistor is $I^2 R t$
= $1^2 \times 4 \times 60$ or 240 J

An Interesting Application

An interesting, purposeful application of a "resistance in series" is connection of a 60 W or 100 W bulb in series with the usual regulator of a ceiling fan for control of fan speed. The connections for this are shown in Fig.1.18.

If the fan is found to rotate too fast even at the lowest setting of the regulator, its speed can be reduced appreciably by letting the bulb to be in series with the regulator. This is achieved by opening switch S. This would result in the current passing through the bulb, making it light dimly. The bulb can be disconnected from the circuit by closing the switch when the fan speed would revert to as set originally by the regulator[27].

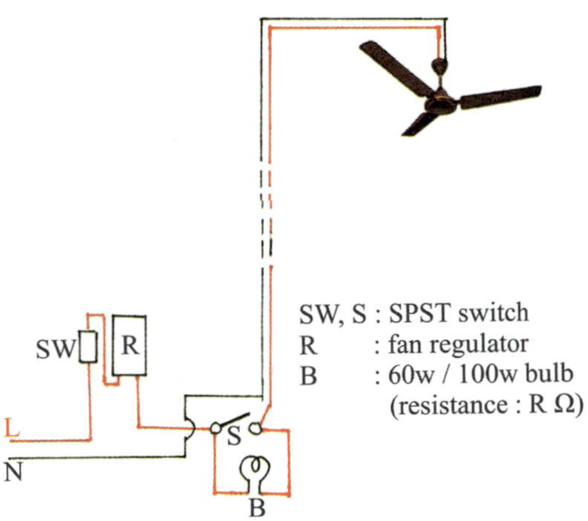

Fig.1.18 : Speed control of a fan by connecting a bulb in series

B. Resistances in parallel

Similar to above, resistances in any number can be connected in parallel as shown in Fig.1.19.

To derive an expression for equivalent resistance of the resistances connected in parallel, consider the circuit of Fig.1.20.

---

[27]When the bulb is in series and dimly lit, it can function as a "night lamp" in the room.

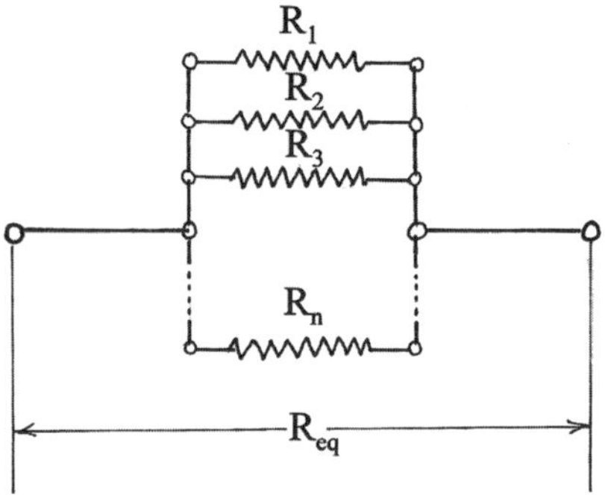

Fig.1.19 : n resistances connected in parallel

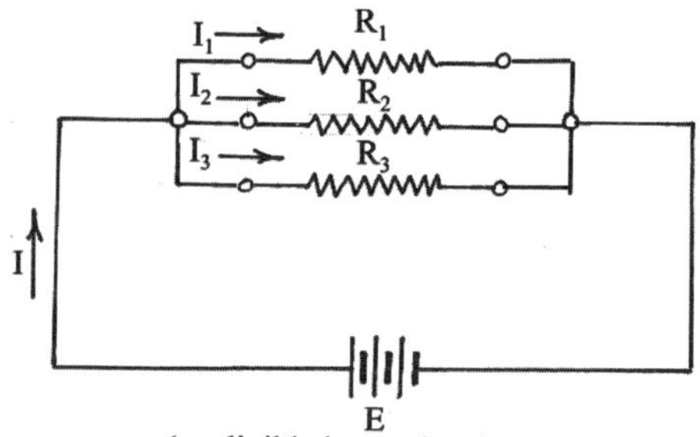

Fig.1.20 : A circuit comprising three resistances in parallel

In the circuit shown in Fig.1.20. three resistances R_1, R_2 and R_3 are connected in parallel across a battery of EMF E and negligible internal EMF. The battery supplies a net current I which divides into individual resistance as shown in the figure such that the potential drop across each resistance equals the EMF E.

Or $$E = I_1 \times R_1 = I_2 \times R_2 = I_3 \times R_3, \text{ that is}$$
$$I_1 = E/R_1, \quad I_2 = E/R_2, \quad I_3 = E/R_3$$

and the net or total current given by

$$I = I_1 + I_2 + I_3$$

Substituting for I_1, I_2 and I_3

$$I = \frac{E}{R_1} + \frac{E}{R_2} + \frac{E}{R_3}$$

If the equivalent resistance of the circuit were R, the total current would be given by

$$I = \frac{E}{R}$$

Equating the two expressions for I

$$\frac{E}{R} = \frac{E}{R_1} + \frac{E}{R_2} + \frac{E}{R_3}$$

whence

$$\frac{1}{R} = \frac{1}{R_1} + \frac{1}{R_2} + \frac{1}{R_3}$$

Clearly, if there are n resistances connected in parallel in a circuit (see Fig.1.19), their equivalent resistance will be given by

$$\frac{1}{R_{eq}} = \frac{1}{R_1} + \frac{1}{R_2} + \frac{1}{R_3} + ... + \frac{1}{R_n}$$

If all the n resistances were equal, each R, the equivalent resistance will simply be

$$\frac{1}{R_{eq}} = \frac{n}{R}$$

Or

$$R_{eq} = \frac{R}{n}$$

Comparison of Series and Parallel Circuits

The main features associated with series and parallel circuits can be compared as listed in the table below.

Series Circuits	**Parallel Circuits**
• The current is the same in every resistor; this current is equal to that in the battery • The sum of the voltage drops across the individual resistors is equal to the voltage rating of the battery • The overall resistance of the resistors is equal to the sum of the individual resistance values, $R_{tot} = R_1 + R_2 + R_3 + ...$	• The voltage drop is the same across each parallel branch • The sum of the current in each individual branch is equal to the total current supplied by the battery • The equivalent or overall resistance of the resistors is given by the equation $1/R_{eq} = 1/R_1 + 1/R_2 + 1/R_3 ...$

Example

1. Given three resistors shown below,

Find the total resistance across A and B.

The three resistances are connected in series, so the total resistance is equal to the sum of the resistances across A and B.

$$R_T = 2 + 3 + 6 = 11 \text{ ohm}$$

2. Find the total resistance for three resistors as connected

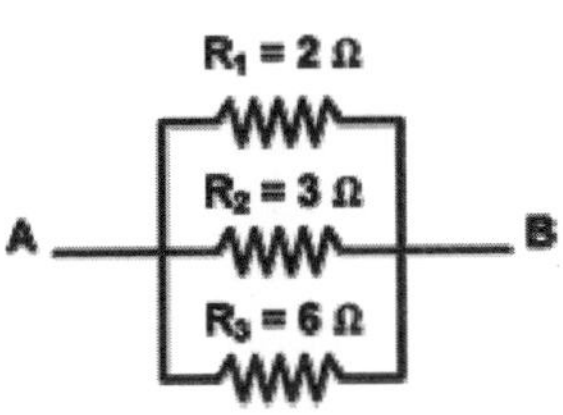

The three resistances are connected in parallel so using the formula for resistances in parallel

$$\frac{1}{R_{AB}} = \frac{1}{R_1} + \frac{1}{R_2} + \frac{1}{R_3}$$
$$\frac{1}{R_{AB}} = \frac{1}{2} + \frac{1}{3} + \frac{1}{6} = \frac{3}{6} + \frac{2}{6} + \frac{1}{6} = \frac{(3+2+1)}{6}$$
$$R_{AB} = \frac{6}{(3+2+1)} = 1\,\Omega$$

C. Resistances in series and parallel

This presents a more common combination of resistances in practice where numerous resistances may be connected in different form of series *and* parallel links or configuration.

A simple configuration of two resistances, R_1 and R_2, in series is followed by two resistances, R_3 and R_4, connected in parallel as shown in the circuit of Fig.1.21. The combination is fed from a battery of EMF E and a net or total current ensues as indicated.

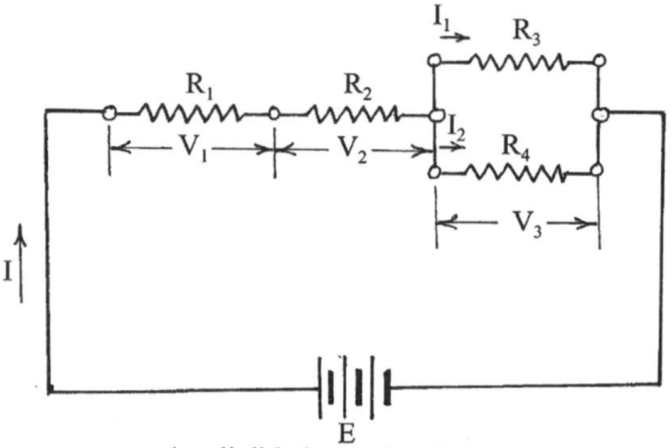

Fig.1.21 : A circuit showing combination of resistances in series and parallel

The various potential drops across the resistances are marked in the figure such that

$$V_1 = I\,R_1, \quad V_2 = I\,R_2$$

and $$V_3 = I_1\,R_3 = I_2\,R_4$$

where I is the net current common to both R_1 and R_2 that divides into I_1 and I_2, flowing through R_3 and R_4[28].

Then

$$I_1 = V_3/R_3 \text{ and } I_2 = V_3/R_4$$

and
$$I = I_1 + I_2$$
$$= V_3/R_3 + V_3/R_4$$

Now
$$E = V_1 + V_2 + V_3$$
$$= I\,R_1 + I\,R_2 + V_3/R_3 + V_3/R_4$$

If R_{eq} is the equivalent resistance of the combination in the circuit, then

$$E = I \text{ x } R_{eq}$$

Or
$$I\,R_{eq} = I\,(R_1 + R_2) + I\,(1/R3 + 1/R_4)$$

[28]Note that currents I_1 and I_2 through parallel combination of resistances R_3 and R_4 divide in *inverse proportion* of the magnitudes of resistances.

whence

$$R_{eq} = (R_1 + R_2) + \left(\frac{1}{R_3} + \frac{1}{R_4}\right)$$

that is, the total or sum of the resistances in series and equivalent of resistances in parallel.

A Simple Model[29] to Demonstrate Series, Parallel or Series-parallel Connection of Resistances

The schematic of the model is illustrated in Fig.1.22.

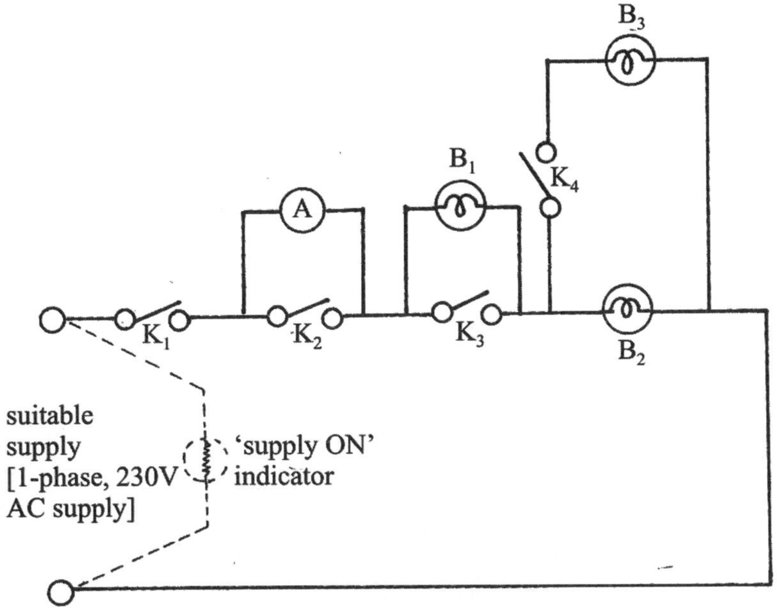

Fig.1.22 : Schematic of the model showing operation of resistances in series/parallel

The device comprises

- three bulbs B_1, B_2 and B_3; for example, of rating 15 W each and of different colours, and four keys K_1, K_2, K_3 and K_4 organised as shown and connected across a suitable supply; this could be normal AC supply at 230 V

- an ammeter[30] that may be connected across key K_2 to measure the circuit current in a given case as discussed later.

---

[29]The model, designed and fabricated by the author, can be easily rigged using various listed parts or items for demonstration in the class room.

[30]Of appropriate range, say 0-1 A and preferably AC/DC or digital type.

Operation

Single resistance

- Close keys K_2 and K_3 and open key K_4. Energise the circuit by closing main key K_1. This allows only bulb B_2 (representing a single resistance, R_1, say) to be in the circuit, making it to glow with full intensity. Keep key K_2 open if it is desired to measure current in the circuit[31].

Two resistances in series

- Open key K_3. This brings bulbs B_1 and B_2 in series (representing two resistances in series). Observe the glow of the two bulbs and the current read on the ammeter.

Two resistances in parallel

- Close key K_3, short circuiting bulb B_1. Close key K_4. This results in bulbs B_2 and B_3 connected in parallel. Observe glow in the two bulbs and current indicated in the ammeter.

A Resistance in Series With Two Resistances in Parallel

- Open key K_3. This leads to bulb B_1 connected in series with bulbs B_2 and B_3 in parallel. Again, observe the glow of all the three bulbs, *relative to one another*, and total circuit current.

A photographic view of an actual model is shown in Fig.1.23[32].

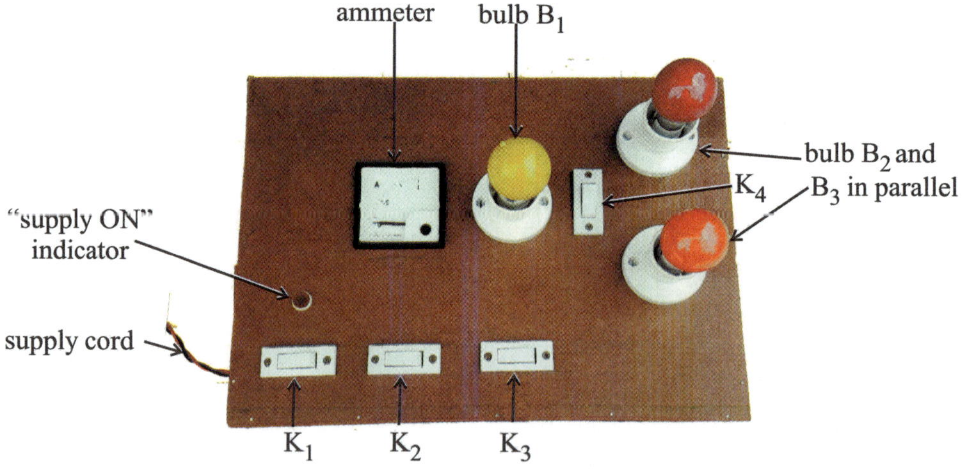

(a) series-parallel resistance model

---

[31]Note that the resistance of the bulb would be much less when cold than when it is carrying (normal) current and hot.

[32]Clearly, the model can be modified to include more bulbs, appropriately controlled by keys, to represent more intricate combinations of resistances in series/series parallel.

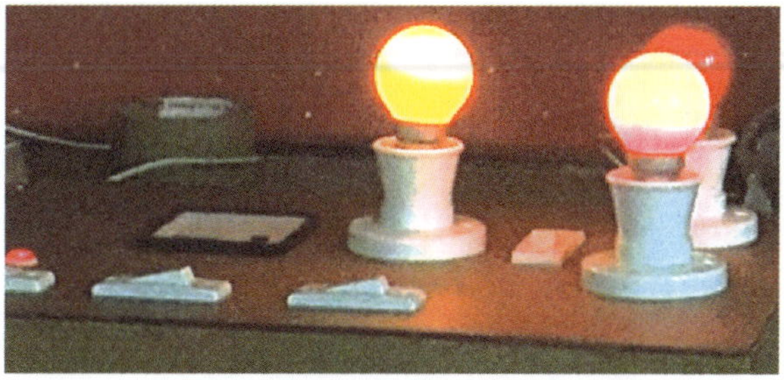

bulbs in series
[observe the position of switches]

bulbs in parallel

(b) resistances or bulbs in series and parallel

Fig.1.23 : Photographic view of the model

Example

1. Calculate equivalent resistance in the network as shown between points
(a) A and B, (b) B and C, (c) A and C

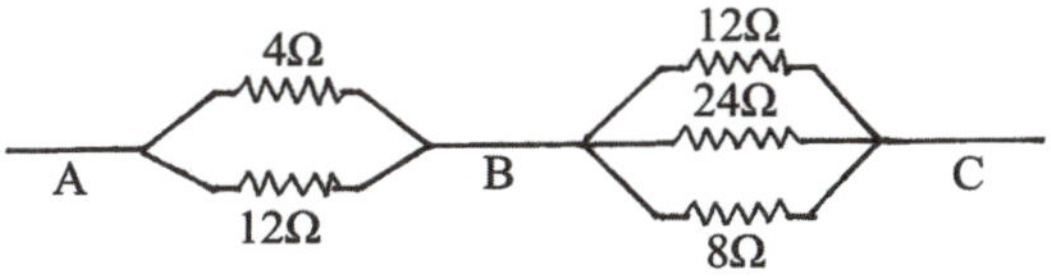

Resistance between A and B is $\frac{1}{4}+\frac{1}{12}=\frac{1}{3}$

$\therefore \qquad R_{AB} = 3\ \Omega$

Resistance between B and C is $\frac{1}{12}+\frac{1}{24}+\frac{1}{8}=\frac{1}{4}$

$\therefore \qquad R_{BC} = 4\ \Omega$

Since R_{AB} and R_{BC} are connected in series in the given circuit

$$R_{AC} = 3 + 4 = 7\ \Omega$$

2. In the network shown , calculate the value of resistance X.

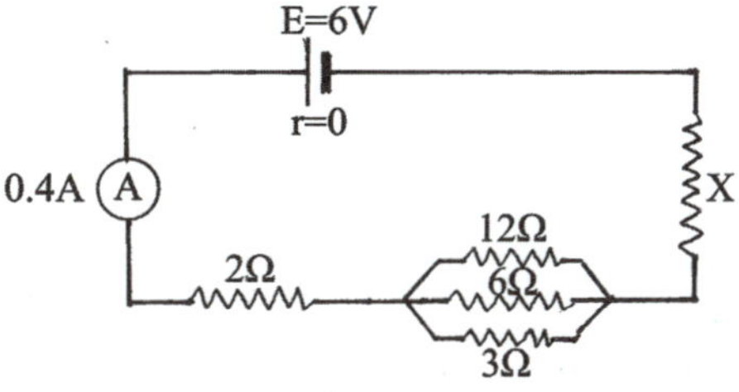

The equivalent resistance, R, of the three resistances in parallel is given by

$$\frac{1}{R} = \frac{1}{12} + \frac{1}{6} + \frac{1}{3}$$

whence

$$R = 12/7 \ \Omega$$

Total resistance in the circuit

$$R_{total} = 12/7 + 2 + X = 1.7 + 2 + X$$

or

$$= 3.7 + X \ \Omega$$

Since the net current in the circuit when fed from a battery of 6 V is 0.4 A

$$\frac{6.0}{3.7+X} = 0.4$$

whence

$$X = 11.0 \ \Omega$$

3. Two cells, each of EMF 1.5 V and internal resistance 1.0 ohm are connected in parallel to form a "battery". The battery is connected to an external resistance of 0.5 ohm and two resistances of 3.0 ohm and 1.5 ohm in parallel.

(i) draw the circuit diagram as above

(ii) calculate the current in the 'main' circuit

(iii) the current in the 1.5 ohm resistance

(iv) the drop in potential across the terminals of the battery

(i) The circuit is drawn below

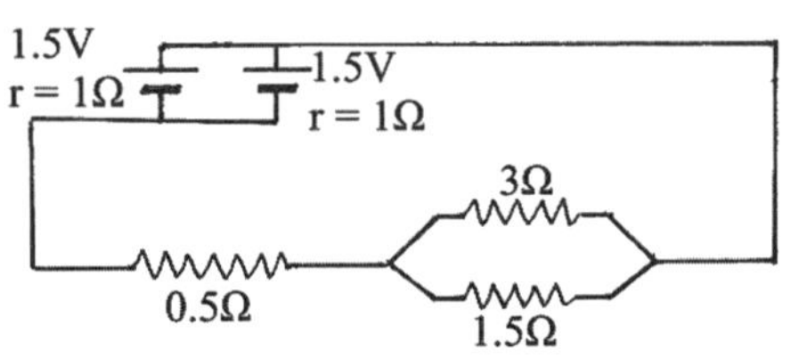

(ii) Since the cells are connected in parallel, the *net* terminal voltage of the battery so-formed will be 1.5 V and its equivalent internal resistance will be 1.0/2 or 0.5 ohm.

With the resistances of 3.0 ohm and 1.5 ohm connected in parallel, and in turn connected in series with the resistance of 0.5 ohm in series, the total resistance of the main circuit will be

$$R_{total} = \frac{1}{3.0} + \frac{1}{1.5} + 0.5 + 0.5 \ \Omega \quad \text{[battery resistance]}$$

$$= 1.0 + 0.5 + 0.5$$

$$= 2.0 \ \Omega$$

∴ current in the main circuit $= V/R_{total} = 1.5/2.0 = 0.75$ A

(iii) The PD across resistances in parallel is

$$= 0.75 \times 1.0 = 0.75 \text{ V}$$

∴ current through the 1.5 ohm resistance will be

0.75/1.5 or 0.5 A

(iv) The drop in potential across the battery terminals will be

main current x internal resistance of the battery

that is 0.75 x 0.5 or 0.375 V

Note that the terminal voltage of the battery will be

(open circuit) EMF of the battery - internal voltage drop

That is 1.5 - 0.75 x 0.5 V

or 1.125 V

Also, if only one cell were used to supply the *same* circuit current, its terminal voltage will drop to 1.5 – 0.75 or 0.75 V!

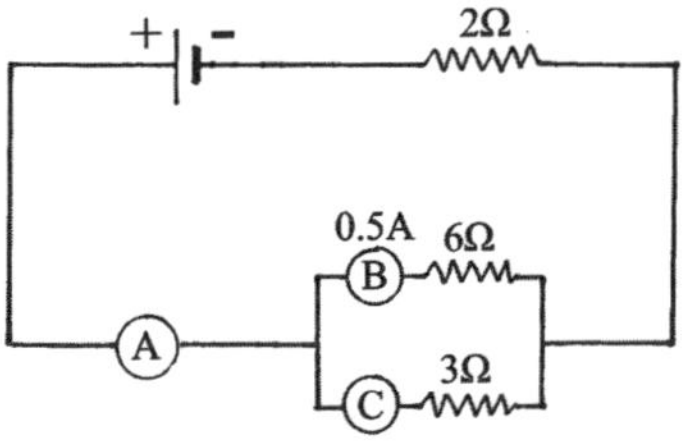

4. The diagram shows three ammeters A, B, and C as connected. The ammeter B reads 0.5 A. Calculate the readings of ammeters A and C and total resistance of the circuit. Assume the internal resistance of the battery to be negligible.

The equivalent resistance of the two resistances in parallel is

$$R_{eq} = \frac{1}{6} + \frac{1}{3} = 2\Omega$$

and total circuit resistance $R_{total} = 2 + 2 = 4\ \Omega$

The current through resistance B being 0.5 A, the PD across this resistance is $V_B = 0.5 \times 6$ or 3 V. This is also the PD across resistance C.

∴ current through resistance C = 3.0/3.0 or 1.0 A, the reading of ammeter C

The current through ammeter A will be the sum of currents through resistances B and C or the sum of current readings of ammeters B and C, that is

$$I_A = 1.0 + 0.5 \text{ or } 1.5 \text{ A}$$

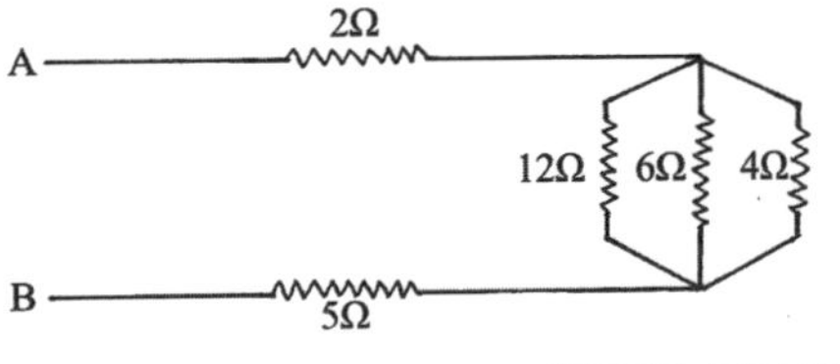

5. Find the equivalent resistance between points A and B.

[ICSE, 2015]

Resistances 4Ω, 6 Ω and 12 Ω are in parallel

∴ equivalent resistance of these,

$$\frac{1}{R} = \frac{1}{4} + \frac{1}{6} + \frac{1}{12}$$

$$= \frac{1}{2} \text{ or } R = 2\Omega$$

R is connected in series with resistances 2Ω and 5 Ω

∴ equivalent resistance between A and B = 2 + 2 + 5 or 9 Ω

6. Find the equivalent resistance between points A and B.

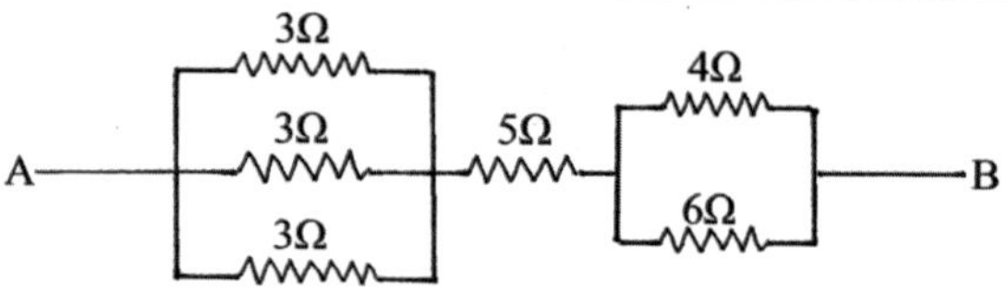

[ICSE, 2014]

(a) The equivalent resistance of resistors 3Ω, 3 Ω, 3 Ω connected in parallel is
$1/R = 1/3 + 1/3 + 1/3$ or $R = 1\ \Omega$

(b) The equivalent resistance of 4Ω and 6 Ω connected in parallel (on the right) is
$$1/R' = 1/4 + 1/6 \quad \text{whence} \quad R' = 2.4\ \Omega$$

∴ total resistance between points A and B
$$R_{AB} = R + 5 + R'$$
$$= 1 + 5 + 2.4 \quad \text{or} \quad 8.4\ \Omega$$

7. Two resistors of 4Ω and 6 Ω are connected in parallel to a cell to draw 0.5 A current from the cell.
(i) Draw a labelled circuit diagram showing the above arrangement
(ii) Calculate the current in each resistor

[ICSE, 2014]

(i) The circuit diagram is drawn below

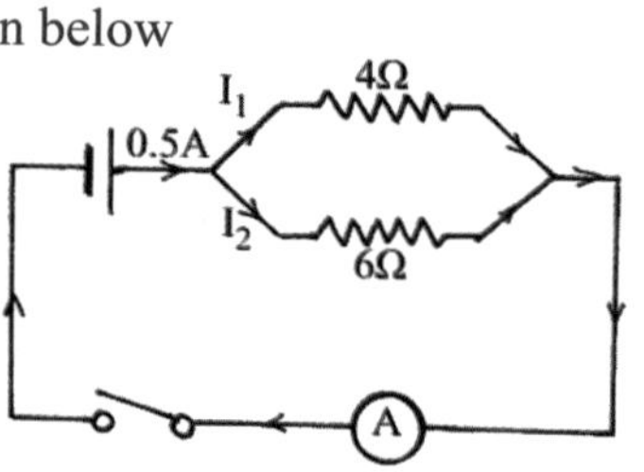

(ii) The equivalent resistance of resistors 4Ω and 6 Ω connected in parallel is
$1/R = 1/4 + 1/6$ whence $R = 2.4\ \Omega$
The current drawn from the cell is 0.5 A

∴ potential drop across R or the resistors in parallel, PD = 0.5 x 2.4 or 1.2 V
Then, current through 4Ω resistance = PD/4 = 1.2/4 or 0.3 A
current through 6Ω resistance = PD/6 = 1.2/6 or 0.2 A

8. Calculate the equivalent resistance between the points A and B for the following combination of resistors.

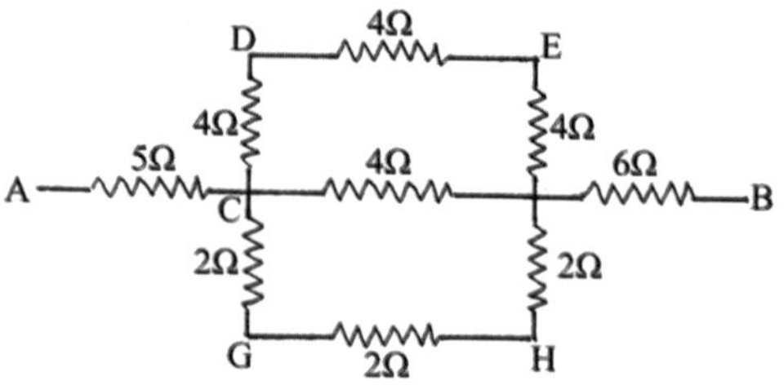

The resistances 4Ω, 4 Ω, 4 Ω in the upper half of the circuit are in series

∴ equivalent resistance of these,
$$R_1 = 4 + 4 + 4 \quad \text{or} \quad 12\ \Omega$$
The resistances 2Ω, 2 Ω, 2 Ω in the lower half of the circuit are in series

∴ equivalent resistance of these, $R_2 = 2 + 2 + 2$ or $6\ \Omega$

Now R_1 and R_2 together are connected in parallel with the resistance of 4 Ω of the circuit in middle

∴ equivalent resistance of these in parallel is

$$\frac{1}{R} = \frac{1}{R_1} + \frac{1}{R_2} + \frac{1}{4}$$

Or $1/R = 1/12 + 1/6 + 1/4$

whence $R = 2\ \Omega$

Now the resistance R is in series with resistances 5 Ω and 6 Ω to the left and right in the circuit

∴ Total, equivalent resistance between the points A and B of the combination is

$$R_{eq} = 5 + 2 + 6 \quad \text{or} \quad 13\ \Omega$$

9. Five resistors of different resistances are connected together as shown in the figure. A 12 V battery is connected to the arrangement. Calculate
(i) the total resistance in the circuit
(ii) the total current flowing in the circuit

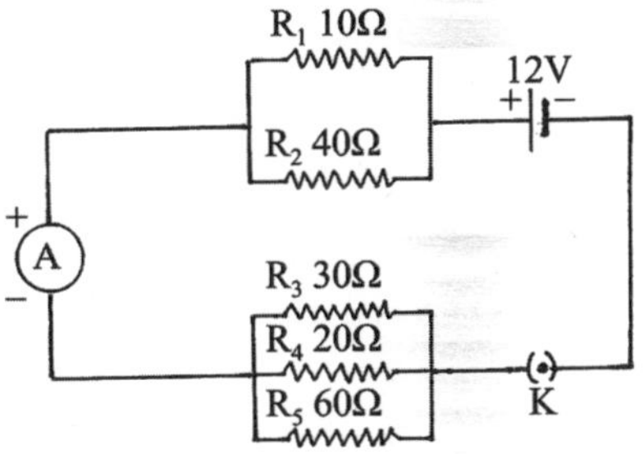

[ICSE, 2010]

(i) Combined resistance of R_1 and R_2 in parallel is

$$R'' = \frac{10 \times 40}{10 + 40} = 8\ \Omega$$

Similarly, combined resistance of R_3, R_4 and R_5 in parallel is

$$R'' = \frac{1}{\frac{1}{30} + \frac{1}{20} + \frac{1}{60}}$$

$$= 60/6 \quad \text{or} \quad 10\ \Omega$$

Since R′ and R″ are now in series, the total circuit resistance is given by

$$R = R' + R'' = 8 + 10 \quad \text{or} \quad 18\ \Omega$$

(ii) Then, total current flowing in the circuit

$$= V/R = 12/18 \quad \text{or} \quad 2/3\ A\ (0.67\ A)$$

10. Find the total resistance of eight resistors connected as shown in the circuit below

$R_1 = 10\ \Omega$
$R_2 = 2\ \Omega$
$R_3 = 3\ \Omega$
$R_4 = 17\ \Omega$
$R_5 = 20\ \Omega$
$R_6 = 20\ \Omega$
$R_7 = 8\ \Omega$
$R_8 = 10\ \Omega$

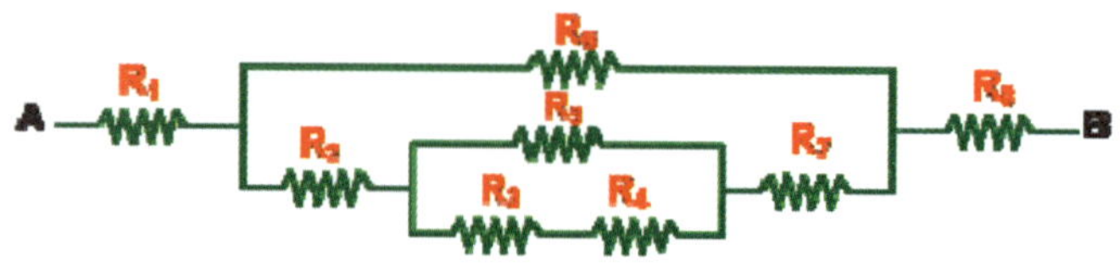

→ Series for R_3 and R_4 → R_{34}
$R_{34} = R_3 + R_4 = 3 + 17 = 20\ \Omega$
→ Parallel for R_5 and R_{34} → R_{35}
$R_{35} = 10\ \Omega$
→ Series for R_2, R_{35} and R_7 → R_{27}
$R_{27} = 2 + 10 + 8 = 20\ \Omega$
→ Parallel for R_{27} and R_6 → R_{276}
$R_{276} = 10\ \Omega$
→ Series for R_1 , R_{276} and R_8 → R_{AB} or R_{total}
$R_{AB} = 10 + 10 + 10 = 30\ \Omega$

11. 8 resistors, each having a resistance of 10 Ω, are arranged as shown

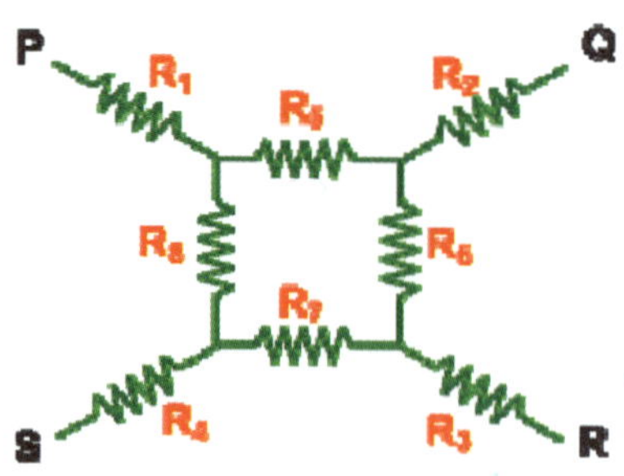

Find the total resistance across terminals P and R

There will be no current flow along R_2 and R_4, being out of circuit considering P and Q, so these can be disregarded
→ Series for R_5 and R_6 :
$R_{56} = 20\ \Omega$
→ Series for R_7 and R_8 :
$R_{78} = 20\ \Omega$

→ Parallel for R_{56} and R_{78} :

$$R_{58} = 10\ \Omega$$

→ Series for R_1 , R_{58} and R_3 to find $R_{PR} = R_{total}$:

$$R_{PR} = 10 + 10 + 10 = 30\ \Omega$$

12. Find the equivalent resistance, R_{EQ} for the following resistor combination circuit.

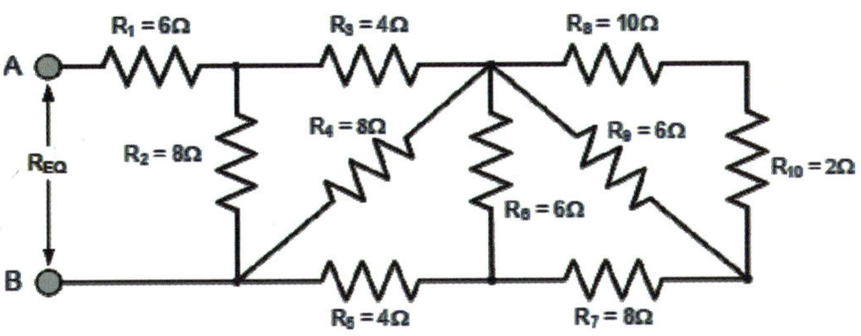

Step-by-Step solution

At first glance, this resistor ('ladder') network may seem a complicated one, but this too is just a combination of series and parallel resistors connected together. Starting from the right hand side and using the simplified equation for two parallel resistors, obtain the equivalent resistance of R_8 to R_{10} in series, in parallel with R_9. Call it R_A.

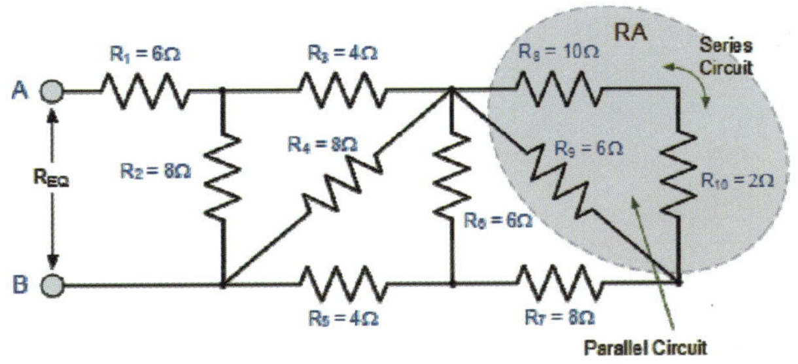

Then

$$R_A = \frac{R_9 \times (R_8 + R_{10})}{R_9 + R_8 + R_{10}} = \frac{6 \times (10 + 2)}{6 + 10 + 2} = 4\Omega$$

R_A is in series with R_7 therefore the total resistance of R_A and R_7 in series will be
$R_A + R_7 = 4 + 8 = 12\ \Omega$
$R_7 = 4 + 8 = 12\ \Omega$
This resistive value of 12 Ω is now in parallel with R_6 and can be calculated as R_B.

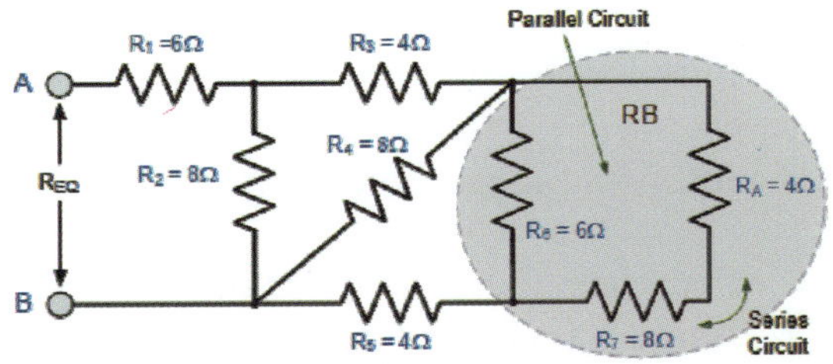

$$R_B = \frac{R_6 \times (R_A + R_7)}{R_6 + R_A + R_7} = \frac{6 \times (4 + 8)}{6 + 4 + 8} = 4\Omega$$

R_B is in series with R_5 therefore the total resistance will be $R_B + R_5 = 4 + 4 = 8\ \Omega$ as shown.

This resistive value of 8 Ω is now in parallel with R_4 and can be calculated as R_C as shown.

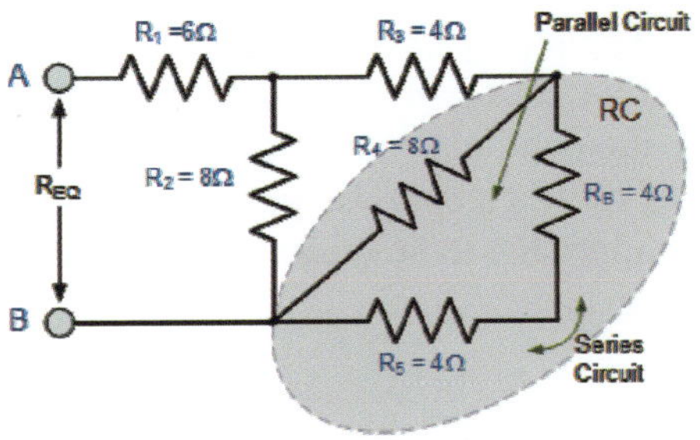

$$R_C = \frac{R_4 \times (R_B + R_5)}{R_4 + R_B + R_5} = \frac{8 \times (4+4)}{8+4+4} = 4\Omega$$

R_C is in series with R_3 therefore the total resistance will be $R_C + R_3 = 8\ \Omega$.

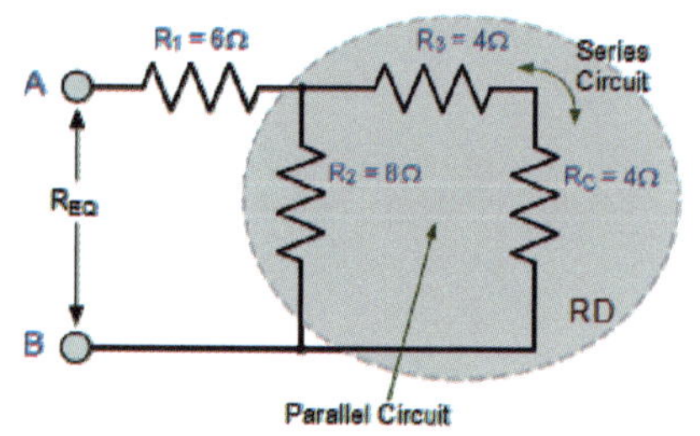

This resistive value of 8 Ω is now in parallel with R_2 from which R_D can be calculated as

$$R_D = \frac{R_2 \times (R_C + R_3)}{R_2 + R_C + R_3} = \frac{8 \times (4+4)}{8+4+4} = 4\Omega$$

R_D is in series with R_1 therefore the total resistance will be $R_D + R_1 = 4 + 6 = 10\ \Omega$

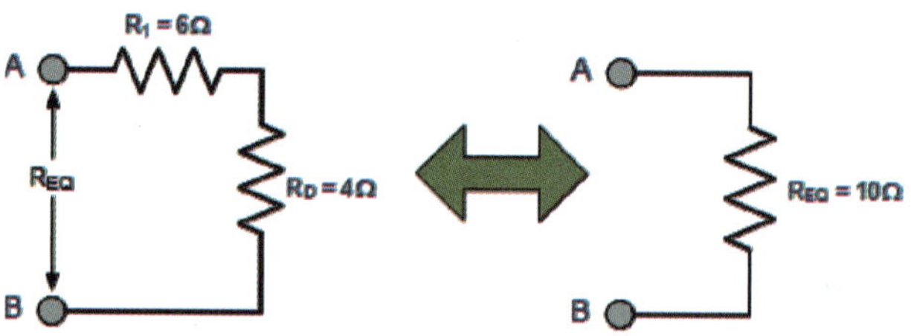

Thus, comprising ten individual resistors connected together in series and parallel combinations can be replaced with just one single equivalent resistance (R_{EQ}) of value 10 Ω.
[Note the sequence of simplification, highlighted in the steps, to the right]

Strategy for Solving Series-parallel Combination Problems

The following suggestions for approaching combination circuit problems are offered

- If a schematic diagram is not provided, take the time to construct one. Use 'schematic' symbols such as adopted in the examples above.
- When approaching a problem involving a combination circuit, take time to organise the given values and marking them with appropriate symbols such as R_{12}, R_{AB} etc.
- Recall and use the appropriate formulae for the equivalent resistance of series- and/or parallel-connected resistors.
- Transform a combination circuit into a series circuit by replacing the parallel section first with a single resistor having a resistance value equal to the equivalent resistance of the parallel section.
- Proceed step by step as, for example, in example "3" or "12" above to systematically approach to arrive at the final single equivalent value of the circuit resistance.

Appendix A

Construction of a lead acid battery

Main Parts

- Container

Container is a box of vulcanized rubber, moulded rubber, moulded plastic, glass or ceramic. On the base of this box there are supports blocks on which the positive and negative plates are fitted.

- Separator

It is most important part of a lead acid battery which separates the positive and negative plates from each other and prevents their short circuit. The separators must be porous so that the electrolyte may circulate between the plates. They should also have high insulating resistance and mechanical strength. The material used for separators are wood, rubber or PVC.

- Electrolyte

In a lead acid battery dilute sulphuric acid (H_2SO_4) is used as an electrolyte. For this purpose one part concentrated sulphuric acid is mixed with three parts of *distilled* water; also used for top-up of the battery periodically as required.

- Cover of cell

It is also made of the same material which is used for making the container and is used to cover the complete cell assembly after the installation of the plates in it. It protects the cell from the dust as well as other external impurities.

- Vent plugs

The vent plugs are provided in the cover plate of each cell and are used to fill up the electrolyte or add distilled water in the cell or the inspection of internal condition of the cell. The vent plugs also serve for exhaust of the gases generated in the cell to the atmosphere.

- Connecting bar

It works like a link and used to connect the two cells in series: positive terminal of one cell to the negative terminal of another cell.

■ Terminal posts

These are the terminals of the battery which are connected to charging circuit as well as the load. For identification the diameter of the positive terminal is designed to be bigger as compared to the negative terminal.

Working

[See the figure below]

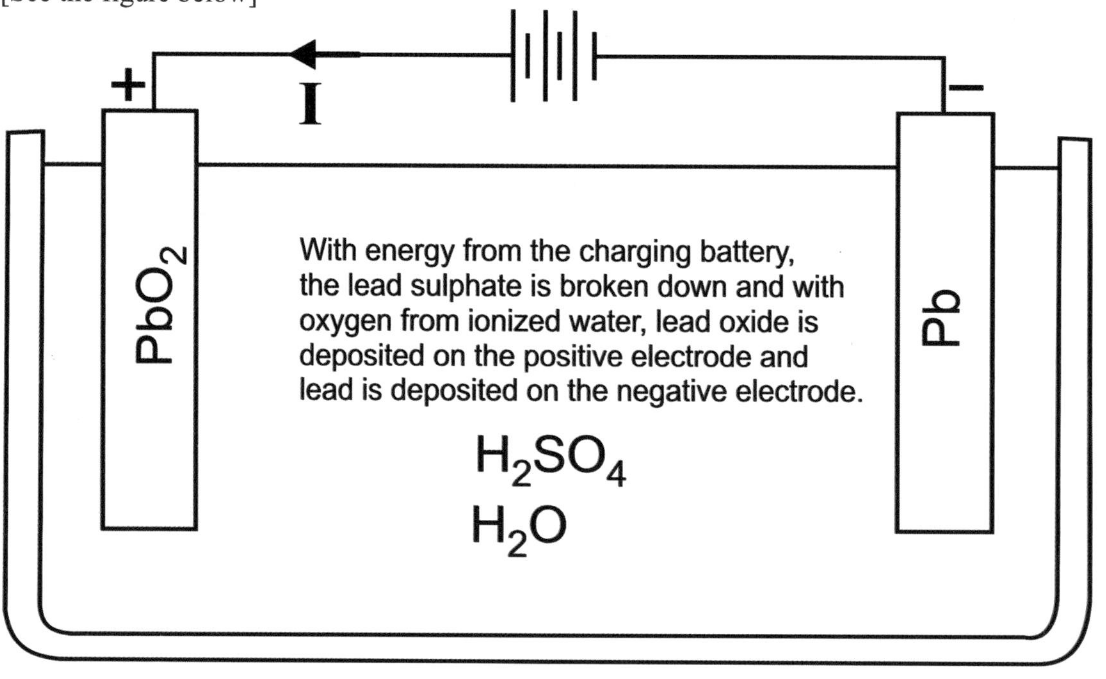

A load is connected externally between the plates of the battery. In diluted sulphuric acid, the molecules of the acid split into positive hydrogen ions and negative sulphate ions. The hydrogen ions when reach the lead (PbO_2) plate, they receive electrons from it and become hydrogen atom which again attack PbO_2 and form PbO and H_2O (water). This PbO reacts with H_2SO_4 and forms $PbSO_4$ and H_2O (water).

EXERCISES

1. The equivalent resistance of the following circuit is 6.0 Ω. Find the value of resistance X.

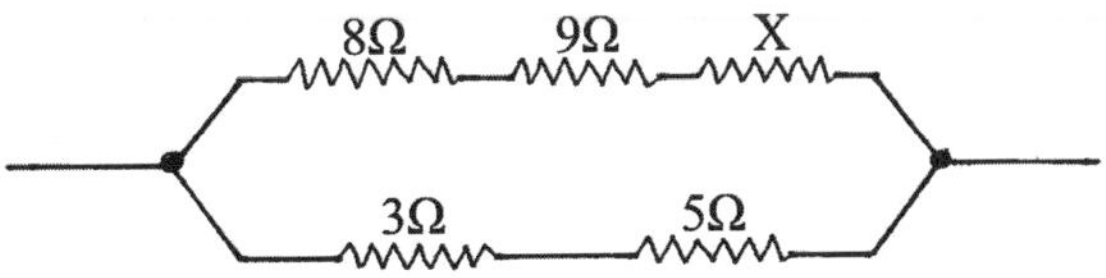

[7.0 Ω

2. Four resistances of values 0.5 Ω, 1.5 Ω, 4.0 Ω and 6.0 Ω, respectively, are connected in series across a battery of 6.0 V and negligible internal resistance. Calculate (i) the current drawn from the battery, (ii) PD across each of the resistances.

[0.5 A; 0.25 V, 0.75 V, 2.0 V, 3.0V

3. A cell of EMF 1.8 V is connected to an external resistance of 2.0 Ω, the potential difference across the resistance being 1.6 V. Calculate the internal resistance of the cell.

[0.25 Ω

4. In the circuit shown, ammeter A reads 0.3 A. Calculate (i) total resistance of the circuit, (ii) value of resistance R, (iii) current flowing through R

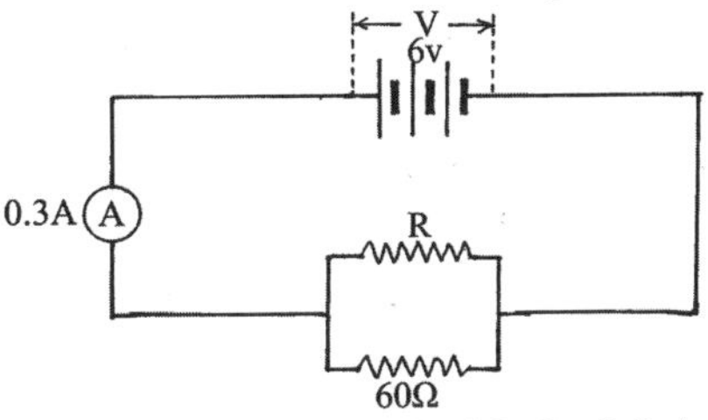

[20 Ω; 30 Ω; 0.2 A

5. In the circuit shown, calculate (i) the main current, (ii) current in each of the parallel circuits.

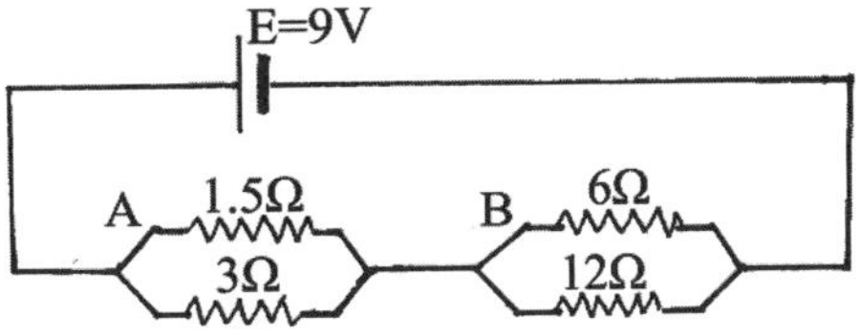

[1.8 A; 0.6 A. 0.6 A

6. A metal wire of resistance 6 Ω is stretched so that its length is increased to twice its original length. Calculate its new resistance.

[ICSE, 2013]

[24 Ω

7. Calculate the equivalent resistance between P and Q from the diagram shown:

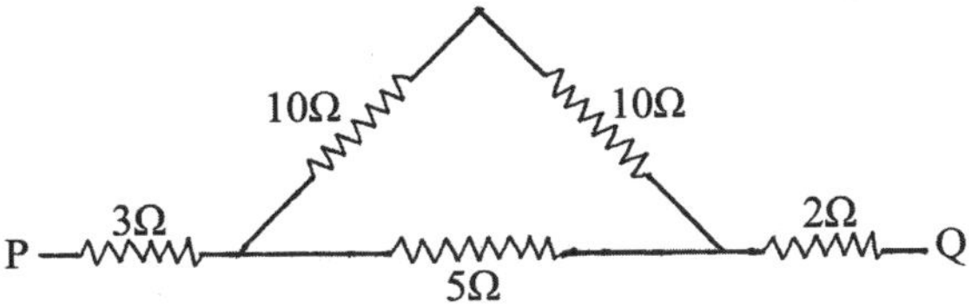

[ICSE, 2012]

[9 Ω

8. Three resistors are connected to a 6 V battery as shown in the figure.

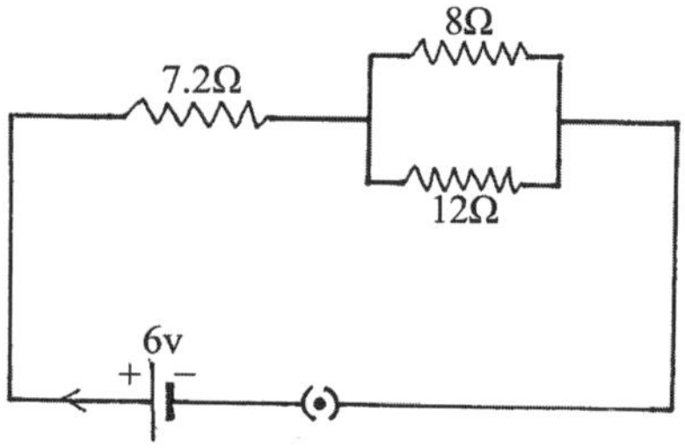

Calculate:
(i) the equivalent resistance of the circuit
(ii) total current in the circuit
(iii) potential difference across the 7.2 Ω resistor.

[ICSE, 2012]
[12 Ω; 0.5 A; 3.6 V

9. Calculate the equivalent resistance between A and B from the diagram shown.

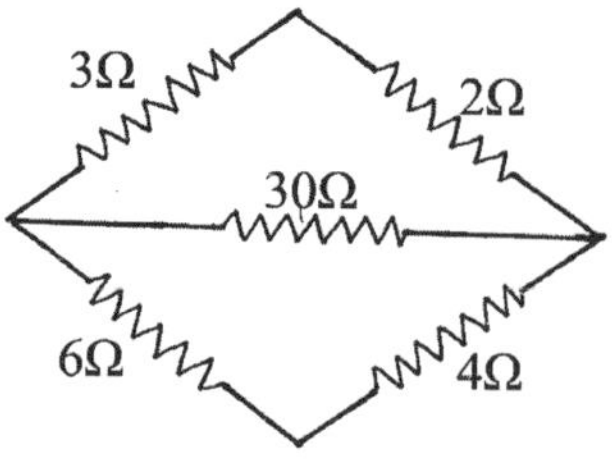

[ICSE, 2011]
[3 Ω

10. Three resistors are connected to a 12 V battery as shown in the figure.

(i) What is the current through the 8 ohm resistor?
(ii) What is the potential difference across the parallel combination of 6 ohm and 12 ohm resistor?
(iii) What is the current through the 6 ohm resistor?

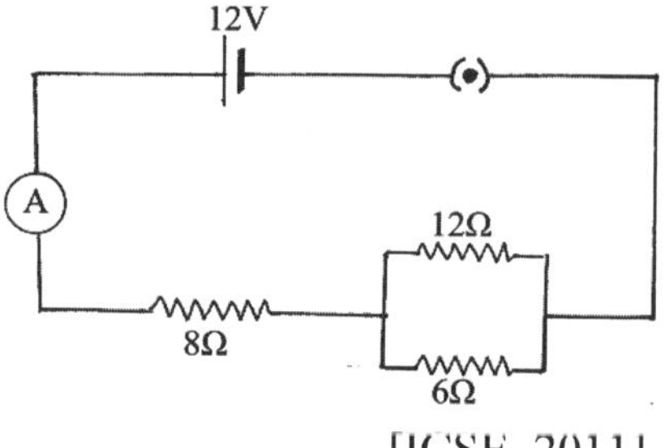

[ICSE, 2011]
[1 A; 4 V; 2/3 A

11. Six resistances are connected together as shown in the figure. Calculate the equivalent resistance between the points A and B.

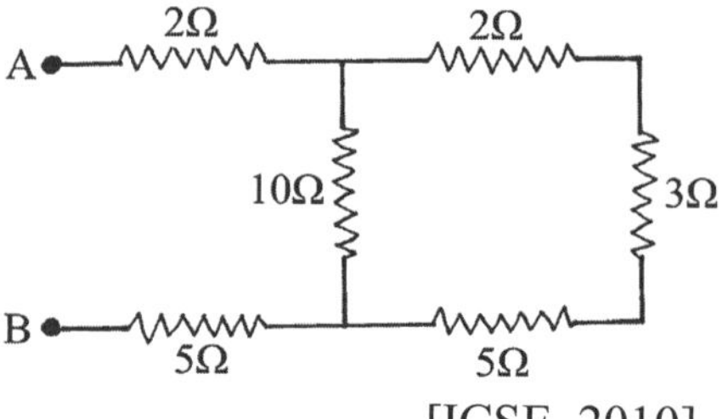

[ICSE, 2010]
[12 Ω

12. The equivalent resistance of the circuit diagram shown is 4 Ω. Calculate the value of x.

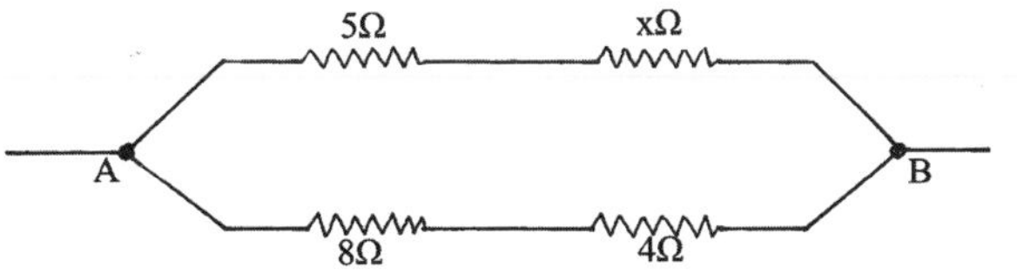

[ICSE, 2009]

[1 Ω

13. Three resistors of 6.0Ω, 2.0 Ω and 4.0 Ω, respectively, are joined together as shown in the figure. The resistors are connected to an ammeter and to a cell of EMF 6.0 V. Calculate:

(i) the effective resistance of the circuit

(ii) the current drawn from the cell

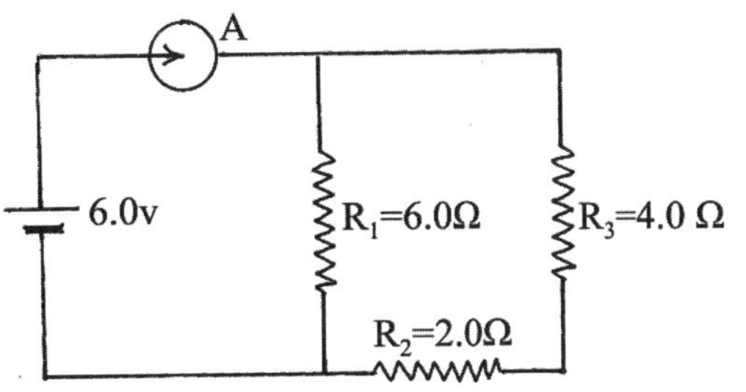

[ICSE, 2008]

[3 Ω; 2 A

14. A cell of EMF 1.5 V and internal resistance 1.0 Ω is connected to two resistors of 4.0 Ω and 20.0 Ω in series as shown in the figure:

Calculate:

(i) current in the circuit

(ii) potential difference across the 4.0 ohm resistor

(iii) voltage drop when the current is flowing

(iv) potential difference across the cell

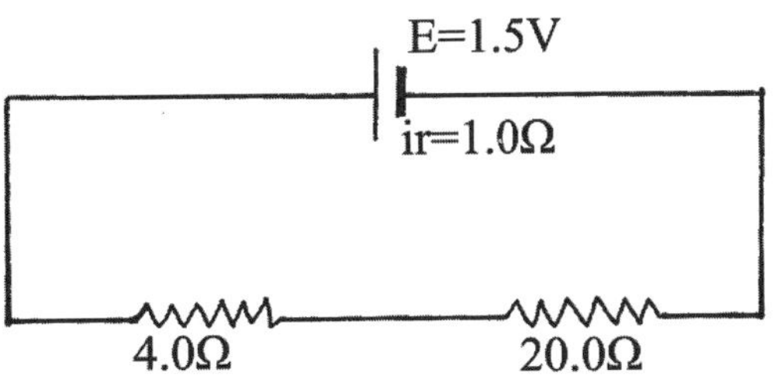

[ICSE, 2007]

[0.06 A; 0.24 V; 0.06 V; 1.44 V

15. Suggest the value of the resistance R and the method of connection to result in an equivalent resistance of (i) 2 Ω; (ii) 10 Ω from a resistance of 6 Ω.

[(i) resistance of 3 Ω in parallel,

(ii) a resistance of 4 Ω in series

II Electrical Power and Energy

Electrical energy; examples of heater, motor, lamp, loudspeaker, etc. Electrical power; measurement of electrical energy, $W = QV = VIt$ from the definition of pd. Combining with ohm's law $W = VIt = I^2Rt = (V^2/R)\,t$ and electrical power $P = (W/t) = VI = I^2R = V^2/R$. Units: SI and commercial; power rating of common appliances, household consumption of electrical energy; calculation of total energy consumed by electrical appliances; $W = Pt$ kilowatt x hour = kWh), [simple numerical problems].

ELECTRICAL POWER

The concept of power, mainly in context of mechanical power was discussed at length in part III of Unit 1. Electrical power, like mechanical power, is likewise the rate of doing work. It is denoted by the letter P, occasionally by p (for example power at an instant) and is the product of voltage, V and current I, being related as

Power, P = voltage, V x current, I

When the voltage is expressed in **volt** and current in **ampere**, in SI units, the power is obtained in **watt**, denoted by letter W.

Source(s)

The two 'components' of electrical power being voltage and current, a source of electrical power must provide a voltage across its terminals from where an electric current can be drawn into an external circuit, 'consuming' power (= VI W).

A basic source of electrical or electric power for supplying small quantity of power, frequently used even today, is a battery of suitable (Ah) capacity; the battery producing electric power through chemical reaction. On a larger scale, electric power would be supplied from machines called electric generators[1] where power conversion is essentially from mechanical to electrical.

[1] A generator producing direct-current power is known as a "dynamo" whilst an alternating current (AC) generator is identified as an "alternator".

Electric Power *into* a Circuit

It follows that if a voltage, V volt, is applied to a 'circuit', as shown in Fig.2.1, having an (equivalent) resistance of R ohm, a current given by

$$I = V/R \ \text{A}$$

will be drawn by the circuit and the electric power supplied, or 'consumed' by the circuit, will be

$$P = V \times I \ \text{W}$$

$$= V \times V/R \ \text{W}$$

$$= V^2/R \ \text{W}$$

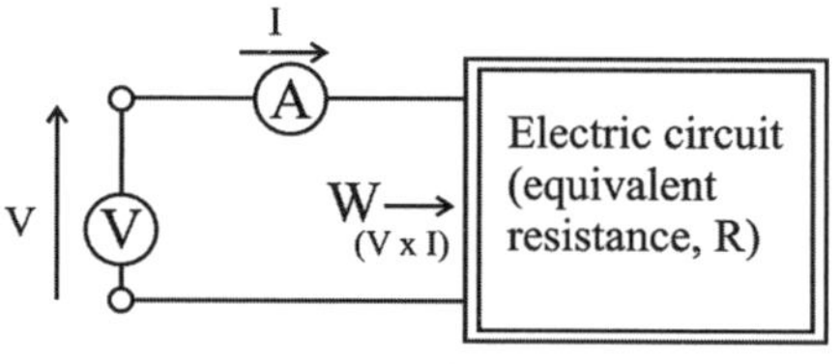

Fig.2.1 : Electric power into a circuit

Alternatively, since V = I R, power can be expressed as

$$P = (I\,R)^2/R$$
$$= I^2 \times R \quad \text{W}$$

Unit of electric power

As mentioned above, the SI unit of power is watt, denoted by W.

Multiples and Sub-multiples of Unit of Power

Multiples

The frequently used multiples of unit of power are

one kilowatt, 1 kW = 1000 W = 10^3 W
one megawatt, 1 MW = 1,000,000 W = 1000 kW = 10^6 W
one gigawatt, 1 GW = 1,000,000,000 W = 1000 MW = 10^9 W
one terawatt, 1 TW = 1,000,000,000,000 W = 1000 GW = 10^{12} W

Sub-multiples

In numerous applications, esp. in electronic circuits, very small values of electric power are encountered. It is thus customary to use sub-multiples of unit of power. Some of the common small units are

a milliwatt, 1mW = 1/1,000 W = 10^{-3} W
a microwatt, 1μW = 1/1,000,000 W = 10^{-6} W
a picowatt, 1 ρW = 1/1,000,000,000,000 W = 10^{-12} W
Of these, the first two are in commonest use

Power Rating of Common (domestic) Appliances

Domestic appliances in common or infrequent use are usually rated for their *power* consumption in terms of watt. Some of these appliances and their *typical* power rating are listed in the table below.

Appliance	Power rating[2], W
• electric bulbs[3]	
- incandescent	10 - 500
- LED	0.5 - 20
• fluorescent tube	
- 4 ft.	40
- 20 ft.	20
• toaster	800-1500
• blender	300
• cooking range (cooker with oven)	1000-2500
• microwave	600-1500
• hot plate	1200
• electric kettle	750-1000
• dish washer	1200-1500
• washing machine (automatic)	500
• vacuum cleaner (upright)	200-700
• electric mower	1500
• pumping motor	200
• sewing machine	100
• electric iron	1000
• geyser	2000
• ceiling fan	100
• television	
- CRO type	150
- LED type	25
• VCR	40
• CD player	35
• cassette player	8
• electric clock	3

[2]The figure listed in the table are only *indicative* and NOT exact ratings; variations are possible depending on the design and manufacturer. In fact, the competition amongst manufacturers is very keen, to evolve energy-efficient and cost-effective products.

[3]Electric bulbs and illumination devices are now rated in terms of their "lumen" output, in addition to, or in place of, their power rating in watt. For example, a 60 W incandescent bulb may have a lumen output of about 1000 lumens; a 12 W LED lamp may provide a lumen output of nearly 1200 lumen.

ELECTRICAL ENERGY

In general, energy involves the 'dimension' time and is best comprehended in terms of work done of some kind[4] in time t. The work is performed by en entity called power so that the corresponding energy is expressed as

energy, E = power, P x time, t

In SI units, power is expressed in watt and time in second, the energy is then obtained in joule so that

joule = watt x second

and

1 joule = 1 watt x 1 second

which defines the SI unit of energy as

1 joule = 1 watt-second

joule being denoted by the letter J.

Electrical Energy and Electric Power

Using the above concept, *electric* power is the rate at which **electrical** energy is transferred to, or consumed by, an electric circuit. Since the current can be interpreted in terms of electric charge flowing in a circuit, the electric power in watt produced by an electric current I consisting of a charge of Q coulomb every t second passing through an electric circuit having a potential difference of V is

$$P = (Q \times V)/t$$

Or

$$P = V \times I$$

since

$$I = V/t$$

where

Q is electric charge in coulomb
t is time in second
I is electric current in ampere
V is electric potential (difference) or voltage in volt

---

[4]Recall Unit 1, Chapter III.

Therefore, electric power is electric energy per unit time, expressed in joule/second, that is watt where the energy can be interpreted as "*work required to move a charge, Q, through a potential (difference), V, in an electric circuit*".

Or electrical energy, $E = Q\,V = V\,I\,t$ J

Since, by ohm's law, $V = I\,R$

$\therefore$ $E = I^2\,R\,t$ J

Or $E = (V/R)^2 \times R\,t$ J

$= (V^2\,t)/R$ J

Example

A current of 2 A is passed through a coil of resistance 75 ohm for two minutes. Calculate (i) the heat energy 'produced', (ii) the charge passed through the resistance.

(i) Heat produced in the resistance $= I^2\,R\,t = 2^2 \times 75 \times (2\times60)$

$= 36{,}000$ J

(ii) Since charge is equal to current x time,
the charge passed through the resistance $= 2 \times 120$ or 240 C

Electrical Energy: Generation and Delivery

The two important aspects related to electrical energy are its "generation" and "delivery" considering its common usage in practice. Thus, electric power or energy is generated in electric power plants, followed by its appropriate delivery to the users or consumers in domestic, commercial or industrial sphere.

Examples of Usage of Electrical Energy

Some of the common, important forms in which electrical energy is used are

- **illumination**: the most common purpose for which electricity is used in variety of ways to provide "general" or "localised" illumination employing bulbs and other light emitting gadgets
- **mechanical power**: electric power used to drive vast variety of motors in a number of applications or machinery
- **heating**: for providing heat energy in various forms, from heaters, toasters, ovens and geysers etc. in houses to different type of furnaces in the industry[5]

[5]The basic principle of conversion of electrical energy into heat usually involves heating of a 'resistors' in different forms where the power lost in heat is given by $P = I^2\,R$ where I is the current into resistor R and the heat energy is obtained to be $I^2\,R\,t$ in time t.

- **sound**: a common example is the conversion of electrical energy into sound using loudspeakers of different types and design[6]

- **electric traction**: uses AC or DC electricity for electric locomotives to haul goods and passenger trains at very high speeds[7]

- **electrolysis**: this depends largely on DC electricity for processing of ores etc. in industry, for example use of 'pots' in aluminium industry for processing bauxite; now for electrolysis of water to produce hydrogen as fuel for cars and vehicles in place of petrol or diesel

- **electromagnets**: huge electromagnets are used in machine shops in industries to lift ferrous metal pieces to move them around

Measurement of Electrical Energy

As brought out, electrical energy is the product of electric power and time. Thus

$$\text{electrical energy} = \text{electric power x time}$$

When electric power is expressed in watt and time in hours, electrical energy is obtained in watt-hour – also the SI unit of electrical energy – else if power is in joule per second from the concept of work done, and time is specified in second, the energy is obtained in joule. Accordingly, measurement of electrical energy is best carried out by measuring the power input to a circuit, or an electric appliance, with the help of a wattmeter, as illustrated in Fig.2.2 and measuring the time separately using a clock for which the power is ON. The product of the two quantities would then provide the amount of electrical energy consumed[8].

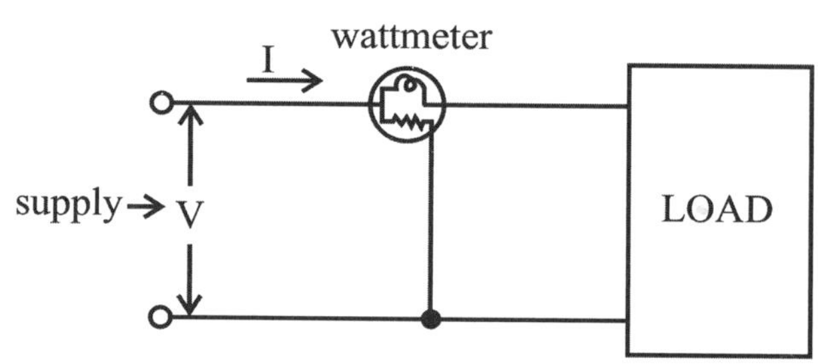

Fig.2.2 : Use of a wattmeter to measure power in a load

Household Consumption of Electrical Energy

The consumption of electrical energy in a typical household would clearly depend on the use of various types of appliances as listed in the table above. This would involve "switching

---

[6]See Chapter III, Unit 3.

[7]Trains at incredible speeds (> 300 kmph) are now increasingly run using "linear motors" and "magnetic levitation", incorporating superconductors.

[8]In practice, and in universal usage, a single instrument or an "energy meter" is incorporated suitably to measure electrical energy directly in appropriate unit(s). This is discussed further later.

ON" of a particular, or assorted, gadget(s) for desired number of hours[9]. Note that all the appliances or gadgets in the house[10] will *not* be switched on at the same time and for the same length of time during a period of 24 hours. For example, some lights using given type of devices may be singly or simultaneously on for some hours; at the same time a fridge may be on for a few hours[11]; a mixer may be on for 10/15 minutes; a TV may be on for a couple of hours at a time; an electric clock may be on all the time, that is 24 hours, and so on. The consumption of energy can then be worked out item-wise and added to obtain total "average" consumption over the 24-hour cycle. And similarly for the entire month[12]. This is demonstrated by the following calculations.

Calculation of total energy

Assume that in a given household various appliances are switched on for certain number of hours as indicated below during a 24-hour period.

Appliance/load, W*	Duration of operation, hours(h)	Consumption, Wh
1. 4 fluorescent tubes, 4x40 W	4 hours [7- 11 pm]	640
2. 4 CFL lamps, 4x25 W	2 hours [7-9 pm]	200
3. 6 LED bulbs, 6x12	4 hours [6.30-10.30 pm]	288
4. Ceiling fans, 4x100 W	10 hours in all	4000
5. 1 conventional TV, 1x100W	4 hours [7-11 pm]	400
1 LED TV, 25 W	- do -	100
6. Fridge, 125 W ON/ 30 W OFF	8 hours ON, 12 hour OFF	1360
7. Mixer, 800 W	1 hour in all	800
8. Microwave, 1200 W	½ hour in all	600
9. "Box" oven, 1200 W	½ hour [on 'average]	600
10. Geysers, 2x1200 W	½ hour each	1200
11. Electric iron, 1000 W	½ hour	500
12. Toaster, 700 W	½ hour	350
13. Exhaust fan (in the kitchen), 100 W	6 hours in all	600
14. Water pumping motor, 200 W	1 hour	200
15. Electric kettle, 1000 W	½ hour	500
16. Electric clock, 3 W	24 h	72
17. Assorted (cell phone)"chargers", 1 W [all considered together]	2 hour	2

*Assumed ratings for calculations

[9]Also known as the "duty cycle".

[10]Also called the "connected load".

[11]Note that the duty cycle of a typical fridge consists of the compressor being on for 'some time' and off for many minutes.

[12]Once again, this will be arrived at automatically by the electricity meter provided by the electricity-supply company to every household.

As arrived at in the table, total energy consumption is 12412 Wh in "24 hours" or per "day"

Or 12.5 kWh[13] (approx.)

Observe that the energy consumption of *six* LED bulbs is less than half of that due to *four* fluorescent tubes whilst providing better illumination in lumen [tubes: 4x1600 or 6400 lm; LED bulbs: 6x1200 or 7200 lm]. Clearly, the LED bulbs (and LED TV) would do much better as compared to incandescent lamps or even fluorescent tubes in terms of lumen output *and* life of bulb/filament, and TV etc.

Average Monthly Consumption and Charges

Assuming average daily consumption of the household to be 12.5 "units" (or kWh) as derived above, the average monthly consumption (for 30 days) would be 30x12.5 or 375 units and assuming the rate of electricity charges to be Rs.5/- per unit, the monthly bill for the consumer will be

375 x 5 or Rs.1875/-

Example

1. Four "tube lights" of rating 40 W each and two fans of rating 100 W each are connected to a supply of 200 V and operate on an average of 8 hours a day. If energy costs Rs.1.50 per kWh, calculate monthly bill of the consumer, assuming 30 days in a month. Calculate also the "fuse rating" for the appliances.

Electric load comprising

(i) tube lights: 4 x 40 = 160 W
(ii) fans: 2 x 100 = 200 W
Total load = 360 W

Energy consumed per day = 360 x 8
= 2880 Wh or 2.88 kWh

Energy consumption per month = 2.88 x 30
= 86.4 kWh

∴ monthly bill of the consumer = 1.5 x 86.4 or Rs.129.60

The current drawn by the load is 360/200 or 1.8 A. The fuse should be rated at 1.8 A or 2.0 A as the nearest 'standard' value.

2. An electric motor rated at 2 HP and two cookers of 500 W each operate on a supply of 250 V for 4 hours per day. If the energy costs Rs.1.80 per kWh, calculate (i) weekly bill of the consumer, (ii) minimum fuse rating. Assume one HP to be equal to 750 W.

---

[13]kWh or "kilowatt-hour" is the actual unit of electrical energy consumption in practice as discussed later; also commonly known as a "unit" equal to one kWh.

Power consumed by

(i) the motor:	2 x 750	= 1500 W
(ii) cookers:	2 x 500	= 1000 W

Energy consumed per week = (2500 x 4) x 7 (days)
= 10,000 x 7 Wh or 70 kWh

Weekly bill of the consumer = 1.8 x 70 or Rs.126.00

Fuse rating for the load = total load in watts/supply voltage
= 2500/250 or 10 A

EXERCISES

1. Calculate the current through a 60 W bulb rated for 250 V. If the supply voltage falls to 200 V, how is the power consumed by the bulb affected, still drawing the same current?

[0.24 A; 48 W

2. An electric heater is rated for 1000 W, 200 V. Calculate resistance of its heating element and current passing through it when connected to the supply.

[40 Ω; 5 A

3. An electric heater rated for 4 kW is connected to a supply of 220 V. Calculate the cost of using the heater for 12 hours if one unit (kWh) of electricity costs Rs.3.25.

[Rs.156.00

4. An electrical appliance is designed to consume 1000 kWh (that is 1 kW in an hour) when connected to a supply of 220 V. If the appliance is operated for 2 hours, calculate the energy consumed by the appliance in (i) kWh, (ii) joule.

[2000 kWh; $7{,}200 \times 10^6$ J

5. An electric oven having a resistance of 20 ohm draws a current of 10 A. If it is operated for 3 hours daily, calculate weekly bill of its operation if electricity costs Rs.1.50 per kWh.

[Rs.63.00

6. Calculate the quantity of heat that will be produced in a coil of resistance 75 Ω if a current of 2 A is passed through it for 2 minutes.

[ICSE, 2010]
[8571.4 J

7. An electric heater is rated 1,000 W – 200 V. Calculate
(i) the resistance of the heating element
(ii) the current flowing through it

[ICSE, 2009]
[40 Ω; 5 A

8. Calculate the electrical energy consumed when a bulb of 40 W is used for 12.5 hours every day for 30 days.

[ICSE, 2009]
[15 kWh

9. An electrical heater is rated 4 kW, 220 V. Find the cost of using this heater for 12 hours if one kWh of electrical energy costs Rs.3.25.

[ICSE, 2008]
[Rs.156

III Household Circuits

Main circuit, switches, fuses; earthing; safety precautions; three-in plugs; colour coding of wires.
House wiring (ring system), power distribution; main circuit (3 wires – live, neutral, earth) with fuse/MCB, main switch and its advantages – circuit diagram; two-way switch, staircase wiring; need for earthing, fuse, 3-in plug and socket; conventional location of live, neutral and earth points in 3-in plugs and sockets; safety precautions; colour coding of wires.

HOUSE WIRING

Wiring System

Wiring systems for house wiring involve TWO aspects:

A. Supply
B. Supply voltage

Supply

For the last century or so, the electric supply for ALL usage has been and, continues to be, AC or alternating current supply[1].

Single Phase or Three Phase

The supply to a typical house or building may be "single" phase or "three" phase. The term single-phase implies use of *one* "live" wire, one "neutral" wire and one "ground" or "earth" wire. A three-phase supply would mean use of *three* live wires, one neutral wire and one ground wire. The former is almost invariably employed for wiring in 'small' houses,

[1]Till about the close of nineteenth century, the use of electricity for almost all purposes was in the form of DC or direct current, albeit limited to only a few applications. It was owing to advancements towards generation of AC and its usage with several important advantages, mainly related to transmission of electric power over long distances, that AC electricity gradually replaced DC almost entirely. Noteworthy contributions in this regard were made by the Italian researcher, innovator and 'engineer' Nikola Tesla, a contemporary of Thomas Alva Edison.

dwellings or shops whilst the latter is essentially recommended and used for 'big', multi-storey houses/buildings or commercial complexes and industries.

Supply Voltage

In accordance with the type of supply, whether single-phase or three-phase, the supply *voltage* would be

- 230 V across the (single) live wire and neutral
- 400 V across *any two* live wires of the three live wires whilst 230 V across *any* of the three live wires and the neutral[2]

This is illustrated schematically in Fig.3.1.

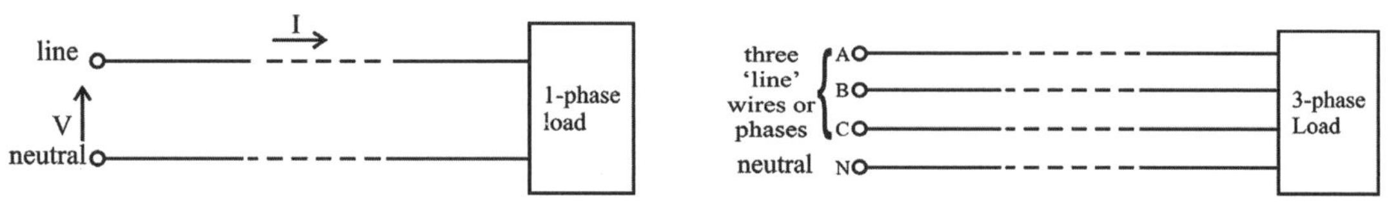

Fig.3.1 : Single- and three-phase supply systems

The Passage of AC Supply: Generation to delivery

As referred to in the previous chapter, an idea of how electricity reaches households from its 'source', that is the power plants where it is generated, to end point, typically called a "power system", can be had from the diagram[3] given in Fig.3.2. The salient feature of the diagram is representation of every 3-phase AC circuit by a "single line"; each circuit comprising three 'live' or "phase" wires with incorporation of transformers, switching and protection devices en route as required, involving all the phases.

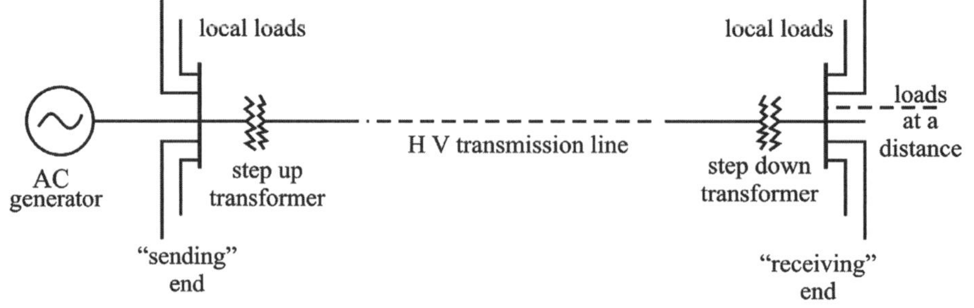

Fig.3.2 : Single-line diagram of a typical three phase power system

[2]The phase-to-neutral voltage in the three-phase supply is line-line voltage/$\sqrt{3}$. For example, 230 V = 400/$\sqrt{3}$ V.

[3]This is known as a "single-line" diagram in electrical engineering practice, each 'line' representing three live wires of the three-phase system.

The schematic of Fig.3.2 illustrates

- a "sending end" of a power system housing one or more generators, synchronised or 'tied together electrically', producing desired power
- local electrical loads such as lights, motors etc.
- a "step-up" transformer that is essential almost invariably in order to transmit electric power to long distances at an appropriate high voltage
- a high voltage, at times several hundred kilovolts, transmission line to carry the (generated) power to long to very long distances, typically several hundred kilometres
- a "step-down" transformer at the end of the transmission line to bring down the voltage to suitable 'level' or value for onward transmission
- the other end of the system where the power is received, known as the "receiving end" from where the power is distributed to local loads or to houses and commercial establishments and industries via distribution network(s).

Stages of a typical power system

The three main stages or parts of a power system are

Generation

This is the first stage that corresponds to generation of electricity in thermal, hydro, nuclear, gas and, now increasingly, in solar-electric power plants. In the former cases, a prime mover in the form of a turbine drives an alternator to generate *AC electricity*, the alternator being typically rated from 100 MW to 1500 MW, the total installed capacity of the plant may be a few thousand MW. Solar electricity is produced from the sun power using photo voltaic cells, the generation being essentially DC that is converted into AC for general usage.

Transmission

This stage follows the generation, characterised by stepping up of the voltage of generation (usually from about 6.6 kV to 24 kV) employing "step up", power transformers[4] at the power station premises itself in what is known as a "switch yard". This is an important step for long to very-long transmission of power with minimal loss of energy and high efficiency. The voltage is usually stepped up to 110 or 132 kV and even to 400 kV[5]. The basic principle involved is that high voltage results in smaller value of current for a given power that is to be transmitted, considerably bringing down $I^2 R$ loss in the transmission line conductors; the longer the distance, higher will be the resistance and higher the loss. Hence, longer the distance of transmission, higher the transmission voltage.

---

[4]Transformer action and topic of transformers is dealt with in detail later in Chapter IV.

[5]The voltage of transmission is being elevated to 765 kV in India as the voltage of future national "grid".

Distribution

At the destination or "receiving" end, the transmission-line voltage is stepped *down* using "step down" (power) transformers in three steps

- from 132 or 400 kV to 33 kV in 33 kV "receiving" station or switch yards, sometimes also identified as the "grid";
- from 33 kV to 11 kV to connect to conductors called "feeders" that form the 3-phase distribution network, and usually supply electricity to bulk consumers or industry directly at this voltage;
- from 11 kV to 400 V (three-phase) lines or conductors that constitute what is known as "secondary" distribution network, using step-down or "distribution" transformers, each of these supplying electricity to a cluster of houses (in a residential colony) or a commercial complex[6].

An image of a typical transmission/distribution network is illustrated in Fig.3.3 whilst Fig.3.4 shows 100 kVA (power rating) distribution transformers, mounted on a pedestal, 'serving' a cluster of about 50 households.

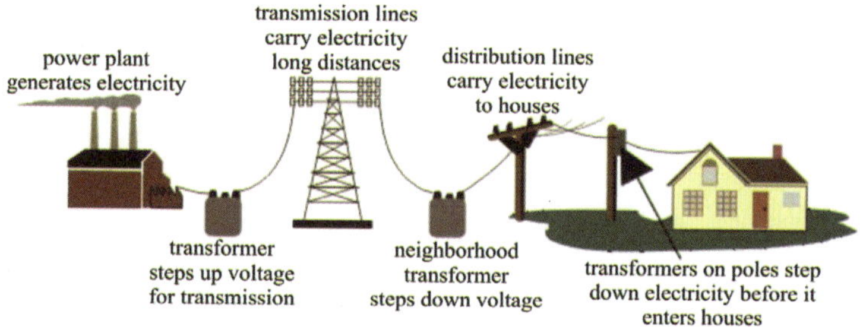

Fig.3.3 : Generation, transmission and distribution of electricity

Fig.3.4: Images of 100 kVA pole-mounted and distribution transformers

---

[6]These transformers can be usually spotted close to residential localities, mounted on a suitable pedestal or support or sometimes on the poles – hence identified as "pole-mounted" distribution transformers - each typically rated from about 50/100 kVA to 500 kVA rating or capacity.

Radial and Ring-main Distribution

Radial electrical power distribution system

In the radial electrical power distribution system, different feeders are directed 'radially' outward from the substation, each feeder being connected to (the primary of) a particular distribution transformer directly. Each of the feeders emanating from the substation would operate independently, feeding all consumers or households connected to the (secondary, at 400 V) of the transformer.

A typical radial distribution network is shown in Fig.3.5.

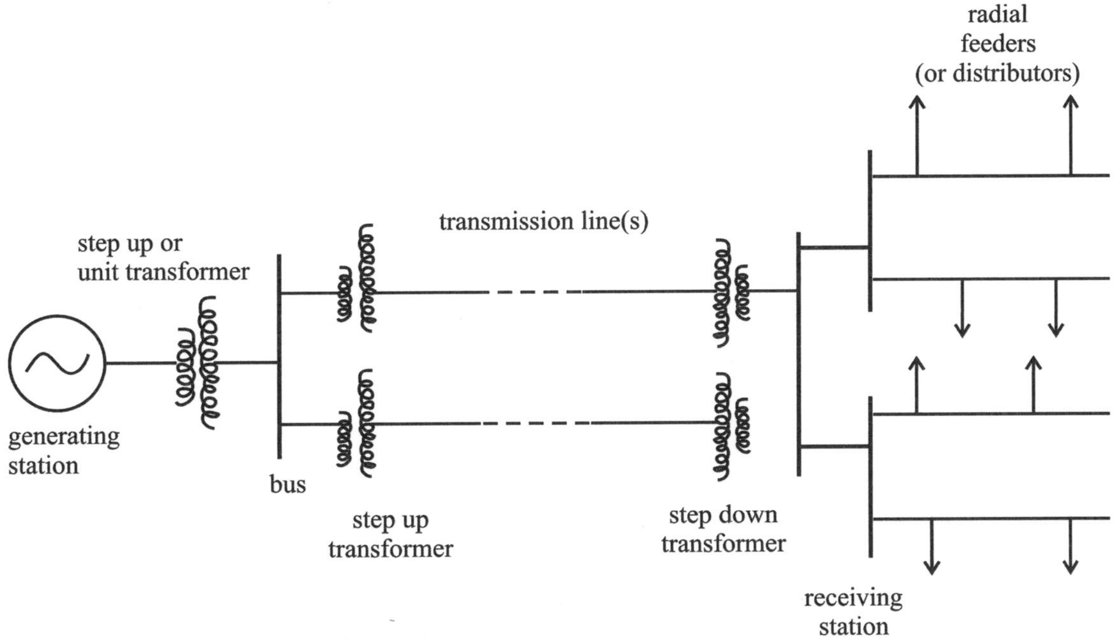

Fig.3.5 : A radial distribution network

However, all radial distribution systems suffer from a major drawback: in the event of a failure or system fault, for example failure of a distribution transformer, that would isolate or disconnect it from the substation[7], all the consumers fed from the feeder would be affected immediately; there be no power supply available to any consumer. All the consumers served by the particular feeder would remain in darkness until the fault on the feeder (or the transformer) is rectified and the feeder re-energised.

[7]Invariably effected by opening of the circuit breaker to eliminate possibility of damage to the system.

Ring main electrical power distribution system

The drawback of radial electrical power distribution system can be overcome by introducing a "ring main" electrical power distribution system. Here, the various feeders from a particular substation, or even involving more than one substation, are laid or organised in the form of a "ring" as indicated in the network shown in Fig.3.6. Each of the feeders would connect to a number of consumers through a distribution transformer as usual.

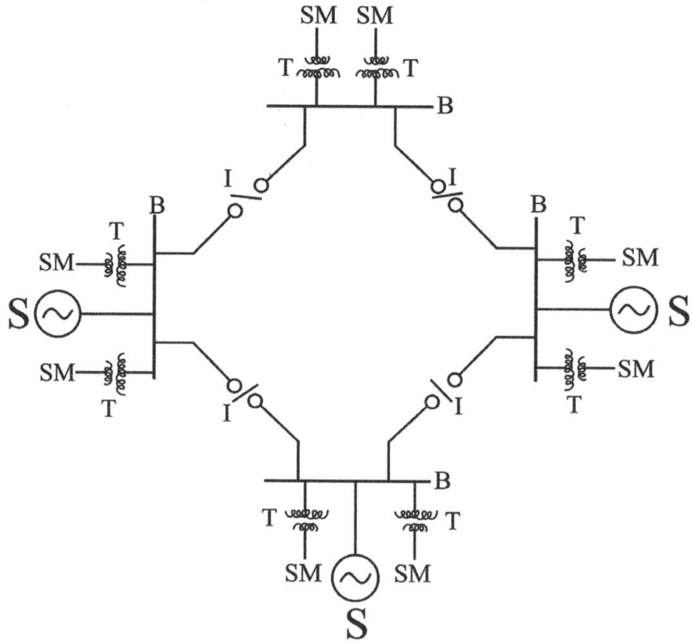

Fig.3.6 : A ring main distribution system

Identification

S	:	Distribution sub station
B	:	11 kV bus
T	:	Step down transformer [11 kV/400 V]
SM	:	Service main [three phase, 400 V]
I	:	Isolator

The important advantage of this system is that in the case of any one feeder being affected by a fault, the supply to the other feeders can still be maintained by switching on appropriate "isolators" as shown in the figure. In this way the supply to the consumers is not affected even when any feeder goes out of service. Also, if a fault occurs on any section of the ring, the same can easily be isolated by opening the associated section isolators *on both sides* of the faulty zone. In this way, supply to the consumers connected to the healthy zone of the ring can easily be maintained even when one section of the ring is under shutdown and rectification of the fault.

Supply to Houses

From the distribution transformer, three live overhead wires with a PD of 400 V across any two of them, along with a neutral wire are commonly routed to individual houses over electric-supply poles along facing roads/streets. The single- or three-phase supply from the pole is connected to a particular house using overhead "service" wires or underground cables, terminating at the house distribution board from where supply is routed appropriately inside the house to various loads.

Wiring Schemes

There are two main schemes of wiring of houses or premises in general

- Radial or "tree" system
- Ring system

Radial scheme

In this scheme or system, the pair of live and neutral wires – and a earth where required –for a given circuit is drawn from the distribution board, branching out to various loads in different parts of the house, comprising lighting fixtures, fans and sockets etc. An appropriate fuse (5, 10 or 15A) is included in *live* wire of each circuit, the wires terminating at the end of the circuit. Since the wires have to carry current equivalent to total, or considering judicial staggering of, connected load, the wire size must be chosen accordingly[8].

A schematic of such a system of wiring, showing various components, is shown in Fig.3.7. The various bulbs and sockets for table fan or lamp etc. are connected in this circuit, each in parallel across L and N, live and neutral wires, respectively, and operated by a separate switch *on the **live** wire side.* For 'power' circuits, that is, loads drawing up to 15 A current, the third wire is the earth wire which is connected to the earth pin of the 3-pin sockets and metal body of the appliance which is earthed.

Similar to radial distribution system, the tree system of wiring has some disadvantages. The wires have to be of thick cross-section to provide for the current requirement of *total* connected load when in practice only a few of them may be in use at any time. If a fuse blows off, two or more appliances have to be checked for a fault. Too many wires of different ratings crowd into terminal bars on the distribution board from different sections of the house. If extra points are required at a later stage, it may not be easy to provide for them.

---

[8]This may be about 1 mm^2 for loads up to 5 A and 2.5 mm^2 for higher currents up to, say, 15A.

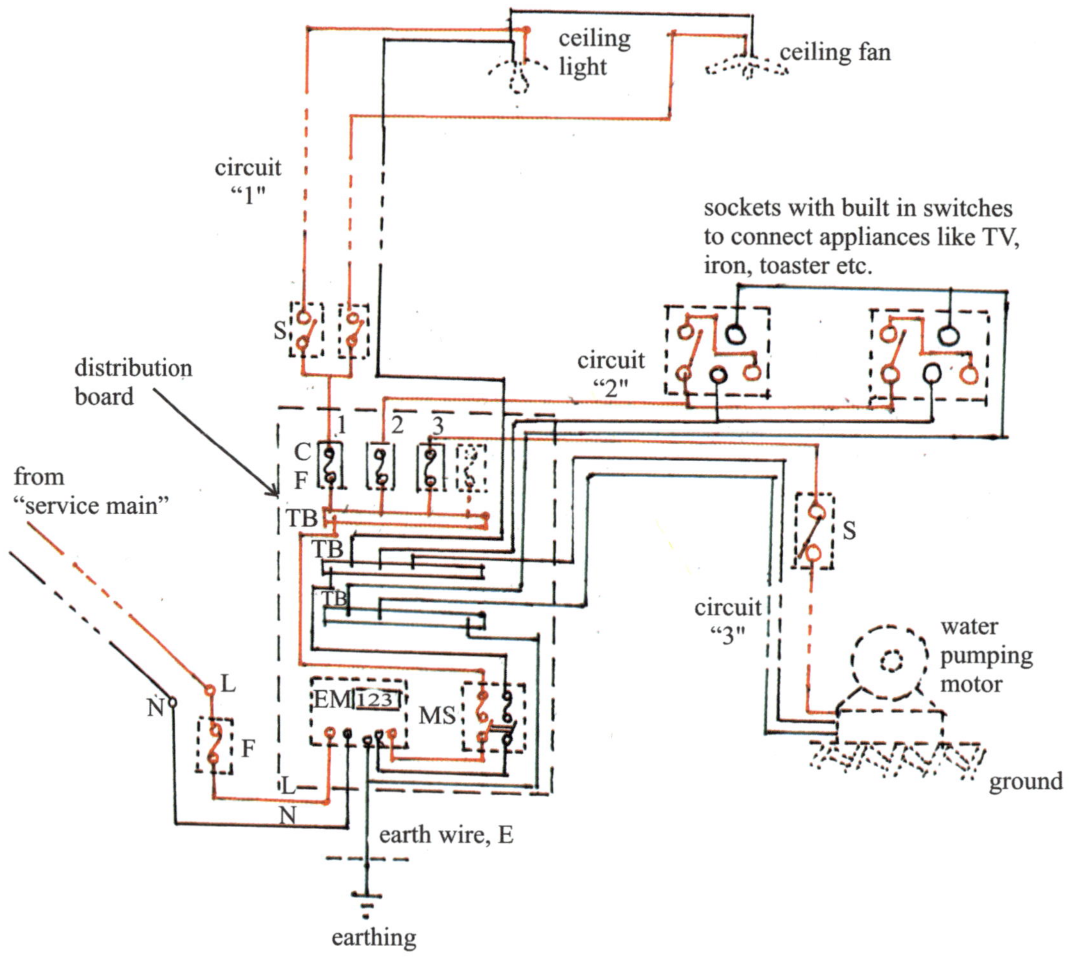

Details

F	: 'input' fuse
EM	: energy meter
MS	: (iron-clad) main switch (with fuses)
CF	: circuit fuse (s)
S	: SPST switch
TB	: "terminal bar(s)" (line, neutral and earth)

Fig.3.7 : Schematic of a radial or "tree" system of wiring

Ring system

This system, originally intended to connect desired number of 3-pin sockets, starting from the terminal bars or distribution box using line, neutral and earth wires, is illustrated in Fig.3.8.

From the distribution box, the three wires connect the sockets *in a sequence* one after the other (or in "series") till all the sockets are connected forming a closed loop or **ring** – hence the name for the wiring – and the wires return to the distribution box as indicated. At the beginning of the distribution box, a fuse or MCB (miniature circuit breaker)[9] of appropriate rating, say 30A, is provided to afford protection of various sockets or loads, the wire also being rated for the same capacity. Provision may also be made for further addition of sockets, marked as A and B in the figure, called "spurs". The wires 'go round' a part of the house or rooms forming a ring for the particular circuit. Other circuits covering the entire house may similarly be wired, resulting in as many rings starting from the distribution box and terminating to it, the live wire of each being provided by a (30A) fuse or MCB. Effectively, a ring wiring system can be interpreted as comprising TWO radial wiring systems starting from the distribution box in opposite directions so that the loads may not be interrupted even when the live wire gets disconnected in a circuit at any point en-route.

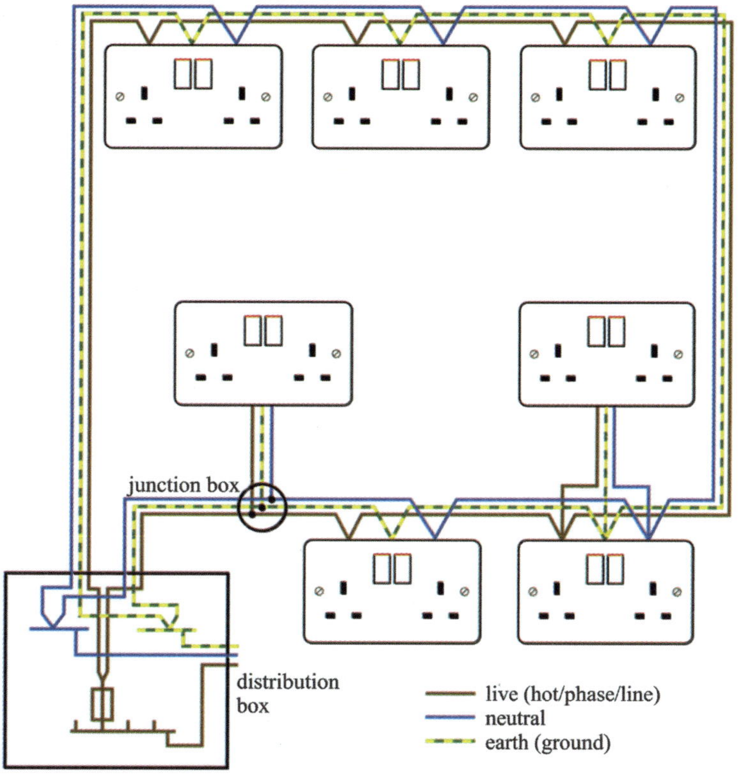

Fig.3.8 : Ring wiring connecting various sockets

The system can be extended or modified to include other loads such as bulbs, fans etc., with the provision of an appropriate fuse in the live wire connecting to each appliance. The basic ring structure of *each circuit* remains the same as marked in Fig.3.8.

---

[9]This is a modern protecting device, in place of a fuse, discussed in detail later.

A modified ring wiring of a typical circuit as discussed above is shown in Fig.3.9.

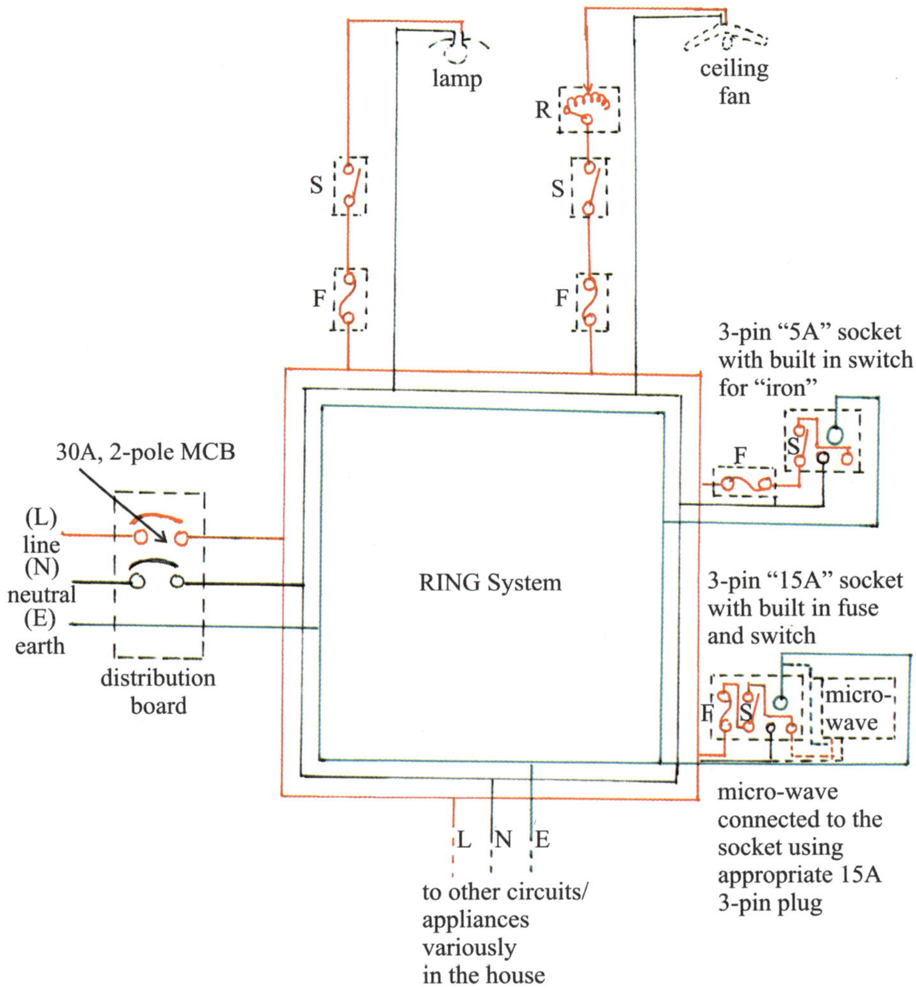

R : fan regulator
S : SPST Switch
F : fuse

Fig.3.9 : Ring wiring of a circuit connecting sockets and appliances

Characteristic features of ring wiring

- The system affords better continuity of supply to every socket or appliance in a given circuit, being connected in 'parallel' from the distribution box.
- Each appliance is operated using separated, independent fuse rated appropriately so that in the event of a fault in a particular appliance only fuse of that appliance is blown, not affecting other appliances (or sockets).

- When required, the supply can be tapped, in the form of additional spur(s), at any convenient point in the circuit inside the house in any room 'served' by that circuit.
- The wire size may be appropriately graded, suiting the rating of a particular appliance, thus resulting in saving of overall cost of house wiring.

House Wiring Using MCBs

A practice that is followed in almost universally these days is to provide an appropriately rated MCB of the "two-pole" type [see Fig.3.10(a)] in a circuit each from the distribution box, along with the neutral and earth wires, terminating at the last appliance or load of the circuit as shown for a typical circuit in Fig.3.10(b)[10].

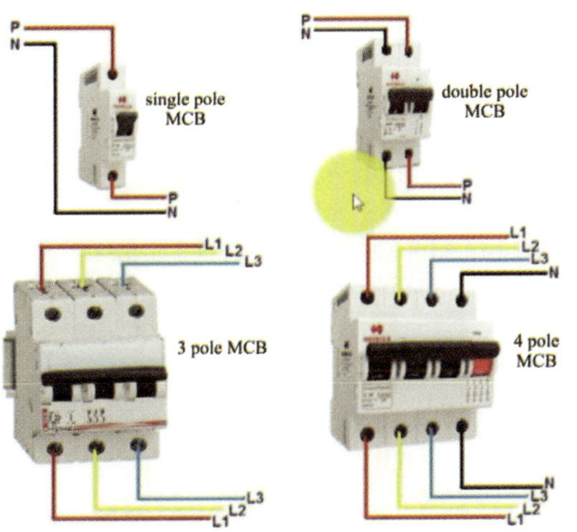

(a) types of MCB for 1-phase or 3-phase circuits

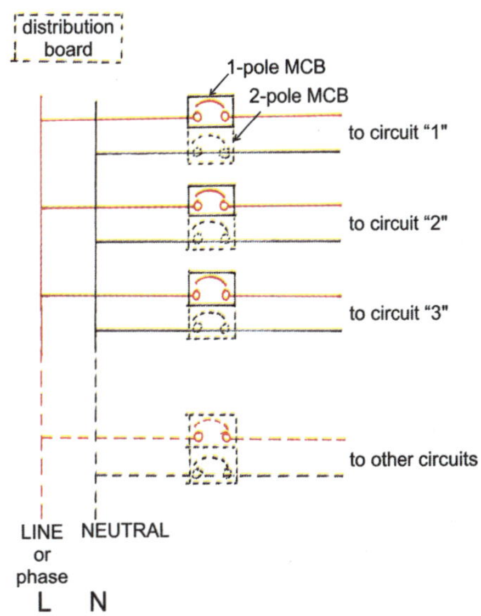

(b) a 1-phase circuit using two-pole MCBs

Fig.3.10 : A circuit showing use of MCBs

Components of the house wiring

With reference to the network shown in Fig.3.5, the various "key" components used in a typical "single-phase" house wiring are

---

[10]This system facilitates the inclusion and connection of an inverter/UPS of suitable rating to energise selected loads in case of a power failure. This is effected by bunching the pairs of live and neutral wires of desired circuits and connecting these to "live" and "neutral" output terminals of the inverter. If this is not warranted, a "single-pole" type MCB would suffice in each circuit in the live wire only; this may lead to relatively economic wiring.

- Line Fuse

A "**line fuse**" is provided in the *live* wire of the service line from the pole *before* the energy meter to protect the latter. This may be rated at, say, 30 A.

Note that a fuse, in simple form of a 'thin' wire, is invariably provided in live wire connecting to any one circuit energising an appliance or load. The wire would safely carry a current slightly more than the circuit current which depends on the rating of the appliance. In the event of an "overload" or a "short circuit", the fuse wire would instantly melt or fuse – hence the name – isolating the appliance from the supply by opening the live wire. The 'standard' ratings of fuse wires in domestic wiring are 1A, 2A, 5A and 10A. Clearly, higher the rating, thicker the wire; although fuses of lower ratings may also be available in directly-replaceable form, enclosed in a tubular glass housing, and generally used for protection of electronic devices by pressing the fuse in a matching recepticle.

The usual type of fuse assembly or housing, also called a "kit-kat" and made of porcelain (it being an electrically insulating material), suitable for 5A usage is shown in Fig.3.11; similar one for 15A usage (bigger in size) would also be available. It consists of two parts: one that is fitted on a board, called a socket, the other which is removable, called a holder carries the fuse wire, fixed between two screws as marked[11]. A symbolic representation of a fuse in an electric circuit is depicted in Fig.3.11(b).

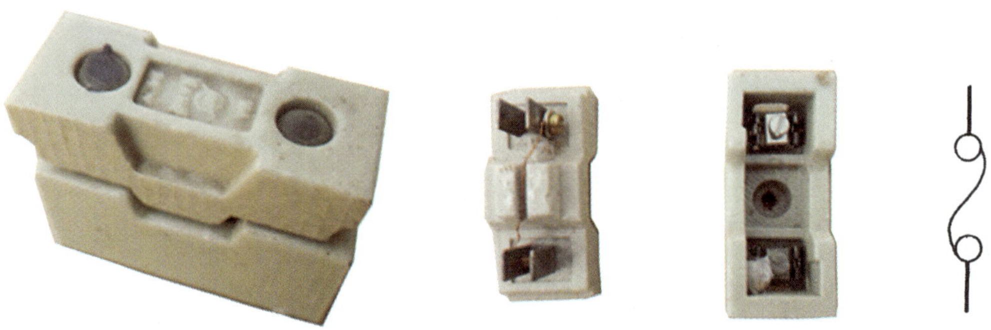

(a) fuse housing ("kit-kat") (b) symbol

Fig.3.11 : A fuse housing

- Energy Meter

This is connected in the circuit immediately following the line fuse. Its purpose is to 'sense' and record energy consumed by the household in kWh over a given period, usually a month. The meter which used to be of electromechanical type, characterised with a rotating aluminium disc, and now increasingly being replaced by "solid-state" (or electronic) type is

[11]For proper fixing of the fuse wire, it is important that the wire end is hooked clockwise round the screw which should be tightened leaving a little slack in the wire.

provided with two terminal pairs: one to the left for connecting incoming live and neutral wires, the other to the right for connecting the live and neutral wire for supplying the loads. A typical meter (of the old type) and schematic of its connections are show in Fig.3.12; also shown in the figure is the front view of a 3-hase solid-state meter.

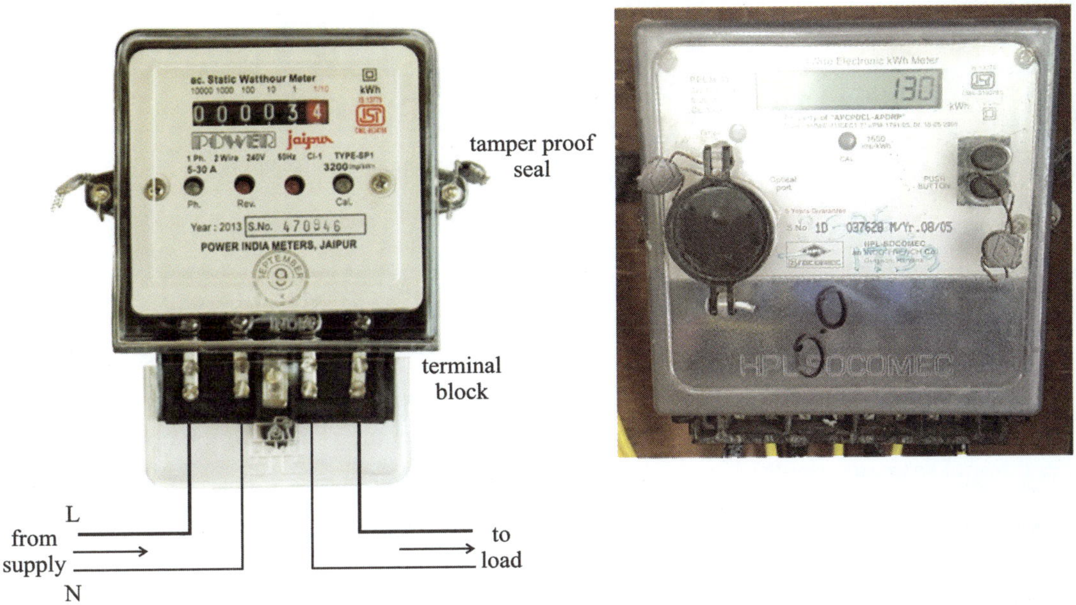

(a) a single-phase energy meter and supply/load connections

(b) a 3-phase, solid-state energy meter, with LCD display

Fig.3.12 : Single- and three-phase energy meter

- “Main” Switch

The purpose of providing this switch, whether in single-phase or three-phase house wiring, is to isolate the internal wiring or the entire load from the supply in the event of a major fault or for carrying out any required changes or modifications in the wiring. This is achieved by *simultaneously* disconnecting both the live and neutral wires by one manual operation. The switch comprises a cast iron casing, fitted with two fuse housings rated for 30A, one each for the live and neutral wire and a specially designed spring-loaded lever to effect simultaneous connection of incoming supply to the load.

An inside view of an iron-clad (IC) main switch and its symbolic representation is shown in Fig.3.13. In modern practice, the main switch is supplemented with an “ earth leakage circuit breaker”, or ELCB, to detect any undesirable ‘leakage’ of current to earth in any part of the wiring and disconnect the supply if there is any.

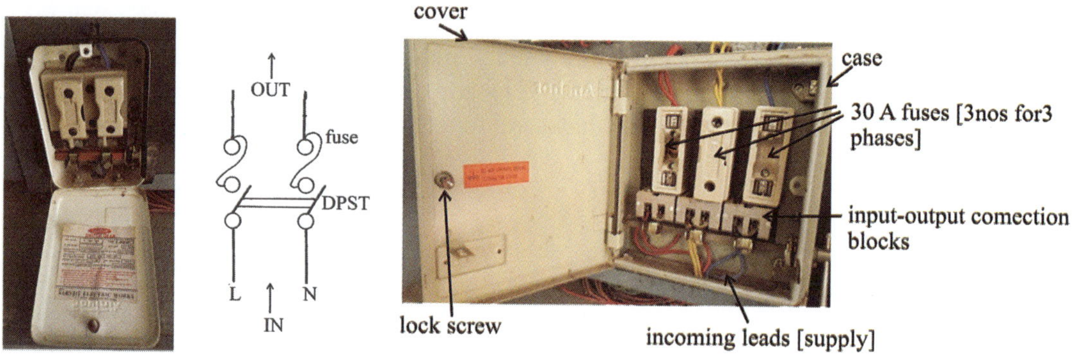

IC switch: single phase — symbolic representation — IC switch: three phase

Fig.3.13 : The interiors of single- and three-phase iron-clad (IC) main switch

Note that the fuses in the switch enclosure *follow* the switching mechanism. This means that for any reason, that is due to a fault in the wiring or otherwise, *the switch is first switched off* before fuse holders are removed from their sockets. This ensures safety to the operator or any personnel from accidental contact to the live part in the wiring.

For the same reason, it follows that the fuses are always provided in the live wire *and never in the neutral wire* as indicated in Figs.3.7 and 3.9.

Current rating of fuse wires for some common appliances

These are listed in the table below.

Appliance	Power rating W* at 230 V supply	Current drawn, A	Current rating for fuse, A
100 W bulb	100	0.434**	5
CRO type TV	125	0.543**	
Fridge (when "ON")	150	0.652**	
Mixie	750	3.26	5
Electric iron	1000	4.38	10
Box oven	1200	5.21	10
Kettle	1500	6.52	10
Geyser	2000	8.69	15
Microwave	2200	9.56	15

* Indicative figures
**All these may usually form one circuit

• Distribution Board

A typical distribution board provided in a house or any premises, prior to wiring circuit(s) *inside* the house, comprises the line fuse, energy meter, main switch and circuit fuses or MCBs, all arranged in a logical order. A photograph of a specially designed and wired distribution board for three-phase wiring of a two-storey building is shown in Fig.3.14. In addition to the various parts listed above for the three-phase wiring, the board provides for monitoring load current in each of the three phases using an ammeter, the voltage across any two phases (with attendant selector switch) using a voltmeter, the frequency of the supply using a frequency meter and the "fuse-blown'' indicators in each circuit that light up un the event of a fault in the particular circuit, followed by blowing of fuse in that circuit. This device[12] is based on the principle of current sharing in a combination of two resistances in parallel and comprises a usual fuse housing fitted with an appropriate fuse wire of 'low' resistance, across which is connected a high-resistance (indicator) lamp as depicted in Fig.3.14(c). During normal operation, the load current flows through the (low resistance) fuse wire and little current through the lamp. In the event of blowing of the fuse due to a fault in the circuit and 'infinite' resistance in the fuse circuit, a small current flows through the lamp which is lit up.

This is a marked improvement on the usual conventional wiring employing "circuit fuses" where in the event of a fault in any circuit, a "trial and error" process would be used to locate the faulty circuit.

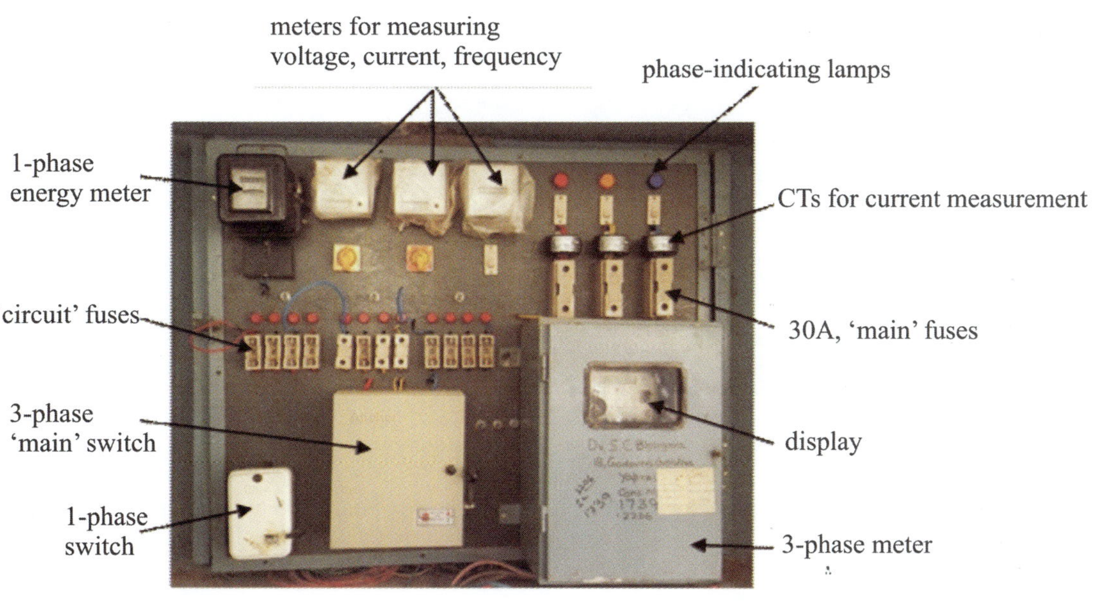

(a) a three phase distribution board

[12]'Invented' by the author.

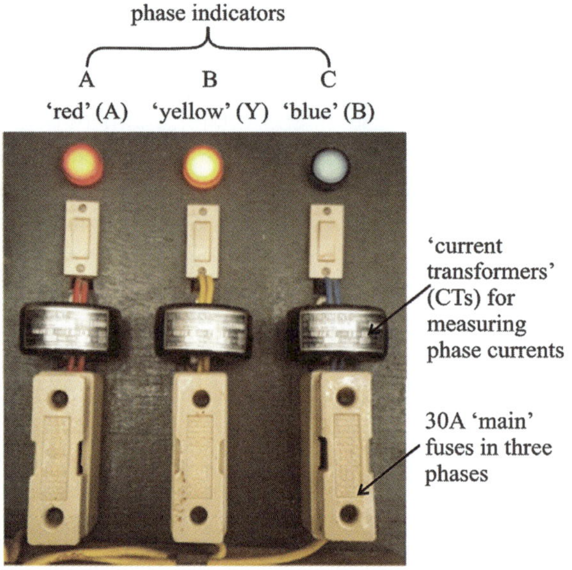

(b) phase indicators, 'main' fuses and CTs for current measurement

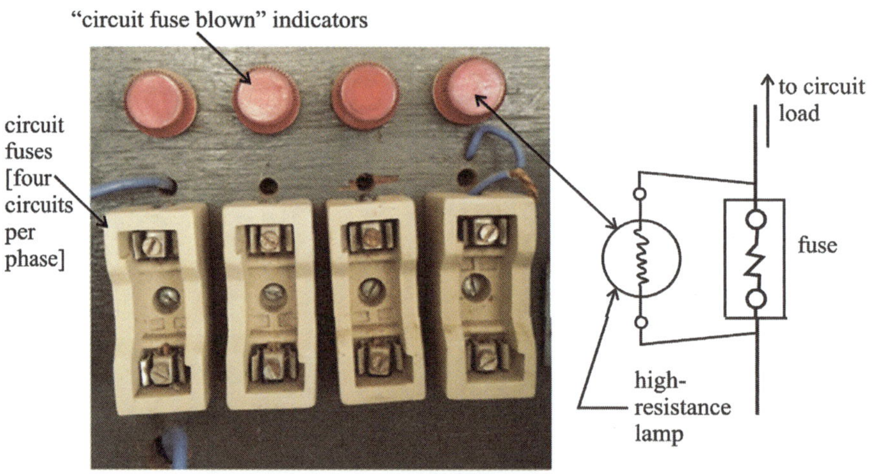

(c) circuit fuses and "fuse-blown" indicators and schematic

Fig.3.14 : Photograph of a three-phase distribution board and circuit details

- Miniature Circuit Breaker or MCB[13]

MCBs or **miniature circuit breakers** are electromechanical devices which protect an electrical circuit from an overcurrent. The overcurrent, in an electrical circuit, may result

[13]Large-rating circuit breakers are essentially provided in power systems, capable of interrupting hundreds to thousands ampere of fault current at high to very high system voltage(s).

from a short circuit, overload or faulty design that may lead to overheating and even fire . An MCB is a better alternative to a fuse since it also does not require replacement once an overload is detected. Unlike a fuse, *an MCB can be easily reset* – this being faster than time taken for rewiring the fuse - and thus offers improved operational safety and greater convenience without incurring appreciable operating cost.

An MCB functions by interrupting the continuity of current flow through the circuit once a fault is detected. In simple terms, an MCB is a switch which automatically turns off (using an "spring" action of moving contacts) when the current flowing through it passes the maximum allowable limit and an indication of the faulty circuit appears on the distribution board. Generally, MCBs are designed to protect against over current as well as overheating caused by a short circuit.

There are two contacts in an MCB: one is fixed and the other moveable. When the current exceeds the predefined limit, a solenoid forces the moveable contact to open (i.e., disconnect from the fixed contact) and the MCB turns off thereby stopping the current to flow in the circuit. In order to restart the flow of current the MCB is manually turned on. The mechanism thus protects an appliance from the damage arising due to over current or over load during a fault.

For its function, a bi-metallic strip is used to sense the overheating. MCBs are generally designed to trip within 2.5 ms when an over current fault arises. In case of temperature rise or over heating, it may take 2 s to 2 min for the MCB to trip.

A disadvantage of MCB over a 'simple' fuse is that this system is somewhat costlier than the system using fuses.

An actual "two-way" MCB unit is shown in Fig.3.15. The connections in the circuit are identical to those for a fuse except that even the neutral wire is also connected in the MCB.

Note that the "ON" position in an MCB is when the control bar is *up*.

Fig.3.15 : A "two-way" MCB unit

- SPST Switch

Single-pole-single-throw, or SPST, switches form an essential requirement in all types of wiring, inserted in the *live wire* of a given circuit to switch on or switch off any appliance such as a bulb, a fan or a socket that may energise a TV, a mixer, a toaster or a microwave etc. A typical switch, of 5 A and 15 rating each, commonly used and its symbolic representation is shown in Fig.3.16. A single-pole, double-throw "knife switch" commonly used in circuit experiments in a physics lab is shown in Fig.3.16(c).

(a) 5A and 15A SPST switches

(b) symbol

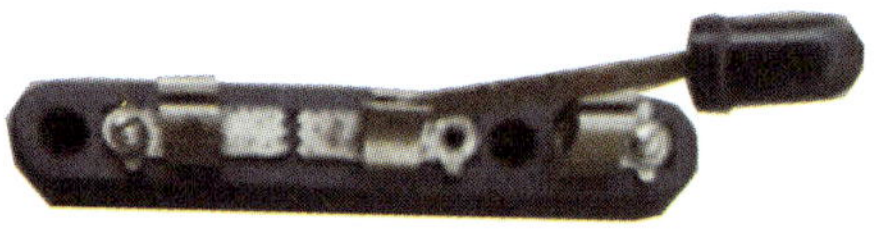

(c) a single-pole, double-throw, SPDT, knife switch

Fig.3.16 : Typical 5 A and 15 A SPST switch for general use and a knife switch

A typical switch board in a (bed) room incorporating SPST switches to control various lights and a fan regulator is depicted in Fig.3.17(a); also shown is a wall thermometer alongside. The location of the board, near the door, is shown in Fig.3.17(b). Note that the location of the board next to the door allows the light(s) to be switched on as one enters the room and switch off on leaving.

Also, the board is positioned at a convenient height, usually about 4 to 5 ft. from the floor, to operate the switch(es) as desired and yet keep them out of reach of children!

Fig.3.17 : Location of a switch board in a room

- Plugs and Sockets

These constitute very common requirements in house wiring to facilitate connection – and disconnection – of an appliance to the electric supply. The appliance is fitted with an appropriate plug at the free end of the wire of the appliance, the plug being inserted into the matching socket for making a connection.

The plugs and corresponding sockets are obtained in the following categories:

- 5A with two prongs or "pins", used for 'small', low rating appliances like TVs, radios, DVDs etc. which draw small current and *do not require earth connection*[14].
- 5A with three prongs or pins, used to connect appliances which draw current less than 5A, but DO require use of earth wire for safety reasons. Electric irons, toasters etc. fall in this category of appliances.
- 15A with (essentially) three pins, used to connect to ('high power') appliances such as geysers, microwaves etc. which are rated above 1000 W, draw over 5A current and for which *earthing connection is a must.*

Typical plugs and sockets of the above categories are depicted in Fig.3.18.

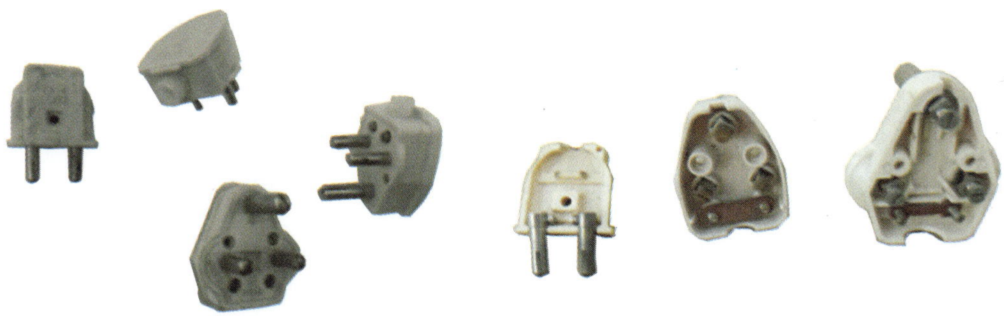

(a) two-pin and three-pin, 5 A and 15 A, plugs

(b) two-pin and three-pin, 5 A and 15 A, sockets

Fig.3.18 : Plugs and sockets of various types

Holding the socket or plug upright, that is with the earth pin at the top as shown in Fig.3.18, note that

- the position of live, neutral and earth pins relative to each other: the live pin is to the left, the neutral pin to the right and earth at the top;
- the projecting length of the earth pin is *more than the other two pins,* more prominently in the case of 15A plug; this is to ensure that as the plug is inserted, or pushed into, the socket, the earth pin makes the contact with ground first for safe operation.

[14]Requirement of earth wire/connection and process of earthing are discussed later.

- Terminal Bars or Distribution Box

These are three bars made of copper and of appropriate cross-section (about 2.5 cm x 1.5 cm) with several threaded holes and matching screws. The bars are positioned and fitted suitably on the distribution board. The live, neutral and earth wires from the main switch are first terminated to the respective bars. The individual circuit connections then ensue from these bars as indicated in Fig.3.7.

- Wires or Conductors

The connections for various circuits are composed of flexible PVC wires, made up of strands of thin copper[15] wires to afford flexibility in handling and drawing of wires through PVC conduits embedded in walls of the rooms

Colour Coding of Conductors

Internationally recommended colour coding of conductors used for wiring is universally adopted to identify live, neutral and earth wires from each other to ensure proper and safe wiring of various circuits. The colour refers to the colour of insulation on the wire.
Colour coding is particularly important in the case of three-phase wiring in buildings where an identification of wires belonging to three phases *and* neutral and earth wires must be maintained for proper wiring.

"Old" and latest convention for single-phase wiring

For single-phase wiring, the colour code for wires in "old" times consisted of using red for live, black for neutral and green for earth. The latest practice is to use brown colour for live wire, light blue for neutral and green (or yellow) for the earth.

Earthing or Grounding and Earth Wire

Earthing at the premises

Earthing, right at the premises of a household, constitutes an important safety requirement of wiring. It consists of providing a terminal, theoretically at ground or "zero" potential from where wires are routed to various 'points' in the wiring as required.

Why Earthing?

Many appliances in a household with high power rating are provided with a "earth" terminal, esp. when enclosed in a metallic casing such as a box oven or a toaster. In the event of a fault or "short" circuit in the wiring, for example breakdown of insulation over time or

[15]The modern practice is to replace copper wires by stranded aluminium wires of appropriate cross-section to economise on the cost of wiring. However, copper wires form a better choice in the long run.

otherwise, the live wire may come in contact with the casing and result in an electric shock to an inadvertent user of the appliance; the shock may at times prove to be fatal. If the casing is connected to ground with the help of a wire from the earth terminal to ground, the live wire will get shorted to ground following the fault thus preventing and untoward consequence.

Earthing Mechanism

The scheme of proper earthing consists of digging a hole or a pit in the ground, close to the location of the energy meter or distribution board, about 2 m deep and 70 cm x 70 cm in section. A hollow GI(galvanised iron: free from rusting) pipe of nearly 5 cm dia. is positioned at the centre of the hole, protruding about 30 cm above the ground level. The lower end of the pipe up to a height of about ¾ metre is suitably perforated; for example by drilling holes diametrically opposite at distances of about 25 cm apart. A suitable funnel is fitted at the top of the pipe for periodically pouring water into the pipe. See Fig. 3.19.

A copper plate, of dimension about 70 cm x 70 cm and 3 or 4 mm thick is positioned at the bottom of the pit as shown. A copper wire of 3 mm dia. is connected firmly (by brazing) at the centre of the plate and brought out through the GI pipe, the length of the wire to be sufficient to reach the distribution board when appropriately routed.

The bottom of the pit, up to a height of about 70 cm, is filled with a mixture of charcoal and common salt, surrounding the copper plate, the latter being firmly packed with the mixture. The latter, ensuring good electrical contact between the plate and the surrounding soil, is essential for effective earthing.

The upper part of the pit is filled with (dug) earth and levelled. It is important that the plate and its surrounding is always maintained moist or damp. For this, water must be poured into the funnel at

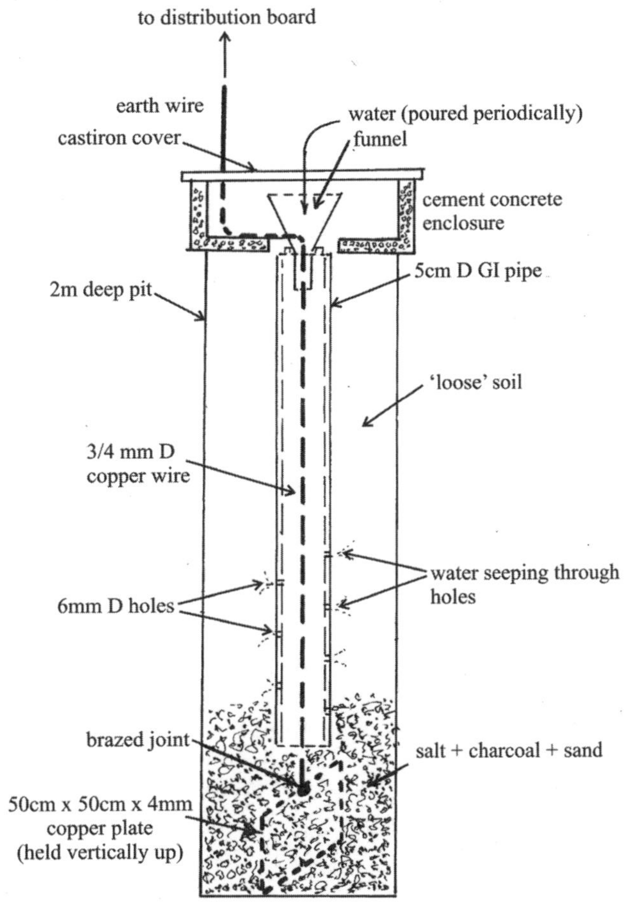

Fig.3.19 : Schematic of earthing at the premises

the top of the pipe at regular interval that would reach the soil surrounding the copper plate through the perforations.

The earth wire is first connected to earthing terminals of the energy meter, the main switch and to the earth terminal bar from where the earth wire is run along the live and neutral wires to various appliances (requiring earthing) and three-pin sockets as indicated in Fig.3.7.

Connections to an Appliance Using Live, Neutral and Earth Wires

As an important example, this is shown for a typical oven in Fig.3.20; also shown is the symbol for earthing used in circuits.

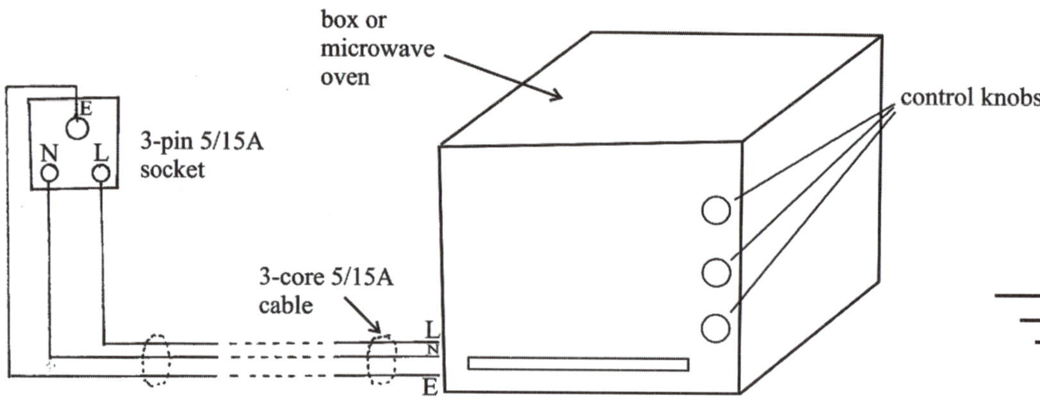

(a) connections to an oven

(b) symbol for earthing

Fig.3.20 : Supply to a typical oven

Two-way (SPDT) Switch and "Staircase" Wiring

A SPDT (single pole double throw) switch is a special switch that has a common terminal and two additional terminals on either side so that the switch can connect the common terminal to either of the other two. A photograph of an actual switch of this type and its symbolic representation is shown in Fig.3.21.

(a) switch

(b) symbol

Fig.3.21 : A two-way switch and its symbol

Staircase wiring

A common use of this type of switch – two are used – is in the "staircase" wiring. The objective of this application is to facilitate a person to switch on staircase light, using one of the two-way switch, as he/she prepares to climb the steps from, say, ground floor to first floor and then switch off the light using another switch after reaching that floor. A simple sketch of the arrangement and schematic of the connections of the two switches, supply wires and the light, under various operating condition, is illustrated in Fig.3.22.

A similar sequence of operation is used for coming down the steps from first to ground floor; and likewise connecting other floors, for example between first and second floor etc.

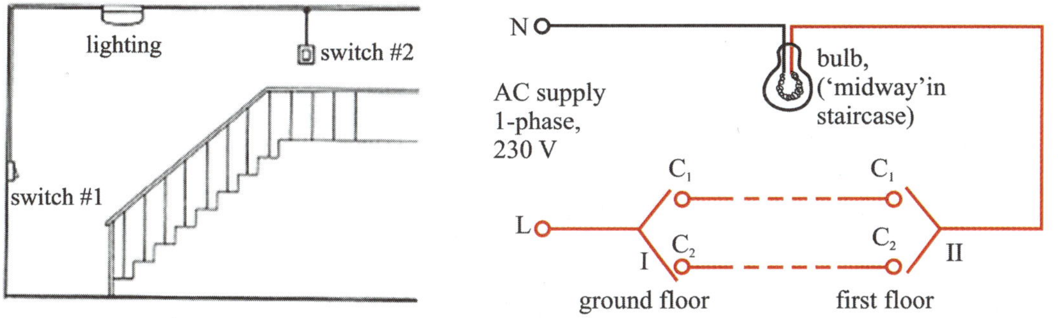

(a) schematic (b) wiring diagram

Fig.3.22 : Staircase wiring

Referring to Fig.3.22 (b), let

(i) the switch I is in position C_2, from an immediate previous operation, whilst switch II is in position C_1, the light B is OFF
(ii) a person climbing up from ground floor operates switch I to close to position C_1, the light going ON
(iii) on reaching first floor, the person operates switch II to close to position C_2, switching OFF the light
(iv) the person coming down after a while would close the switch II to position C_1, switching the light ON, and on reaching ground floor would operate switch I to position C_2 to switch OFF the light, and so on.

Other Applications

- Long corridor

A useful application of 2-way wiring is to control lighting in a 'long' corridor when a person entering the corridor at one end needs to switch on the light and to switch it off after traversing a section of the corridor, till reaching the other end. This may be effected by control or switching on of all the lights along the length of the corridor by providing just two switches at the two ends or multiple pair of two-way switches to switch on lights one after the other as the person keeps walking; both schemes having their advantages and disadvantages.

A schematic of corridor lighting is shown in Fig.3.23.

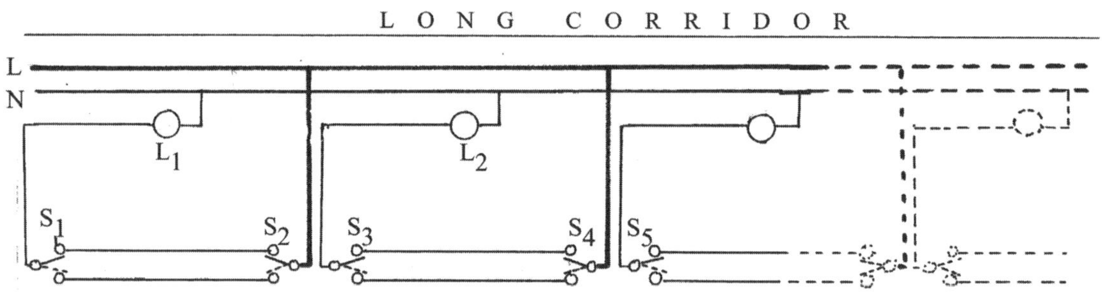

L_1, L_2..............: ceiling lights
S_1, S_2..............: two-way switches

Fig.3.23 : Control of lighting in a long corridor using two-way switches

- Auditorium

Often, a 'central' light in an auditorium needs to be switched off during a PPT or slides presentation or a stage show when near-darkness in the hall is desired. The light initially switched on from a switch board near the dais, can be switched off by an operator (of PA system etc.) from a distant "control" room by means of the other two-way switch. Later, the light may be turned on from either end as desired.

Safety Precautions Whilst Using Electricity

Remember the adage:

Electricity is a good slave but bad master and that **electricity kills**

This means that electricity **must** be treated with 'respect'; do **NOT** take it lightly or for granted!

The following precaution must be observed when dealing with electricity.

- All circuits must be provided with proper fuse or MCB in the live wire.
- The wires used for wiring must be of proper cross-section, more than the total current that may be drawn by an appliance.

- Ensure proper earthing or grounding in the premises all the time and connect earth terminal of all the 'heavy' duty appliances such as ovens etc. to the earth wire.
- Never insert many plugs or wires into one socket; avoid using multi-input socket adapters in a socket to connect many appliances to it.
- Beware touching any "leaky" switches to avoid a shock, esp. in the bathrooms etc. Get these replaced immediately.
- Never touch a wet switch or a switch with wet hand, esp. when barefoot.
- Keep children away from electric appliances, esp. sockets located at 'ground' level. See that they do not attempt to insert a metallic item into pin receptacle of a socket; it is safer to keep the sockets covered when not in use.
- When trying to repair an appliance, disconnect it from the socket; better still, remove the fuse from that circuit or 'switch' "off" the MCB.
- Use well-insulated tools, such as screw drivers and pliers etc., and equipment with non-conducting handles when working on electrical devices.
- When it is necessary to handle equipment that is plugged in, be sure that hands are dry and always wear nonconductive gloves, protective clothes and shoes with insulated soles.
- If an individual is seen to have come in contact with a live conductor, do **NOT** touch the person cord or equipment; first disconnect the power source at once by removing the fuse or pull out the plug using, for example, a leather belt.
- Do NOT attempt to repair a circuit fault, esp. involving a live wire. Call for a qualified electrician or personnel.

This is NOT an exhaustive list; use common-sense in all cases, at all times.

EXERCISES

1. An electrical appliance is rated at 1,000 kVA, 220 V. If the appliance is operated for 2 hours, calculated the energy consumed by the appliance in (i) kWh; (ii) joule.

[ICSE, 2012]
[2,000 kWh; 7.2×10^9 J

2. Two bulbs are marked 100 W, 220 V and 60 W, 110 V. Calculate the ratio of their resistances.

[ICSE, 2011]
[2.4

IV Magnetic Effects of Current

Magnetic effects of a current(principles only, laws not required); electromagnetic induction (elementary); transformer.
Oersted's experiment on the magnetic effect of electric current; magnetic field (B) and field lines due to current in a straight wire (qualitative only), right hand thumb rule – magnetic field due to a current in a loop; electromagnets: their uses; comparison with a permanent magnet; Fleming's left hand rule, the DC electric motor – simple sketch of main parts (coil, magnet, split ring commutators and brushes); brief description and type of energy transfer (working not required); simple introduction to electromagnetic induction; frequency of AC in household supplies; Fleming's right hand rule; AC generator – simple sketch of main parts, brief description and type of energy transfer (working not required); advantages of AC over DC. Transformer – its types, characteristics of primary and secondary coils in each type (simple labelled diagram and its uses).

MAGNETIC EFFECT OF A CURRENT

Oersted's Experiment

Oersted, a Danish physicist[1], performed an experiment on April 21, 1820 and discovered that a steady electric current, that is DC, creates a magnetic field around it.

The set up used by Oersted to demonstrate his experiment to a crowd of onlookers is shown in Fig.4.1.

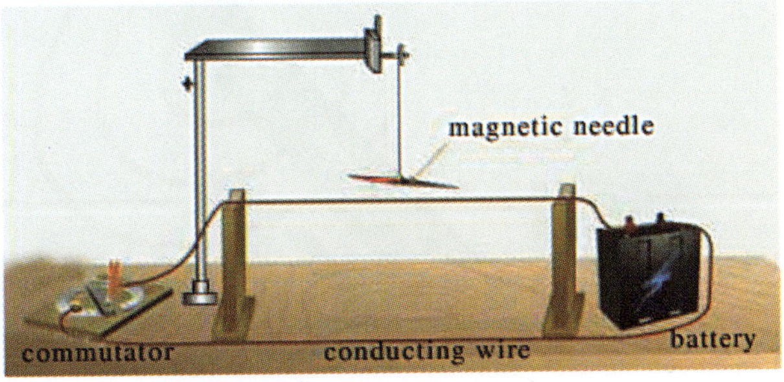

Fig.4.1: Oersted's experiment

It comprised a straight wire supported horizontally, through which a current was passed from a "battery". A compass needle was suspended above the wire – it could be placed below the wire, supported suitably and free to rotate on a vertical axis – at mid-length of the wire. The passing of current through the wire

[1]Hans Christian Oersted (1777-1851).

resulted in setting a magnetic field around the wire, causing the compass needle to deflect, taking a direction perpendicular to the direction of the wire[2].

Oersted's discovery was the first link found between electricity and magnetism and paved the way for many fundamental, historic discoveries involving magnets (or magnetic field) and electric current,

Through his experiments, Oersted established that for a straight wire carrying a steady (DC) current

- The magnetic field lines encircle the current-carrying wire
- The magnetic field lines lie in a plane perpendicular to the wire
- If the direction of the current is reversed, the direction of the magnetic field reverses.
- The strength of the field is directly proportional to the magnitude of the current.
- The strength of the field at any point is inversely proportional to the distance of the point from the wire.

A pictorial view of the current through a conductor and the magnetic field produced by it is depicted in Fig.4.2.

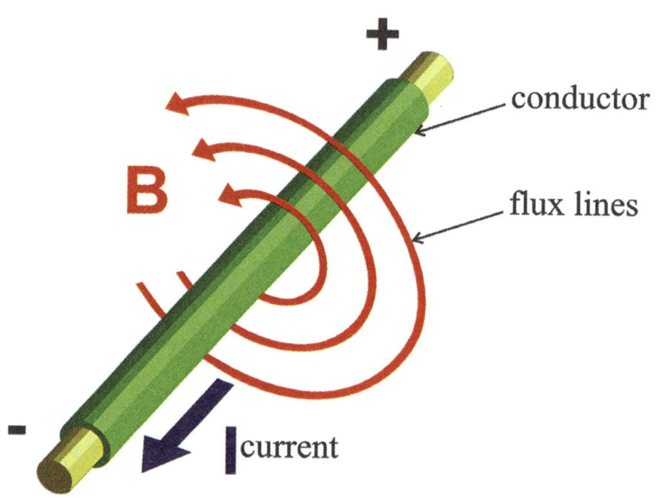

Fig.4.2 : Pictorial view of a current-carrying conductor and its magnetic field

---

[2]It being assumed that the magnetic field produced by the current was much stronger than the (feeble) field of the compass needle.

Performing Oersted's experiment

Oersted's experiment can be easily performed in a laboratory, or even in the class room, as follows:

- Take a 'thick' insulated copper wire (say about 2 mm in dia.).
- Arrange to hold it horizontally on a wooden plank of suitable dimensions or a table between two insulated supports, about 30 cm apart and at a height of about 2 cm.
- Connect the wire in an electric circuit as shown in Fig.4.3.
- Arrange to place a sensitive compass (needle) below the conductor, as close to it as feasible, at mid length. Remember that the strength of the magnetic field decreases with distance.
- Adjust the circuit current to a suitable value using the rheostat.
- Observe the deflection of the compass needle.
- Increase the current and observe the deflection.
- Reverse the direction of current by means of the reversing switch and observe the deflection. Record observations.

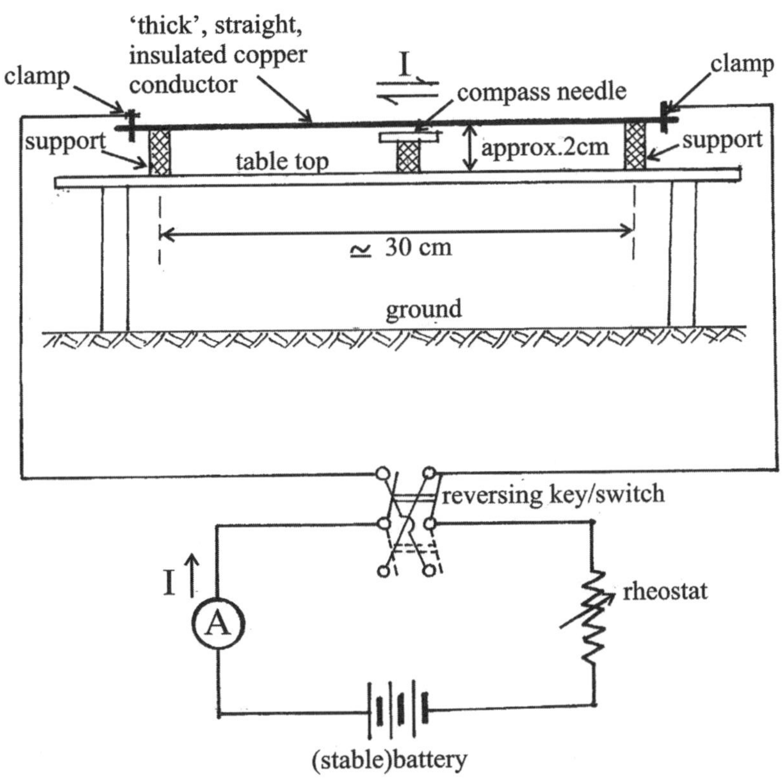

Fig.4.3 : Performing Oersted's experiment

Field lines due to a straight conductor

The magnetic field lines surround a current carrying conductor such that they would form close *concentric* circles in a plane *perpendicular* to the direction of the conductor; strictly speaking, direction of current. This is depicted in Fig.4.4. Note that the actual direction of magnetic field at any point on any field line will be given by a tangent at the point. Clearly, the strength or intensity of the field would depend on the radial distance, r, from the centre line of the conductor or current.

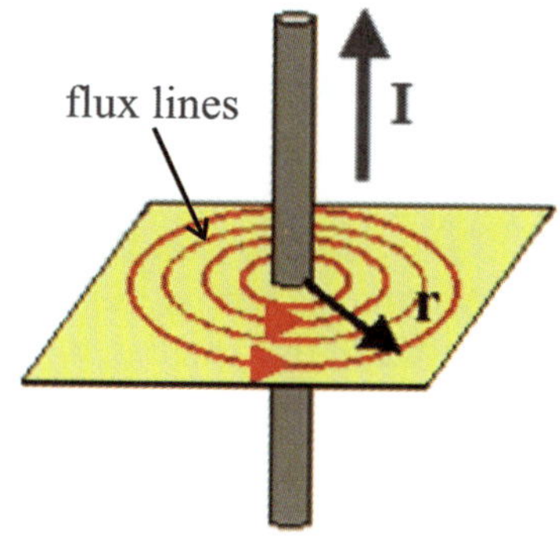

Fig.4.4 : Magnetic field around a current-carrying conductor

Experiments to Observe Magnetic Field of a Straight Current

A visual image of the field lines due to current in a conductor can be had by performing simple experiments.

A. Using iron filings

Iron filings provide a simple yet very effective means to observe patterns of magnetic field due to a magnet (for example a permanent magnet) or any agency producing the fields such as a wire carrying current. To observe the field produced by a current in a straight conductor, take a piece of smooth card board, about 30 cm x 30 cm in area, supported horizontally. At the centre of the board, make a hole through which a straight wire can be inserted, *at right angles to the plane of the board,* that is vertically, to be connected in an electric circuit as shown in Fig.4.5(a).

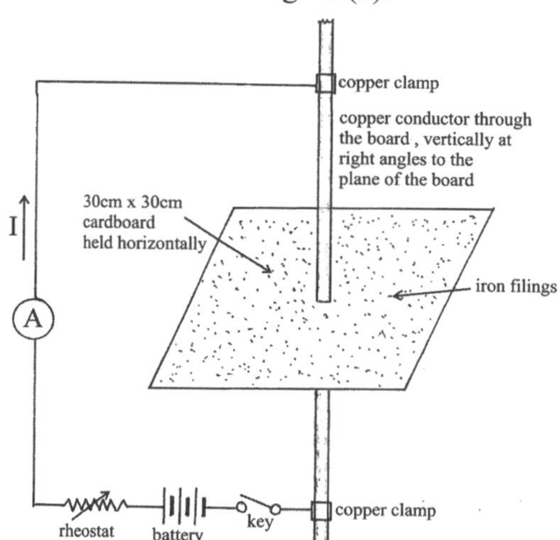

(a) schematic of the experiment

(b) iron filing pattern of the field

Fig.4.5 : Observing magnetic field of current in a straight conductor using iron filings

Allow a suitable current to pass through the wire. A magnetic field will be produced on the plane of the board that would be invisible. To make it visible, sprinkle some iron filings on the board and gently tap it. The iron filings will take the shape of field lines, approximately in the form of concentric circles. See Fig.4.5(b).

By comparison, an iron filing pattern of magnetic field due to a typical permanent magnets is depicted in Fig.4.6.

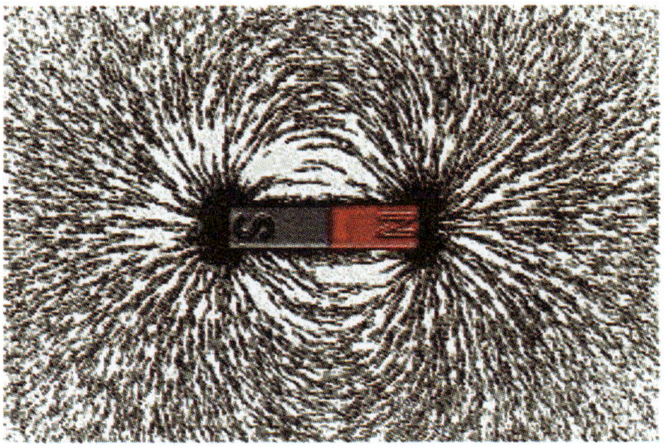

Fig.4.6 : Iron filing patterns of field of a permanent magnet

B. Using a small compass needle

The preparation for this experiment is similar to that using iron filings as shown in Fig.4.7.

In place of (sprinkling of) iron filings, a small compass needle is used to obtain contour of a particular field, remembering that each field *must form a closed circle or path.* Thus, commencing at a suitable point on the card board away from the wire carrying a suitable current in the circuit, two points are marked on the card board [PQRS] coinciding with the tips of the compass needle as indicated in the figure. Joining these points would yield the direction of the field at that point, being tangential to it.

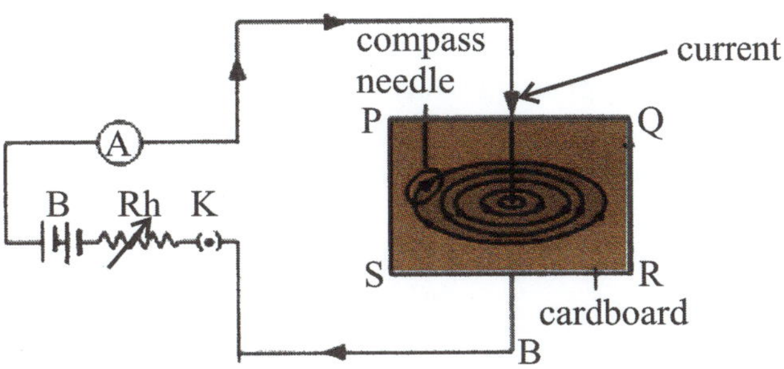

A: ammeter B: battery Rh: rheostat K: key

Fig.4.7 : Observing magnetic field of a current using a compass needle

Move the compass a little further in the anticipated direction of field and mark the two points as before. If the experiment is performed properly, the compass would 'return' to the starting position. Proceed similarly to plot more field lines towards or away from the conductor[3]. Repeat the experiment by changing current in the circuit using the rheostat.

[3]The process can be tedious and time-consuming; however, the experiment would result in 'correct' field pattern.

Right Hand Thumb Rule

This rule provides a quick indication of direction of magnetic field vis-à-vis flow of current in a straight conductor as illustrated I Fig.4.8.

Thus, referring to Fig.4.8, let a conductor be held upright in one's *right* hand enclosing the conductor with fingers, but pointing the thumb up as shown. Then if the flow of current in the conductor is upward, coinciding with the direction of the thumb, the circling fingers point to the direction of magnetic field as viewed on the paper, held in a plane perpendicular to the conductor. Observe that the lines of field are directed anticlockwise *when viewed from the top*[4].

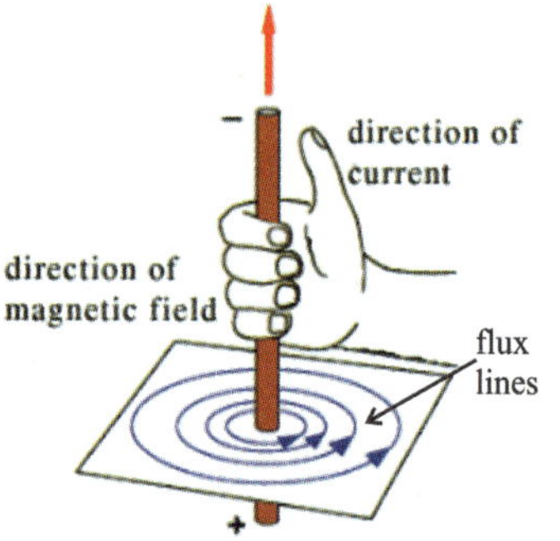

Fig.4.8 : The right hand thumb rule

Magnetic field due to current in a loop

If the straight conductor is assumed to be bent in the form of a loop and a paper or card board is positioned as shown in Fig.4.9, the pattern of the magnetic field would be as marked in the figure. Note that for the left side, the direction of current is upward, that is out of the paper resulting in anticlockwise direction of the field; the reverse is true for the right side current, the field being essentially circular, or concentric circles being relatively dense near the conductor and sparse away from it. Clearly, at exact mid-point between the wire the field would cancel resulting in 'zero' net field.

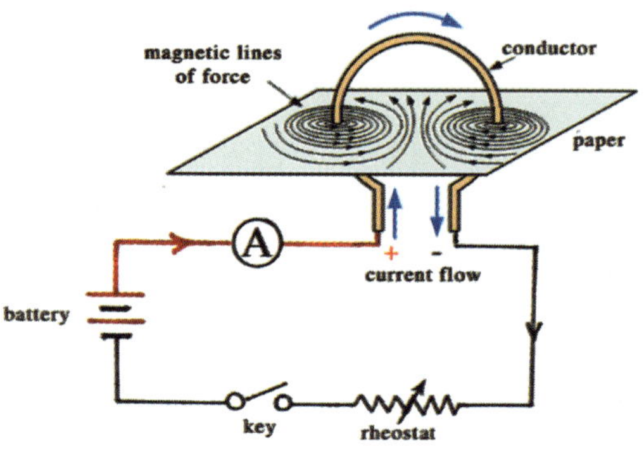

Fig.4.9 : Magnetic field due to a loop current

Magnetic Field due to a Solenoid[5]

A. Air-core solenoid

A **solenoid** is an electromagnetic device having large number of turns, or "loops", of insulated wire wound closely on a tubular support, called "former", with the length of the former (and hence the solenoid) being much greater than its diameter. A schematic of a solenoid is depicted in Fig.4.10[6]. The overall 'arrangement' of turns is known as a "coil"; a

---

[4]This is important since if viewed from below the conductor, the direction of the field would appear to be clockwise.

[5]from the French *solénoïde*, derived in turn from the Greek *solen* "pipe, channel" + combining form of Greek *eidos* "form, shape".

[6]When viewed from the side, a solenoid is usually round in shape; however, there may be solenoids of rectangular or square in shape to suit particular requirements.

coil in general may therefore comprise N number of turns. A current is fed into the coil at one end that would emerge at the other end as shown in the figure.

A longitudinal section of a solenoid is shown in Fig.4.10(b).

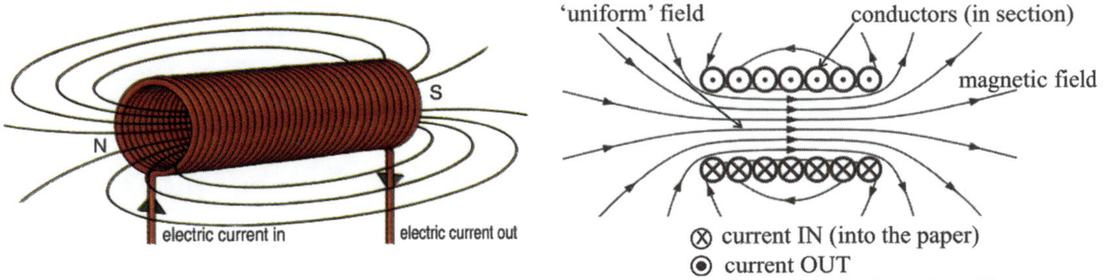

(a) schematic of a solenoid

(b) section of a solenoid

Fig.4.10 : An air-core solenoid

If 'opened', the wire forming the turns could be imagined as a 'straight' conductor, carrying a current. This would result in a magnetic field, the combined effect of current in all the turns manifesting in a pattern shown in Fi.4.10.

Characteristic features of a solenoid

- For a "long" solenoid, that is one having very large number of turns, the field inside at the middle for a short axial length, is uniform, the flux lines being parallel to the axis of the solenoid; meaning that if the field were to be measured using a suitable device it would indicate same magnetic field strength.
- The field is typically strongest in the middle section, reducing in strength and 'spreading' out towards the ends, the field lines closing outside the solenoid as shown[7]. [See Fig.4.10].
- The strength of the field is proportional to the
  a) current fed into the coil
  b) number of turns wound on the former
- Reversing the direction of current flow would result in reversal of direction of field of the solenoid.

A polarity of magnetic field, similar to a permanent magnet, is often assigned to a solenoid. This is illustrated in Fig.4.11.

The concept of polarity emerges from the direction of current input to the coil; for example IN at the left end or entering the beginning turn of the coil, and the field produced, being *outward* at the left end. By comparison with a permanent magnet, this end therefore

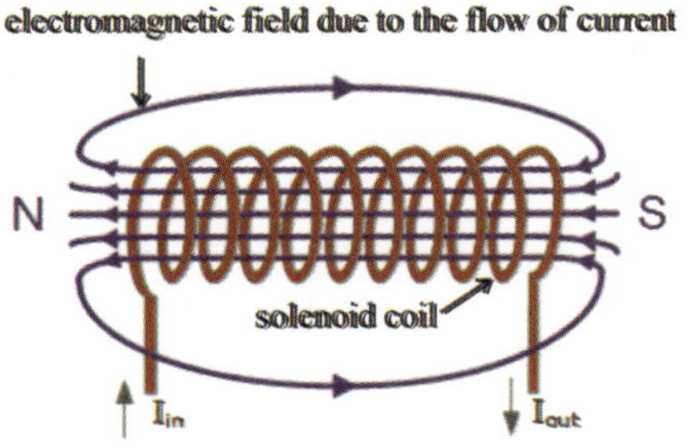

Fig.4.11 : Field of a solenoid with north and south poles

[7]This field or flux of no practical use is termed as "leakage" flux.

is interpreted to represent a *north* pole; the right end thus representing a *south* pole.

B. Iron-core solenoid

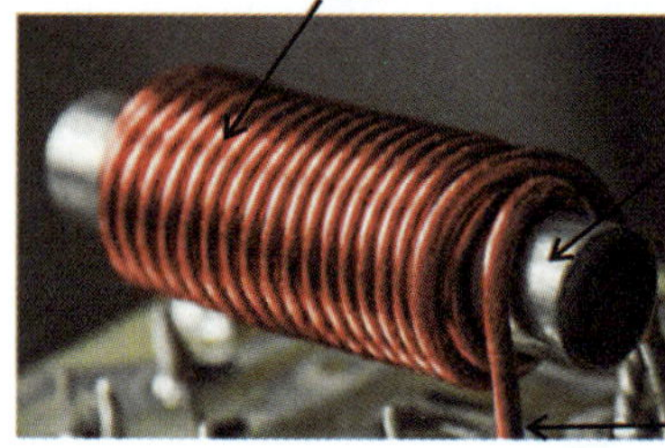

Fig.4.12 : An iron-core solenoid

Fig.4.12 illustrates a typical iron-core solenoid.

Inserting "iron" or a ferromagnetic material inside an air-core solenoid, Fig.4.10, suitably fitting the inside section, can significantly increase the magnetic field produced by the *same* current. The increase in the field strength will be directly proportional to the permeability of the material. Such a solenoid is identified as "iron-core" solenoid and is particularly suited for numerous applications in engineering. Also, the core helps in concentrating the magnetic lines of forces through the solenoid, so that the magnetic field is almost the same as in the middle at the end faces of the core. In the absence of a core, the magnetic lines of forces start to diverge by curving sharply immediately outside the coils as in an air-core solenoid. The smaller the space between the coils and the core, that is closer the fitting of the core inside, the stronger is the magnetism induced in the core.

The "faraday coil"

Magnetic Field Due to a Multi-turn Coil

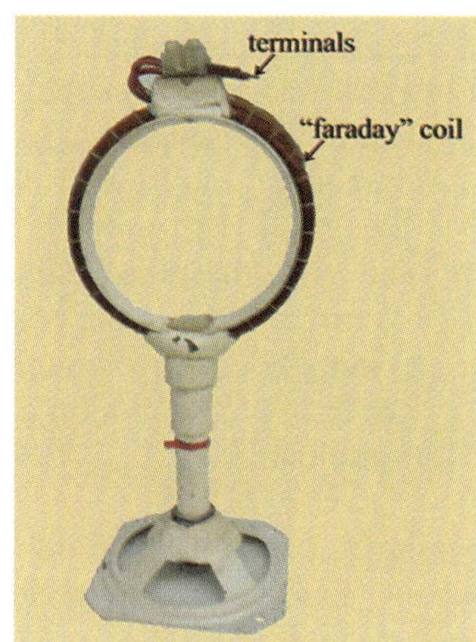

Fig.4.13 : The faraday coil

A special multi-turn coil, to be called "Faraday coil" owing to its resemblance to original Faraday coils used by him for experiments, is shown in Fig.4.13. It consists of several *hundred* turns of 'fine', enameled copper wire wound round a PVC arbor (or circular support), and fitted to a spindle held vertically on a rigid base. This results in the plane of the coil to be vertical. the terminals of the coil are brought out for external connections.

Magnetic effect of current in the coil

The coil forms the example of an air-core solenoid with a very short axial length and relatively large diameter. The magnetic effect of a current passed through the coil is demonstrated by reference to Fig.4.14.

A. In the first instance, Fig.4.14(a), a sensitive compass needle is positioned at the centre of the coil such that the needle is aligned with the plane of the coil, pointing "north" – "south" under the influence of horizontal component earth's magnetic field.

B. A dry cell of EMF 1.5 V is connected to the coil terminals, resulting in a small current to flow, producing a magnetic field that is *directed axially*. The needle is deflected in a horizontal plane, settling in a direction being resultant of the fields due to earth and current in the coil as shown in Fig.4.14(b).
Clearly, the reversal of direction of current flow in the coil would result in reversal of direction of deflection of the needle.

The action is depicted graphically in Fig.4.14(c).

(a) compass needle aligned to earth's field

(b) needle deflected by resultant magnetic field

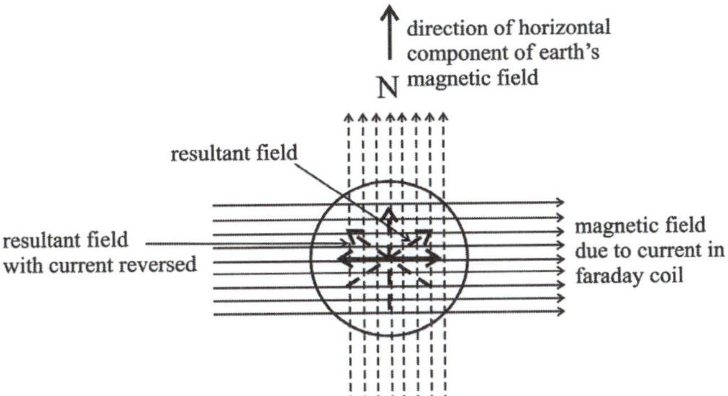

(c) graphical representation of fields

Fig.4.14 : Magnetic effect of current in the faraday coil

Electromagnet

An **electromagnet** is a type of magnet in which the magnetic field is produced by an electric current[8]. The magnetic field disappears when the current is turned off. Electromagnets

[8]The current, called the "excitation" current, can be either DC or AC.

usually consist of a large number of closely spaced turns of wire that create the magnetic field. The turns of wire are often wound around a magnetic core made from a ferromagnetic material such as iron or mild steel; such a core concentrates the magnetic flux and makes a more powerful magnet[9]. With the inclusion of the iron core and being of straight length, an electromagnet can be compared to a permanent magnet having well-defined magnetic poles at the two ends.

The main advantage of an electromagnet over a permanent magnet is that the magnetic field can be quickly changed by controlling electric current in the winding. However, unlike a permanent magnet that needs no power, an electromagnet requires a continuous supply of current to maintain the magnetic field.

The principle of operation of an electromagnet is depicted in Fig.4.15.

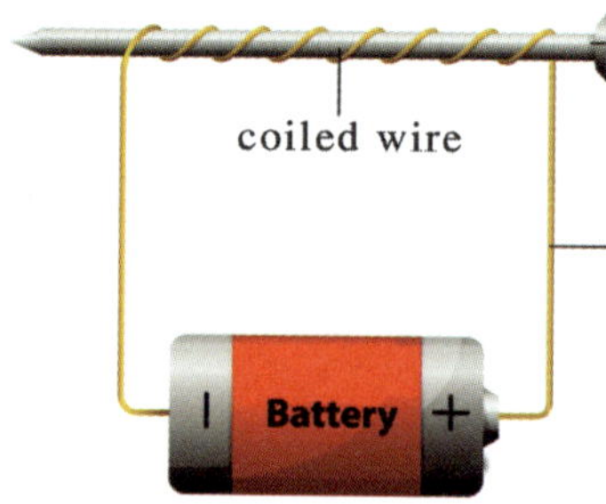

Fig.4.15 : Principle of operation of an electromagnet

Uses of electromagnets

Electromagnets are extensively used in electric and electromechanical devices, including

- Motors and generators[10]
- Transformers[11]
- Relays, including reed relays originally used in telephone exchanges
- Electric bells and buzzers
- Loudspeakers and earphones
- Actuators
- Magnetic recording and data storage equipment: tape recorders, VCRs, hard disks
- MRI machines
- Scientific equipment such as mass spectrometers
- Particle accelerators
- Magnetic locks
- Magnetic separation equipment, used for separating magnetic from nonmagnetic material, for example separating ferrous metal from other material in scrap. See Fig.4.16.

---

[9]When so constructed an electromagnet is similar to an iron-core solenoid.

[10]Used for "field" magnets in DC motors and dynamos and excited rotor in an AC generator or alternator.

[11]Using AC to excite transformer cores.

- Industrial lifting magnets. See Fig.4.17.
- Magnetic levitation
- Induction heating for cooking, manufacturing, and hyperthermia therapy

Fig.4.16 : A separating electromagnet

Fig.4.17 : A lifting electromagnet

Comparison of an electromagnet with a permanent magnet

Although the basic behavior of the two magnets follows the same pattern, there are some differences between an electromagnet and a permanent magnet. A comparison of the two is given below.

Electromagnet	Permanent magnet
• An electromagnet invariably uses a soft iron core	A permanent magnet is made from steel or alloys such as ALNICO or Barium Ferrite
• Its magnetic field comes into play as long as current flows in the winding	Its magnetic field is permanent
• Its field strength can be varied by varying the excitation current	The field of a permanent magnet cannot be altered
• The polarity of an electromagnet can be reversed by reversing excitation	The polarity of a permanent magnet cannot be reversed
• An electromagnet becomes "demagnetised" as soon as the excitation is switched off	Demagnetising a permanent magnet involves difficult process(es)
• An electromagnet does not exhibit hysteresis	Possessing hysteresis of varying degree constitutes an important property of a permanent magnet

ELECTROMAGNETISM

The Electric Motor

Fleming's left hand rule

Fleming's left-hand rule forms the fundamental basis for operation of **ALL** types of motor – DC or AC – and indeed to explain the operation of a variety of electromagnetic devices involving electric current and magnetic field. It was originally evolved by Fleming[12] in the late 19th century as a simple way of working out the direction of motion in an electric motor.

Fleming observed that when a current flows in a wire, and an external magnetic field is applied across that flow, the wire experiences a force *perpendicular both to that field and to the direction of the current flow*. This led to the enunciation of Fleming's left-hand rule.

The phenomenon is depicted in a simple pictorial form in Fig.4.18.

The rule can be stated as follows

> "Hold out your *left* hand with forefinger, second finger and thumb at right angle to one another. Then, if the **forefinger** represents the direction of the *field* and the **second finger** that of the current, then **thumb** gives the direction of the *force*[13]".
> This is depicted in Fig.4.18(a).

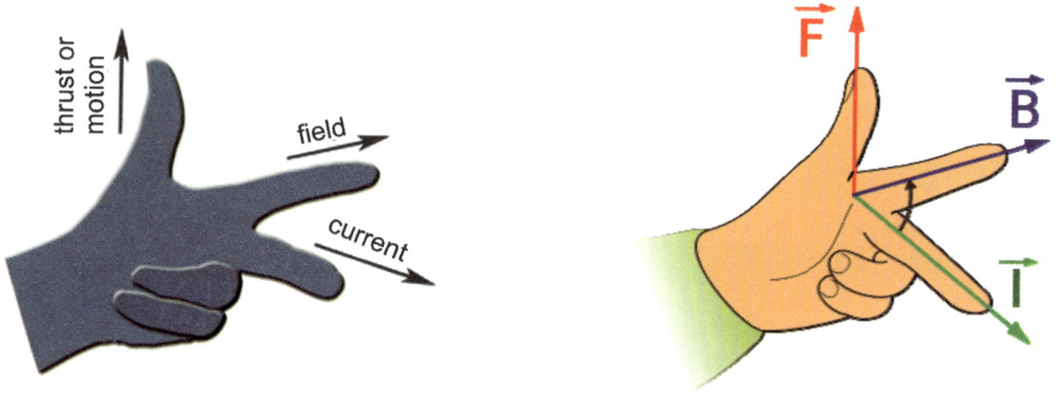

(a) (b)

Fig.4.18 : Images to illustrate Fleming's left hand rule

In Fig.4.18(b), B is the field, I the current and F is the direction of developed force.

---

[12]Sir John Alexander Fleming – a Scottish physician, biologist, pharmacologist and botanist: 1881-1955.

[13]Or the motion.

It is important to note that the phenomenon is essentially *three* dimensional as depicted in Fig.4.19 in Cartesian coordinates.

Mathematically, if a conductor of length l m carrying a current I A is held at right angles to a magnetic field of flux density B T, it would experience a force given by

$$F = B I l \ \text{N}$$

F being at right angles to *both* B and I.

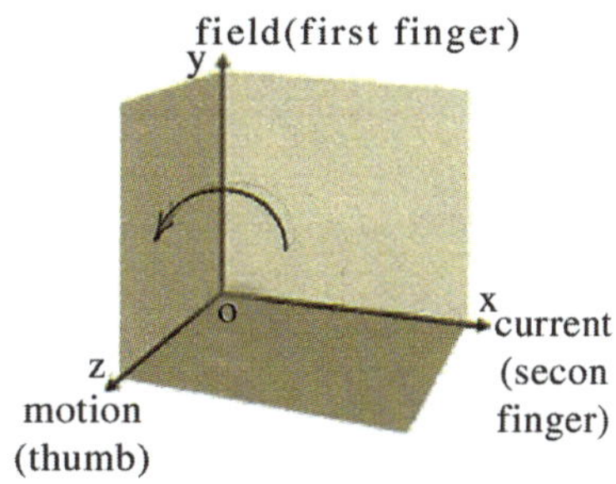

Fig.4.19 : Three dimensional representation of Fleming's left hand rule

A Model to Demonstrate Fleming's Left Hand Rule

A simple yet very effective working model to demonstrate Fleming's left hand rule is illustrated photographically in Fig.4.20(a), its schematic being shown in Fig.4.20(b)..

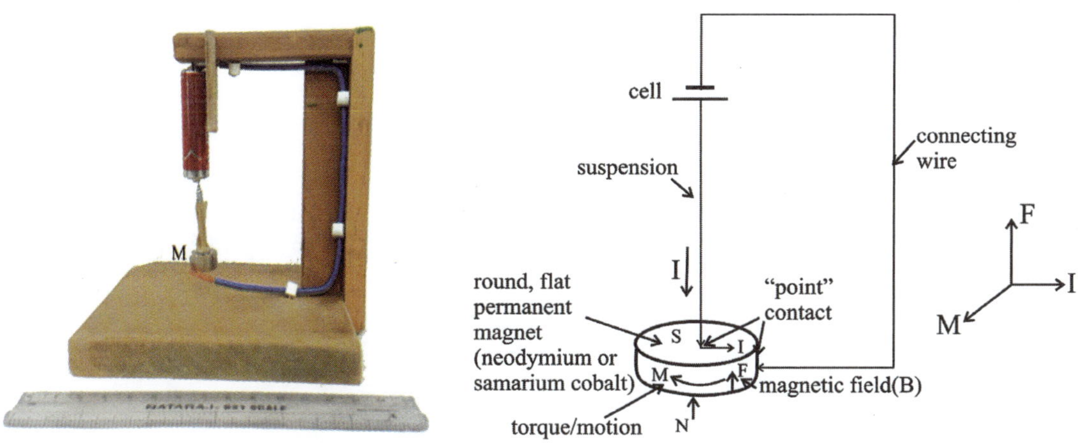

(a) electromagnetic model (b) schematic

Fig.4.20 : A model to demonstrate Fleming's left hand rule

Working

In the photograph and schematic, M is a strong, round magnet, made of modern permanent-magnet material like NEODYMIUM or SAMARIUM (cobalt) with the flat surfaces being magnetised as north and south poles such that the flux due to the magnet is directed vertically up or down depending on its mode of suspension. The current from the

(1.5 V) cell, flowing downward through the suspension screw (made of steel), 'spreads' *radially outward* in the magnet and interacting with the field at every point of the peripheral surface of the magnet results in a force *tangential* to the surface. The force acting at a distance R, the radius of the magnet, gives rise a torque that sets the magnet into a circular motion.
If the direction of current is reversed by connecting the cell up-side down, or the polarity of magnetic field is reversed by flipping the magnet, the direction of rotation is also reversed, conforming to Fleming's left hand rule[14].

The DC motor

A motor is an electric *machine* which converts electrical energy into mechanical energy. A DC motor, energised from a DC supply, works based *directly* on Fleming's left hand rule.

Small DC motors are used in tools, toys, and appliances. Larger DC motors, with ratings from a few hp to several hundred hp, are used in propulsion of electric vehicles, elevator and hoists, or in drives for steel rolling mills and many more industrial applications. The motors possess the important advantage of *easy and effective speed control* and therefore ideally suited to all applications where smooth control of speed of the drive forms a crucial requirement.

Construction of a Typical DC Motor

A **DC motor** essentially comprises two main parts, namely

1. **Stator** – The static part that houses the field windings wound around projecting or "salient" poles constructed from "laminated" magnetic material and receives the DC supply[15], and

2. **Rotor** – The rotating part that brings about the mechanical rotation of the shaft in the motor

The stator and rotor can be further sub-divided into various pars as follows

Stator
comprising a
yoke
poles

[14]Note that when viewed in perspective, the device can be interpreted as a simple, miniature DC motor that is based on Fleming's classic rule. The experiment can be easily repeated in the lab or class room.
[15]When so provided the "field system" in the stator constitutes an electromagnet so that the strength of the field from the poles can be controlled by varying current called "excitation" through the winding.

Rotor

comprising

an armature

provided with a set of coils made from insulated copper wire known as winding

a commutator

a 'cylindrical' structure comprising copper bars or segments, insulated from each other, fitted at one end of the motor shaft and playing key function in the motor operation; see Fig.4.26b) later

a set of brushes

made of special carbon, these are enclosed in specially designed receptacles or (brush) holders and rest under pressure of an spring on the commutator to make 'firm' contact with its segments

ventilating fans

mounted suitably on the motor shaft at the "non-drive" end for cooling the motor from inside

All the parts are enclosed in an appropriately designed case or housing with a provision for mounting the motor on a suitable foundation as required.

An "exploded" view of a DC motor showing all the main parts is illustrated in Fig.4.21.

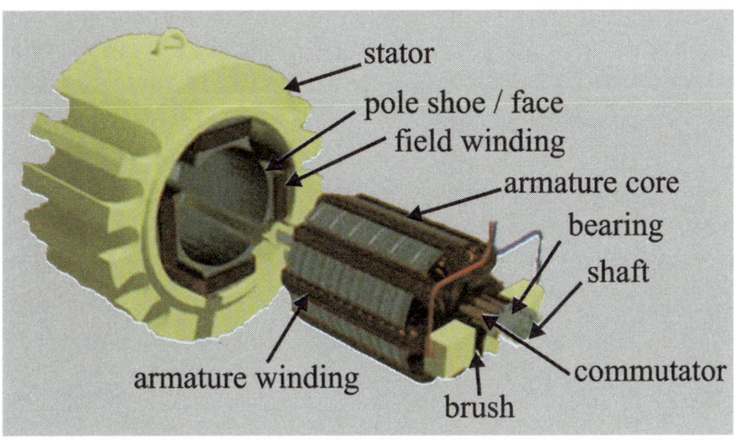

Fig.4.21 : Exploded view of a DC motor

Principle of Operation

A simplified sketch of a motor consisting of just *one* coil, terminating on a *pair* of commutator 'segments', identified as a "split-ring" commutator, with brushes is given in Fig.4.22. Note the relative position of the two conductors constituting the coil with respect to commutator segments and brushes[16].

[16]This forms an important constructional feature of a *DC* motor 'repeated' many times over in an actual motor.

The coil is arranged to rotate about the horizontal axis between the poles of a magnet[17], the field of the magnet being directed horizontally.
A source of DC supply of suitable EMF is connected across terminals T, T that appears in turn across the brushes B, B resulting in a current, I, to flow in the coil, say onward in the left conductor and return in the right conductor as marked. These currents reacting with the field result in producing a force in each conductor according to Fleming's left hand rule, the developed forces in the two conductors being *oppositely directed*, thus forming a couple that causes the coil to rotate in a particular direction.

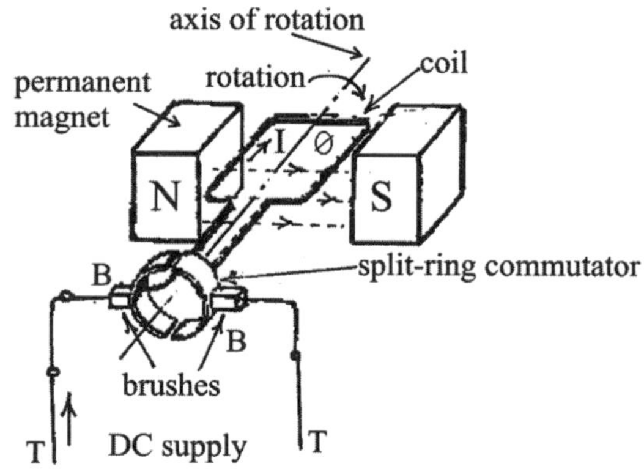

Fig.4.22 : A "basic" DC motor

Function of the Commutator

The commutator plays an important role in a DC motor in that it ensures the rotation of the coil in the *same* direction – clockwise or anticlockwise – *without changing the polarity of electric supply* at terminals T, T, Fig.4.22. In Fig.4.23 are shown various relative positions of the coil 'cross section' as it rotates about its axis. The two commutator segments are shown as a and b.

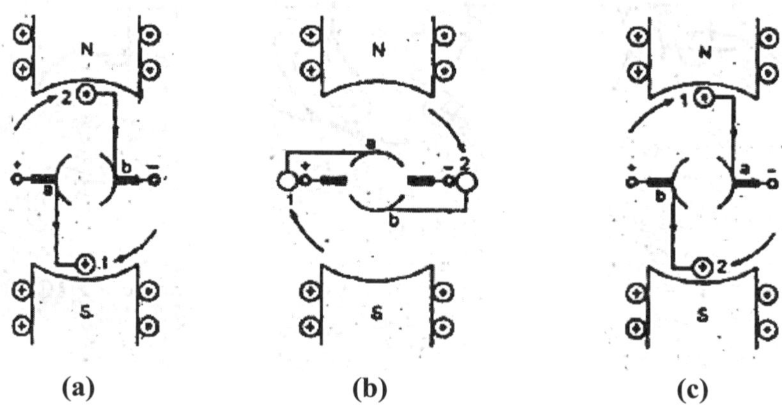

Fig.4.23 : Different stages of a coil undergoing commutation

The step-by-step action of the commutator in a DC motor is explained by reference to Fig.4.24.

---

[17]Shown as a permanent magnet; in an actual motor, in practice, this would be an electromagnet as mentioned above.

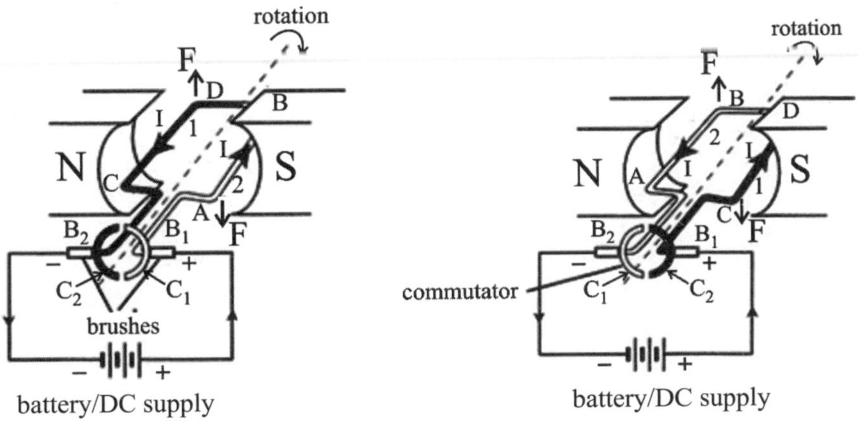

Fig.4.24 : Production of force in conductors of the coil corresponding to cases "A" and "C"

A. Coil in horizontal plane

In this position, shown in Fig.4.24(a), the side conductors AB and CD, each of length L m and carrying current I A as marked are directly under the influence of magnetic field of the magnet[18], positions "1" and "2". If the flux density of the field *at that position and instant* is B T, conductor AB will experience a force F N that will be directed *downward* according to Fleming's left hand rule. Similarly, conductor CD will experience force F N, but directed *upward,* that is opposite to that in conductor AB since the direction of current is opposite in conductor CD[19] whilst the direction of magnetic field is un-changed. The force F in each conductor is given by $F = B\,L\,I$ N and the two forces acting together form a couple (or torque) given by

$$T = F \times W = B\,I\,A$$

A being the area of the coil and where W is the breadth of the coil in metre. The couple or the resulting torque results in an *clockwise* rotation of the coil when viewed from front or commutator side.

B. Coil in vertical plane

After an anticlockwise rotation of 90°, conductors AB and CD occupy a position as shown with conductor AB at the top and conductor CD at the bottom. The commutator segments would also have move as shown, with the brushes *being stationary* and closing on the commutator gaps, thereby short-circuiting the coil and resulting in no current flow in the

[18]The current is fed into the conductor at points 'a" and "b" involving brushes "B_1" and "B_2" and commutator segments "C_1" and "C_2" as shown
[19]It is to be noted that the coil sides AD and BC develop no force since the direction of ('closing') currents in these sections is parallel to the direction of field due to the magnet.

conductors. No forces are developed in the conductors in this position; however, the coil would continue to rotate due to its inertia.

C. Coil again in horizontal plane

The situation in this case is similar to that in "A" except that conductor CD now occupies the position that was held by conductor AB. By virtue of the commutator segment C_2, attached to conductor CD and its connection to brush B_1, it now carries current I in the same direction as AB in "A". Forces F each are therefore developed in the conductors in the same direction resulting in the couple that would continue to rotate the coil in clockwise direction as shown in "A".

D. Coil again in vertical plane

This presents a situation as at "B" with the same reasoning. The brushes short circuit the coil, developing no force in the conductors and no torque. As at "B" the coil continues to rotate for a while along with the commutator segments due to its inertia to occupy horizontal position.

E. Coil again in horizontal plane

The cycle repeats itself as explained in "A".

Speed of Rotation of the Coil

The speed of rotation of the coil in a given direction can be derived from the input power to the coil and torque developed by it. Thus if P is the power input in watt and T the torque in Nm, then

$$P = 2\pi n T$$

where n is the coil speed in revolution per *second* (or rps). That is

$$n = \frac{P}{2\pi T} \text{ rps}$$

Factors affecting operation of a DC motor

There are numerous factors that affect the operation - (mechanical) output and speed - of the motor.

■ Number of Turns

The above discussion pertains to a coil of *single* turn. In practice, the rotating part of a motor, called an "armature", comprises dozens of multi-turn coils connected to one another in a designed manner, the coils being so formed are housed in slots or recesses in the armature, the arrangement of coils known as "winding".

■ Area of Coil(s)

The area of the coil in the armature would control the torque developed by way of affecting the force produced in the conductor – "coil sides" –in the winding and the magnitude of couple, respectively. These parameters constitute two important aspects of motor design and affect the overall size of the rotor and hence the motor.

In an actual motor, the field system comprises salient or protruding poles, fitted to the yoke or outer frame of the stator, a set of field winding on each pole. The field winding when excited is arranged to produce alternate north and south poles in the field system as shown in Fig.4.25. Note that the field system is seldom a two-pole construction; rather may consist of four or six poles.

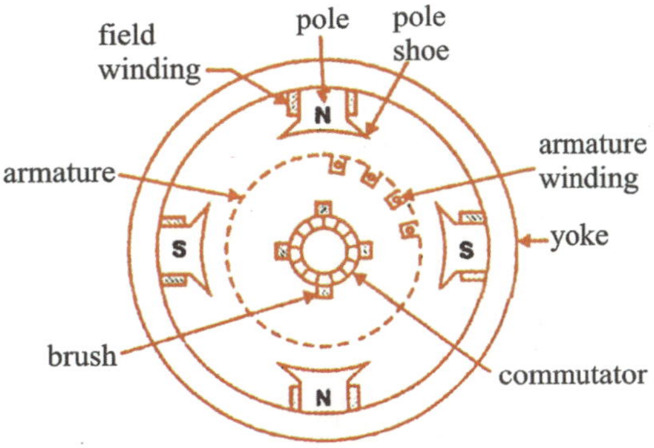

Fig.4.25 : Field system of a four-pole motor

All the above requirement result in a complex construction of an actual motor. Fig. 4.26 illustrates constructional features of a typical DC motor. A dismantled view of a typical DC motor showing all important parts is shown in Fig.4.26(a); construction of a typical armature and commutator is shown in Fig.4.26(b).

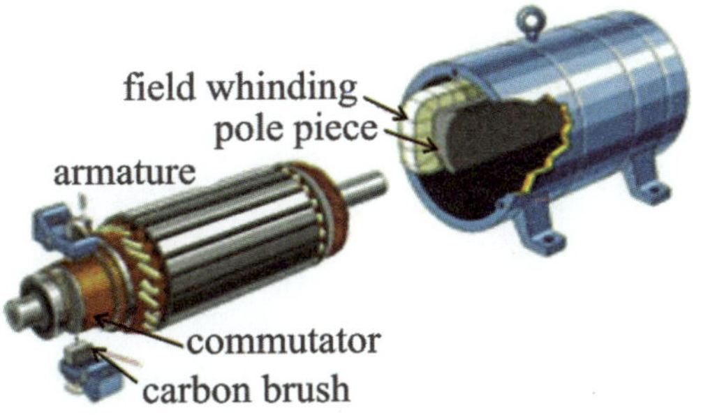

(a) : a dismantled DC motor showing essential parts

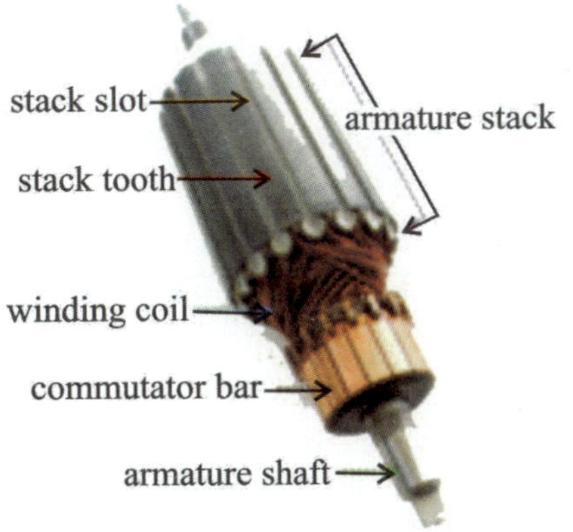

(b) : rotor and commutator of a DC motor

Fig.4.26 : Constructional features of a DC motor

ELECTROMAGNETIC INDUCTION

Faraday's Discovery

Whilst the 19th century was marked by the discovery of magnetic effect of a current and force on a current-carrying conductor, it was the genius of Michael Faraday[20], a British scientist, that led to the discovery of production of EMF in a circuit that linked with a *time-varying* magnetic field. Faraday began his 'research' around 1825, but succeeded only in 1831 to discover now universally known "laws of electromagnetic induction" that form the basis of ALL modes of generation and utilisation of *AC* electricity.

Faraday's experiment

For his experiments, Faraday used a ring-shaped magnetic core of about 25 cm diameter, wound with two overlapping windings, *but not connected in any way*. A view of the historic ring used by Faraday is shown in Fig.4.27(a) whilst a "home-made" replica of the ring, comprising a 'primary' and 'secondary' winding, wound round a circular laminated core, is shown in Fig4.27(b).

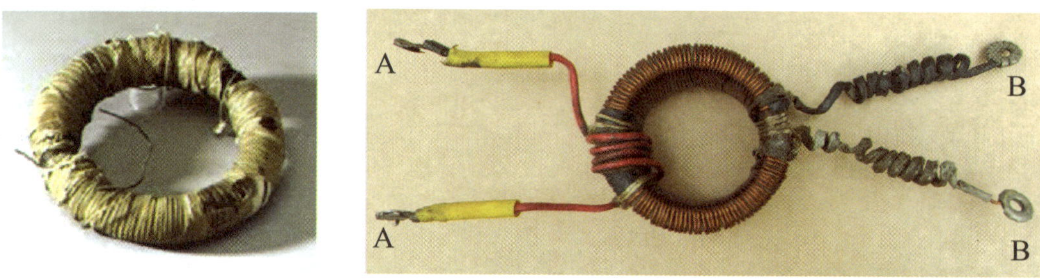

(a) the historic "Faraday ring"

(b) home made replica
[note that either coil can act as 'primary' or 'secondary']

Fig.4.27 : The historic "Faraday ring" and its home-made replica

---

[20]A blacksmith's son, Michael Faraday was born on 22nd September, 1791 at Newington, Surrey, England. He began his 'career' as a book-binder, later to be engaged as a laboratory assistant at the Royal Institution, London. He was inspired and initiated into 'science' by the noted scientist Humphry Davy and became his assistant.
In 1821, Faraday discovered the principle of (DC) electric motor and built one; two years later he learnt to liquefy chlorine, and discovered the laws of electromagnetic induction on 29 August, 1831 – the laws that gave electricity to the world and changed the destiny of mankind!
Among his other accomplishments, Faraday produced the first dynamo, stated the basic laws of electrolysis, discovered that a magnetic field would rotate the plane of polarisation of light and discovered Benzene.
He died on 25th August, 1867 at Hampton Court, Surrey.

A schematic of the ring and experimental set-up is shown in Fig.4.28. One of the windings, A, is connected to a battery via a key whilst the other, B, is connected to a sensitive "centre-zero" galvanometer.

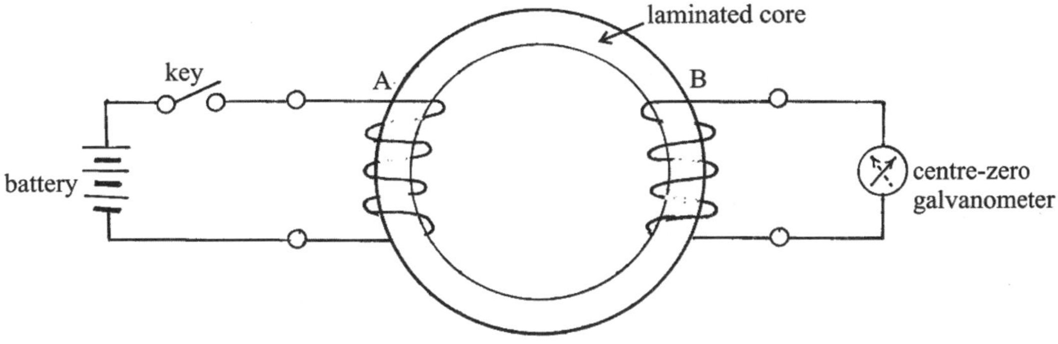

Fig.4.28 : Schematic of Faraday ring experiment

Faraday observed that on sudden closing of the key as the current *grew* in winding A, a throw was observed in the galvanometer connected to winding B *which 'died' as the current became steady*. On opening the key, a throw was again observed in the galvanometer, *but in the opposite direction.* This showed that deflection in the galvanometer occurred only *when current in winding A changed with time*. These observations, based on other similar experiments, led Faraday to establish his laws.

Experiment Using the "Replica Coil"

The above phenomenon can be replicated using the replica coil shown in Fig.4.27(b). A centre-zero galvanometer is connected across terminals AA whilst a battery of 3.0 V is connected across terminals BB with a key connected in the battery circuit. By pressing the key and un-pressing it, same effects as above are observed. Observe that no EMF is induced in the coil and hence no deflection occurring in the galvanometer if the key is kept pressed and a steady current continues to flow in the ('primary') coil (coil BB).

Electromagnetic induction: relative motion of a coil and a magnet

The phenomenon of electromagnetic induction is often couched in terms of motion of a (permanent) magnet with respect to a stationary coil of a given number of turns or *vice-versa*; in other words considering their relative motion. The concept and the experiment involving a magnet and coil, with their axes aligned, is also ascribed to Faraday.

Consider the arrangement of a permanent magnet and an air-core coil with its terminals connected to a galvanometer as shown in Fig.4.29.

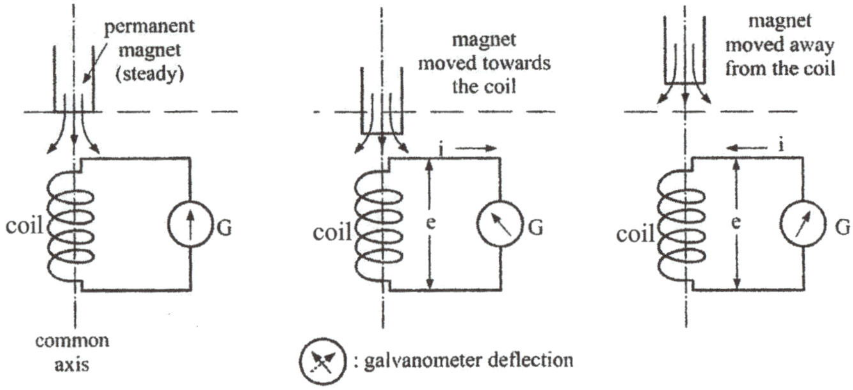

Fig.4.29 : Relative motion of a magnet with respect to a coil and electromagnetic induction

In Fig.4.29(a), when the magnet is stationary, there is no induction of EMF in the coil and the galvanometer points to zero. As the magnet is moved closer to the coil, a throw is observed in the galvanometer in one direction, indicative of induction of EMF, and hence a current, in the coil, Fig.4.29(b)[21]. The induction of EMF is due to change of flux linking the coil owing to the movement of the magnet relative to the coil, that is, due to change of *flux linkages*. If the magnet were pulled away from the coil as shown in Fig.4.29(c), the EMF induced in the coil would be in opposite direction as indicated by deflection of the galvanometer in the opposite direction.

Direction of the Induced EMF

The direction of induced EMF in the coil is governed by Lenz's law, discussed later, so that the flux set up by the induced current in the coil will *oppose* the flux originally responsible for induction of the currents, that is, the flux due to the magnet. This is illustrated in Fi.4.30 where the magnet moving towards the coil results in induction of current in anti-clockwise direction[22].

See also Fig.4.31 for action of magnet moving towards coil and away from it when viewed from left to right.

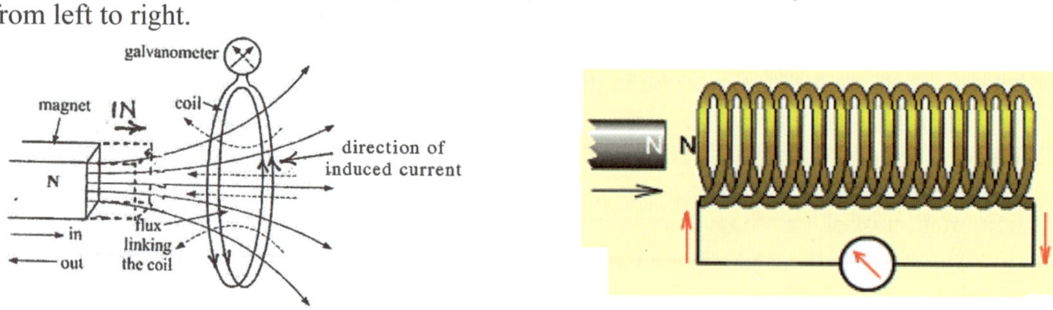

Fig.4.30 : Movement of magnet towards a coil and direction of induced current

[21]The galvanometer pointer returns to zero, however, if the magnet is at rest in its new position.

[22]Note that this is contrary to the direction of flux set up by a steady (or direct) current flowing in the coil.

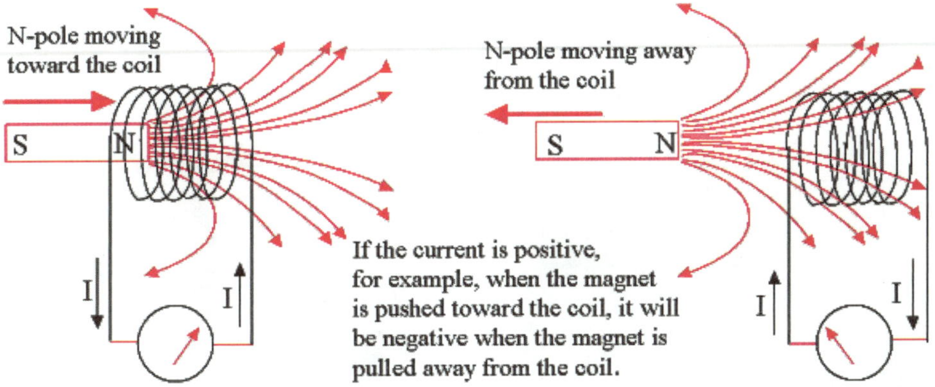

Fig.4.31 : Action of magnet moving towards and away from a coil

Laws of electromagnetic induction

These are stated as

I ***A changing magnetic field* induces *an electromotive force (EMF) in a conductor***

II ***The electromotive force is proportional to the* rate of change *of the field***

III ***The direction of the induced electromotive force depends on the* orientation *of the field***

Here, the change of magnetic field is implied to be relative: a magnet moving relative to the coil or *vice versa*. Also, the strength of the field and speed of relative movement both play a key role.

An experiment to demonstrate electromagnetic induction

This involves the use of the Faraday coil discussed previously, see Figs.4.13 and 4.29, and a permanent magnet moving *relative to the coil.*

- Connect the coil terminals to a sensitive "centre-zero" galvanometer.
- Hold a permanent magnet suitably [see Fig.4.32] and insert it swiftly inside the coil, approximately in the centre, Fig.4.32(a).
- Observe the throw in the galvanometer to left or right due to sudden change of flux linking the coil and an EMF induced in the coil, resulting in a small current through the galvanometer.
- Pull out the magnet in a fast motion, Fig.4.32(b), and again observe throw in the galvanometer, *being in a reverse direction.*
- Reverse the connections of the coil to the galvanometer and repeat the above four steps.
- ▸ If the magnet is inserted 'slowly' in the coil, the deflection in the galvanometer would be small.

Note that the EMF is induced in the coil, and hence throw in the galvanometer occurs, only as the flux linking the coil changes; if the magnet is held stationary anywhere with respect to the coil, no EMF is induced with no deflection in the galvanometer.

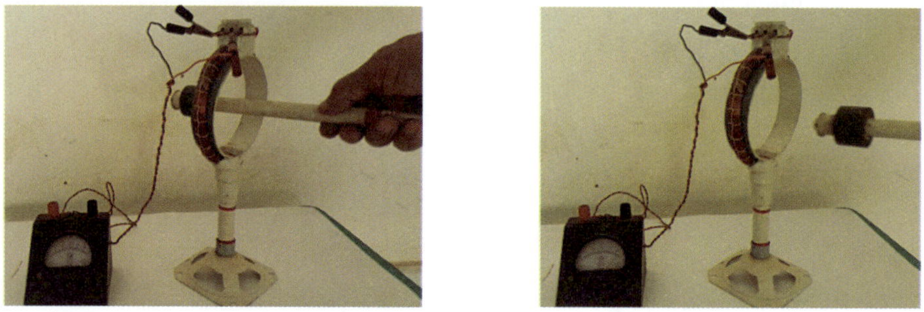

(a) magnet being pushed inside the coil (b) magnet being pulled out

Fig.4.32 : Experiment with faraday coil to demonstrate electromagnetic induction

Expression for Induced EMF

Faraday's laws of electromagnetic induction can be expressed by reference to the phenomenon in a coil by the equation

$$\text{EMF} = N\frac{d\phi}{dt}$$

where N represents number of turns in the coil, ϕ the flux linking the coil at instant t and d/dt expressing *rate of change* of flux, ϕ.

Factors Affecting Magnitude of Induced EMF

As reflected in the above expression, the magnitude of EMF induced in a coil[23] would depend on

- the number of turns in the coil
- the magnitude of flux linking the coil
- the (time) rate of change of flux; faster the rate, greater will be the EMF for the given coil

Direction of induced EMF

Similar to the direction of force produced in a current carrying conductor in a magnetic field according to Fleming's left hand rule, the direction of induced EMF in a conductor can be expressed in terms of change of magnetic field, *effected by the motion of a conductor*, and the direction of magnetic field.

---

[23]In the discussions, reference is made to a *coil* rather a conductor; the latter being unimportant in practice.

Fleming's right hand rule

This rule provides for the *direction* of induced EMF in a conductor moving at right angles to a uniform magnetic field, B T, with a velocity of v m/s. The relationship is expressed in a pictorial form by **Fleming's right hand rule** illustrated in Fig.4.33[24]. See also Chapter IV, Unit 1.

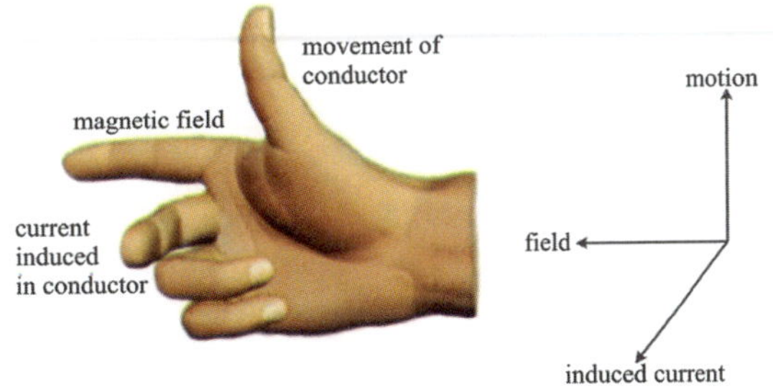

Fig.4.33 : Fleming's right hand rule

The rule can be stated thus

"Hold the right hand stretched out with the forefinger, middle finger and thumb being mutually at right angles with respect to each other as shown. If the forefinger points in the direction of field and the thumb points to the motion of the conductor, then middle finger indicates the direction of induced EMF".

Flux cutting rule

This rule which directly relates to Fleming's right hand rule is fundamental to ALL types of electric generators – DC or AC – where conductors formed as coils, and winding, essentially "cut" the flux provided in the machine by appropriate means – field systems - to generate EMF or voltage at the terminals of the generator.
The concept is illustrated graphically in Fig.4.34. Referring to Fig.4.34(b), if a conductor moves with a velocity v m/s, from its initial position, PQ, to later position, RS, cutting a *uniform* magnetic field at right angles, an EMF (or current) will be induced in it, the direction being given by Fleming's right hand rule as shown. The magnitude of the EMF will be given by

$$E = B\,l\,v \quad V$$

where B is the flux density of the field in tesla and l is the length of the conductor in metre.

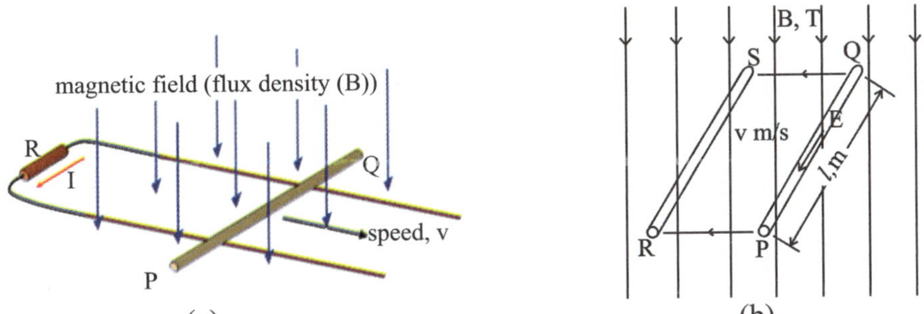

Fig.4.34 : EMF induced in a conductor by flux cutting rule

[24]Cf. Fleming's left hand rule.

Lenz's law

A concept that is sometimes neglected or poorly understood is the basic "cause-effect" relationship in electromagnetism. For example, consider the case of EMF induced in a *stationary* coil due to movement of a magnet to change the flux linkages. If the terminals of the coil are shorted a current would flow that would produce its own magnetic field in a direction that must *oppose* the field due to the magnet, thus 'balancing' the flux linkages with the coil which were zero in the first place. Note that this also follows from the "law of conservation of energy".

These concepts were generalised into an important law by Lenz* in 1835 which states that

"the induced EMF in a (closed) circuit is such that the flux produced by the current in the circuit will oppose the flux responsible for producing the EMF".

Accordingly, the expression for induced EMF given above would modify to

$$E = -N \frac{d\phi}{dt}$$

the minus sign accounting for Lenz's law.

*Russian physicist Heinrich Friedrich Lenz: 1804-1865

AC GENERATOR

A **generator** is a device or machine that converts mechanical energy into electrical energy for use in an external circuit or system. The source of mechanical energy input may vary widely from a diesel or internal combustion engine for a small generator to gas, hydro or steam turbine. See also Chapter IV, Unit 1.

An **AC generator**, also called an **alternator**, is one that generates "alternating" EMF or current; hence acronym "AC". A typical AC generator comprises

- A coil or a set of coils, in practice, known as **winding** in an actual machine
- A magnetic field produced by a **field system**, or a set of electromagnets
- A pair of copper slip rings, to which the coil ends are connected
- A pair of carbon brushes pressing on the slip rings

These parts are shown schematically in Fig.4.35. An essential, additional requirement for a generator is a device (mechanical or otherwise), called a "prime mover", to rotate the coil or winding in either direction, that is clockwise or anticlockwise, about the axis of rotation of the coil. Only anticlockwise direction of rotation is shown in the figure.

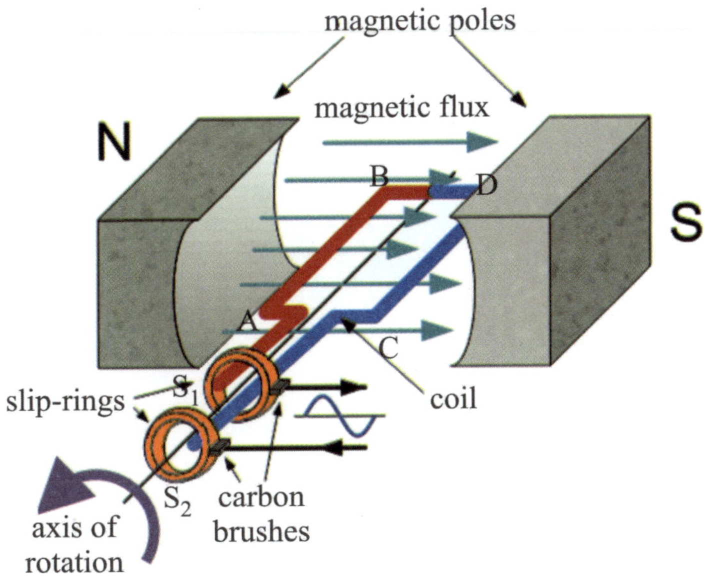

Fig.4.35 : Schematic of an AC generator

Principle of Operation

As provided by Faraday, a conductor moving relative to a magnetic field develops an electromotive force(EMF) in it, the direction of induced EMF being given by Fleming's right hand rule. This EMF would reverse its polarity when the conductor moves across magnetic field of opposite polarity.

In Fig.4.35, N and S are north and south poles of a permanent magnet producing a magnetic field (or flux) in horizontal direction, assumed to be *uniform*, of flux density B tesla. A rectangular coil A B C D, comprising straight conductors AB and CD, the ends of which are connected to a pair of slip rings S_1 and S_2. The coil is free to rotate about a horizontal axis of rotation as shown.

In the position shown, the coil is in horizontal plane with conductor AB cutting the flux from north pole at right angles or 90° from an (arbitrary) datum of 0° as marked, with the peripheral velocity, v m/s, and the flux-cutting being maximum. Since the *magnitude* of induced EMF *at any instant,* e, is given by

$$\boxed{e = Bl\,v}$$

it follows that

$$e \propto v$$

as B and l are constant. If an angle θ is measured from the datum in the axis of rotation, then at θ=90°, the position of conductor AB in Fig.4.35, the peripheral velocity of the conductor will be maximum, v_{max}, and the induced EMF given by

$$(+)e = B\,lv_{max}$$ [25]

and will be a maximum at that instant. At any other instant, for the conductor AB to occupy a position at any angle θ between 0° and 90°, the peripheral velocity of the conductor will be

$$v = v_{max} \sin\theta$$

and the EMF induced in the conductor will be given by

$$e = B\,lv_{max} \sin\theta$$

If the quantity $[B\,l\,v_{max}]$ is expressed as e_{max}, the maximum value of EMF induced in the conductor, the induced EMF at any instant, t, corresponding to that position of the coil, will be

$$\boxed{e = e_{max} \sin\theta}$$

and is called "instantaneous", that is at any instant, (induced) EMF.

As the conductor AB reaches the position corresponding to $\theta = 180^o$, there is no flux cutting and the induced EMF is zero; following another 90° rotation (θ=270°) conductor is once again in the position for maximum induced EMF, given by (-) e *with respect to time*, followed by e=0 at θ=360° or 0°[26]; this cycle being repeated with time.

Similar reasoning would apply to the EMF induced in conductor CD at any instant t, acted upon by magnetic field of south pole, *but in a direction opposite to that in conductor AB.* Since the two conductors are connected in series forming the coil, the two EMFs will be *added* such that the total EMF that would appear across the slip rings will be 2 x e V.
From the above discussion, an important feature of the generator emerges: the induced EMF of the coil configured as in Fig.4.35 varies "sinusoidally" with angular position of the coil, θ[27].
This is illustrated in Fig.4.36 for different position of a coil acted upon by a magnetic field of a magnet, rotating in a clockwise direction about its axis[28].

---

[25]The positive sign is associated arbitrarily to polarity of induced EMF with time, not to be confused with polarity of EMF in DC working.
[26]Once again, the negative sign is indicative of reversal of EMF with time.
[27]Or with time since θ is related to time t expressed by $\theta = \omega t$, ω being the *angular* speed of the coil in rad/s.
[28]Since the EMF alternates with respect to time, the generator is called an "alternator". Note that even though the EMF induced in the coil alternates in sign with respect to time, the *current* in a circuit connected across the slip rings will be 'unidirectional', flowing in the same direction, but varying sinusoidally with time so that in a resistive electric circuit, the current follows the waveform of the induced EMF.

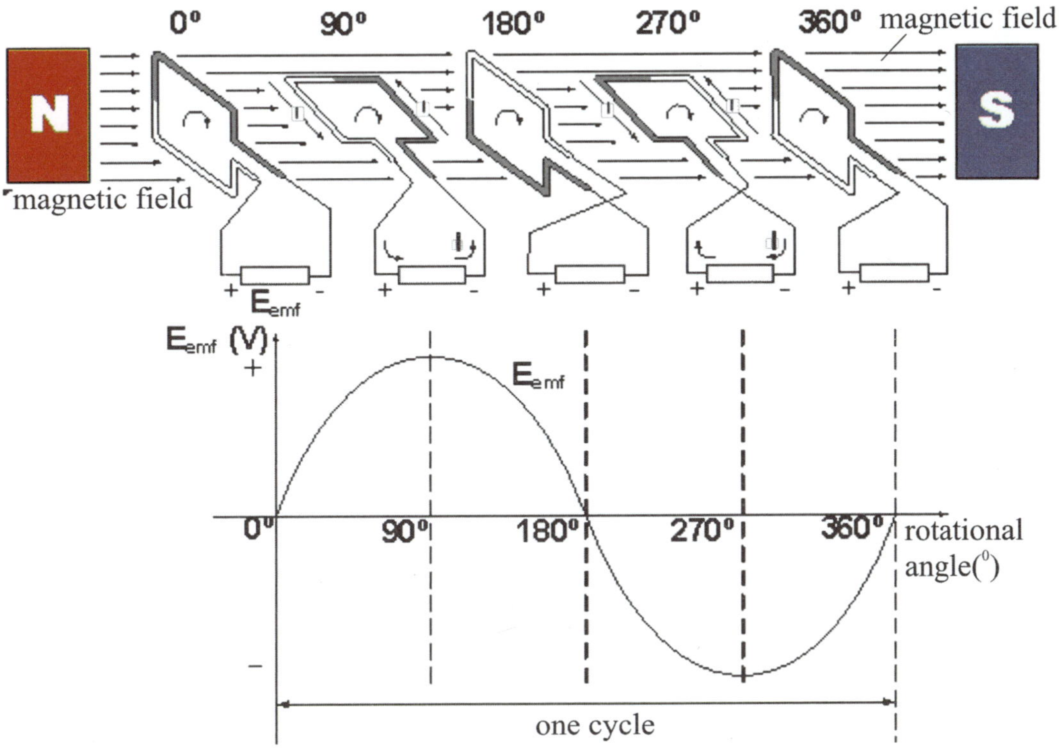

Fig.4.36 : Sinusoidal variation of induced EMF in a coil

Important Differences Between a "Single-coil" and Actual Alternator

An actual AC generator or alternator installed in a power house that generate huge power, feeding into power systems, differs from a simple, schematic 'generator' in the following respects

- the magnetic field is provided by a system of magnets, called the "field system", that *rotates inside* the stationary member of the machine, called the "stator"; the field system being identified as the "rotor"
- the field system invariably comprises suitably designed and constituted form of electromagnets, rather than permanent magnets, the windings of the electromagnets being fed from a controllable source of direct current; thus providing a means to vary magnitude of magnetic field of the electromagnets and hence the induced EMF of the generator
- the single coil is replaced by a set of multi-turn coils arranged to form a "winding" that is housed in several slots in the stationary member or stator of the generator.

The rotating field system turns within the stator at a constant speed[29], cutting the conductors of the winding and generating an induced EMF (electromotive force), as the mechanical input causes the rotor to turn.

The above arrangement is shown in Fig.4.37 for a four-pole "single-phase" generator where the field system is still shown to be permanent magnets. Note the relative connections of stator winding.

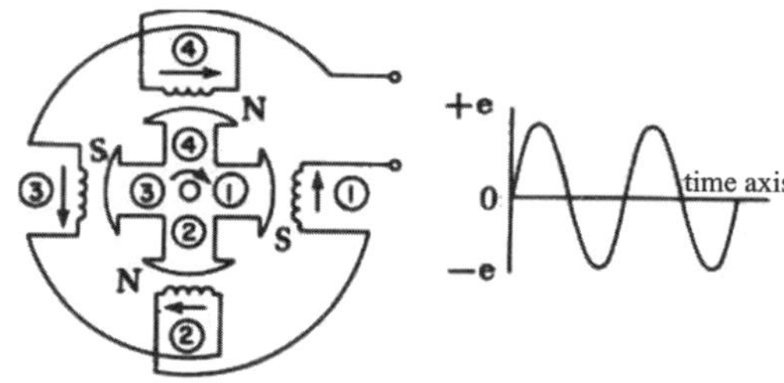

Fig.4.37 : Constituents of a single phase alternator

Frequency of Alternating Current

The AC EMF generated by an alternator changes sign in magnitude with time, related to angular position of the coil as discussed, the variation of magnitude of induced EMF with time following the sine law. The time taken by the variation of EMF to complete one cycle, that is time from beginning of positive loop to end of negative loop is called the "time period" of the sine waveform, measured in time. The inverse of time period is known as **frequency** of the generation of EMF. The standard supply frequency in India being 50 Hz, the corresponding time period of the sine waveform is derived to be 1/50 or 0.02 s (20 ms).

Advantage of AC over DC

As stated earlier, till the close of the 19th century, electricity in all forms – generation, distribution and usage –was confined to use of direct current or DC employing a DC generator[30], even for lighting of street lights. The advent of alternating current system(s), esp. the generation of alternating-current voltage by alternators and use of transformers which made high voltage-transmission possible changed all that. Ever since, AC system has almost completely replaced DC having numerous advantages.

- The generation of AC EMF in an alternator does not involve a commutator, doing away with several problems associated with the commutator, requiring regular inspection and maintenance – a tedious process.

---

[29]The speed of rotation in alternators in practice is linked to the number of poles for which the machine is designed and the frequency of supply which is standardised in a country. Thus, in India and several other countries the standard frequency is 50 Hz. Accordingly, the speed of rotation of the rotor in a generator having P poles is given by N = (120x50)/P; a 2-pole generator will have to be rotated at 3000 rpm and so on. The speed of a given alternator so derived is called the "synchronous" speed and *must* be maintained constant under all circumstances for proper operation.

[30]A machine similar in construction to DC motor with all essential parts except that instead of being supplied by a battery or any source of DC supply to produce torque in the motor, an armature is driven mechanically under the influence of a magnetic field from a stationary field system generating DC EMF at its terminals. A DC generator is usually called a "dynamo".

• Owing to the principle of operation of an alternator, the generation can take place at high to very high voltage resulting in the use of conductors of smaller cross section.

• The AC EMF generated by the alternator can be "stepped up" to very high levels by transformers – electromagnetic devices that can convert voltage from a low value to high value and *vice versa* - right at the power plant premises making *transmission* of electric power to long distances – from a few hundred to thousands of kilometres – most convenient and cost effective as conductors of lesser cross section can be used to transmit a given power.
This is not possible in the case of DC where no such transformers are available to step up the voltage.

• The *distribution* of electric power to consumers, for example domestic or commercial, can be carried out more flexibly and conveniently than it would be possible using DC.

• The usage or utilisation of AC power or energy in various forms is much more varied, convenient and cost effective than would be possible using DC.

TRANSFORMER[31]

A **transformer** is a *static* electromagnetic device used in an electric circuit or system either for raising or lowering the voltage of an AC supply with a corresponding decrease or increase in current in the circuit, if closed. A transformer essentially consists of two windings positioned on a common *laminated* magnetic core.

- Primary Winding – So called as the source of AC supply is connected to this winding

- Secondary Winding – So called as the load is connected to this winding.

Principle of Operation

The principle of operation of a transformer derives *directly* from Faraday's law of electromagnetic induction whereby a *time-varying* magnetic flux induces an EMF in a circuit linking the flux. In terms of cause-effect relationship an AC supply, itself varying sinusoidally with time, when connected to the primary winding of a transformer

▸ sets up a time varying magnetic field in the (magnetic) core of the transformer, also varying sinusoidally with the same frequency

[31]The discussion of transformer in these sections pertains to "power frequency" transformers, that is in which supply frequency is 50 Hz. Slightly modified designs may be used for transformers operating at other frequencies; for example, audio or radio frequency applications. However, the principle of operation remains same.

▸ this flux linking with the secondary winding, wound around the core, induces an EMF in this winding by electromagnetic induction

Thus, If v_{pri} is the AC voltage applied to the primary winding *at any instant*[32] and v_{sec} the voltage induced across the secondary winding, then

$$v_{sec} = -N\frac{d\phi}{dt}$$

where ϕ is the flux set up in the core by current in the primary winding, *common to both the windings*, and N denotes the number of turns in the *secondary winding.*

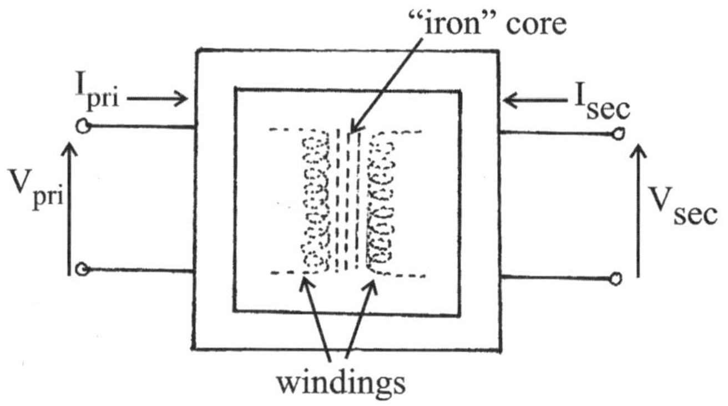

Fig.4.38 : Block diagram of a transformer

The principle is depicted in the block diagram of a transformer in Fig.4.38.

Note that in the above expression the flux ϕ is assumed to be changing with time, but remaining constant in magnitude depending on the value of the primary voltage and corresponding current I_{pri} in the winding.

Characteristic Features of a Transformer

- a transformer is essentially an AC device; it cannot work with DC
- a transformer works on the basis of constant power rating in kW etc. when referred to its primary or secondary side
- the operating frequency of the transformer remains unchanged during conversion of voltage, that is 50 Hz in India, at all time.

---

[32]Although instantaneous values are expressed here, the law of induction will hold for *any* values of the primary and secondary induced EMFs, usually denoted by RMS (root mean square) values encountered in, say, household circuits as depicted in Fig.4.38.

Basics of a Transformer

The basics of a typical transformer are shown in Fig.4.39.

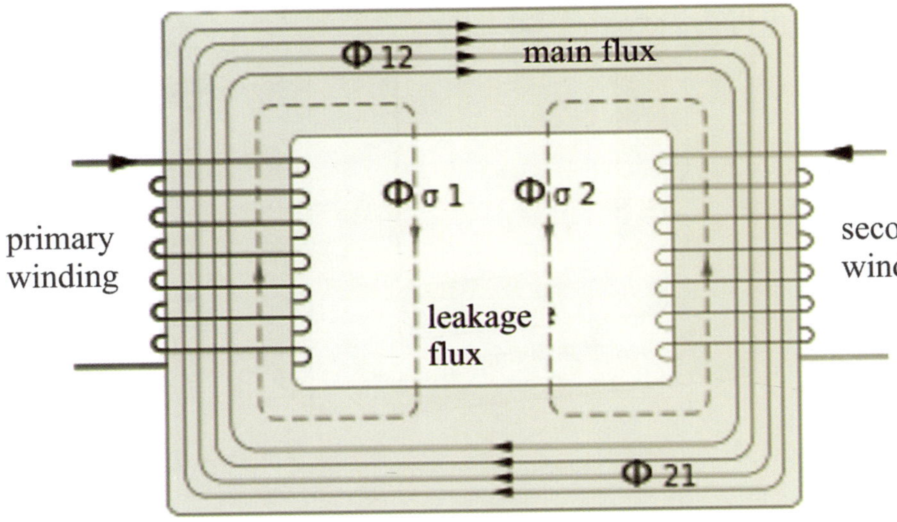

Fig.4.39 : Basic components of a transformer

Construction

The construction of a typical transformer is characterised by three essential components

- an "iron" **core** that is invariably made of "silicon steel", laminated (that is, 'divided' in thin sheet pieces called "stampings") and arranged side by side to form a stack[33]
- a winding of N_1 or N_p turns wound, on one side or 'limb' of the core, called the **primary** winding
- a second winding of N_2or N_s turns, wound on the other side or limb of the core, called the **secondary** winding

The construction of a simple transformer is shown schematically in Fig.4.40 whilst a set of laminations or "stampings" of "E" and "I" type, frequently used to construct a transformer core, is shown in Fig.4.41.

[33]The core is laminated to reduce power loss in the core, and consequent heating, caused by induced "eddy" currents in the core owing to the presence of a time-varying flux.

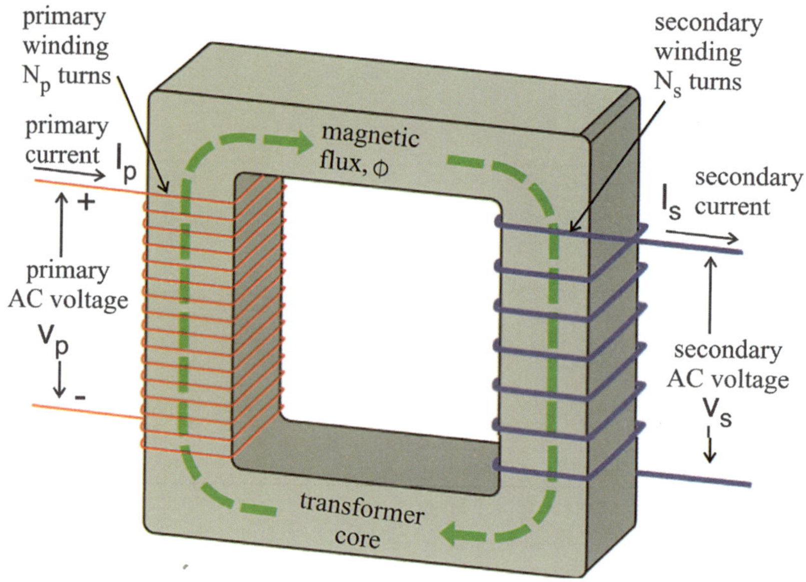

Fig.4.40 : Schematic of a transformer

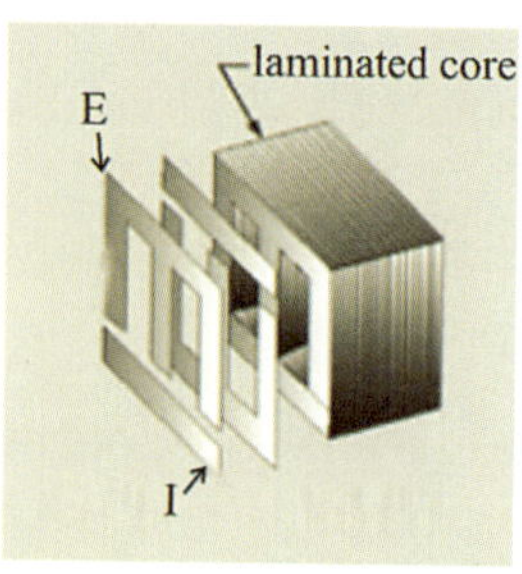

(a) "E" and "I" type

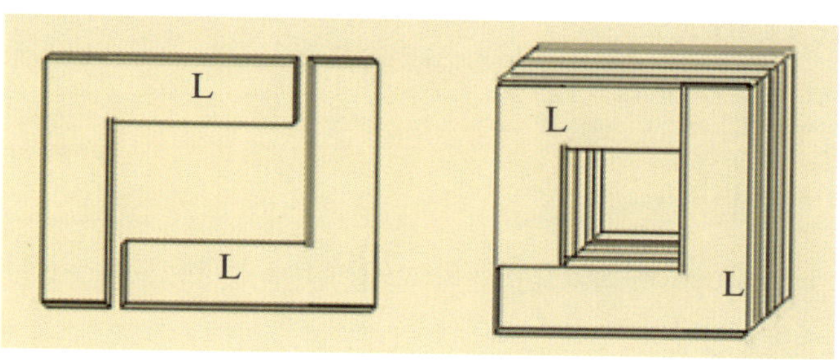

(b) "L" type

Fig.4.41 : Laminations for a transformer core and their stacking

Types of Cores

Two forms or types of core usually employed for the construction of transformer core are

A. Core type

In this case, the core consists of two vertical limbs on two sides, joined at top and bottom by yokes, by arranging the "L" type lamination pieces to obtain overlapping joints by reverse positioning of the two pieces as indicated in Fig.4.41(b). The primary and secondary windings are placed on both the sides, one half turns of each winding on either limb.

B. Shell type

This type core is constructed by stacking "E" and "I" type lamination pieces as depicted in Fig.4.41(a). Both primary *and* secondary windings are placed only on the middle limb, the outer limbs only completing the path of magnetic flux[34].

Typical examples of core and shell type (1-phase) transformers are shown in Fig.4.42(a) whilst constructional features of a *three-phase* core type transformer is shown in Fig.4.42(b),

(i) core type (ii) shell type

(a) single-phase core and shell type transformers

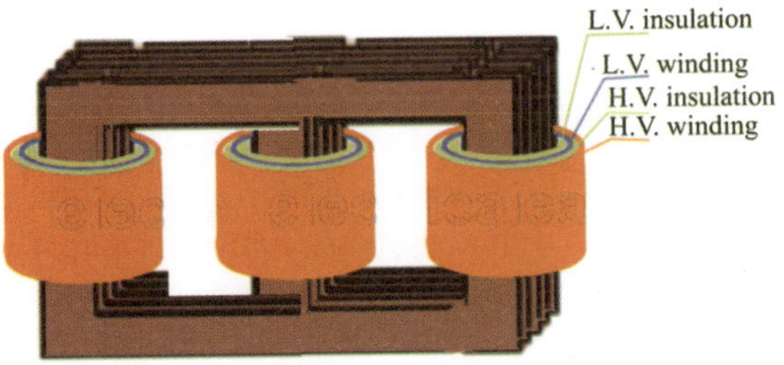

(b) 3-phase core type transformer

Fig.4.42 : Types of transformer

Number of turns and turns ratio

Under 'steady' operating condition, the voltages in the two windings of a transformer are balanced. Applying the principle of electromagnetic induction, it can be written that

$$E_{pri} = - N_p (d\phi/dt)$$

and

$$E_{sec} = - N_s (d\phi/dt)$$

where E_{pri} and E_{sec} represent the induced EMFs in the primary and secondary windings, respectively, and ϕ is the *mutual* flux linking both the windings.

[34]The shell type is commonly used for nearly all "single-phase" transformers.

Dividing the two expressions, it follows that

$$\frac{E_{pri}}{E_{sec}} = \frac{N_p}{N_s}$$

or that the induced EMFs in the transformer are related to the number of turns of the windings.
The ratio of number of turns of the windings, viz. N_p/N_s is called the **turns ratio** of the transformer.

Step Down and Step Up Transformer

In the connection scheme of a transformer if

- the primary voltage is greater than the secondary voltage, that is $V_{pri}>V_{sec}$ or $N_p> N_s$, the transformer is called a **"step down transformer"**

- the primary or applied voltage is less than the secondary voltage, that is $V_{pri}<V_{sec}$ or $N_p<N_s$, the transformer is called a **"step up transformer"**[35]

Special Features

Since power rating of a transformer is the same on either primary or secondary side, that is, the product of $V_{pri} \times I_{pri} = V_{sec} \times I_{sec}$ where I_{pri} and I_{sec} denote the primary and secondary currents, respectively, it follows that in a step down transformer with the primary voltage being higher than secondary voltage, the current (rating) of the primary winding would be proportionately *less* than that for the secondary, requiring wire of lesser cross section; the secondary winding to be wound with relatively thicker wire. And *vice versa* for a step up transformer.

Rectifier Transformers

These represent a special category of transformers designed and constructed to provide DC output when fed from an appropriate AC supply.
As an example, consider a single-phase rectifier transformer with primary voltage being 230 V. The secondary is designed to provide 6 V by selecting appropriate number of turns on the two sides. Let the primary is designed to carry a current of 0.1 A (100 mA). The secondary current will then be 3.8 A (approx.) and the wire size for the windings will have to be chosen accordingly for safe, continuous flow of current(s) without overheating due to I^2R loss.

---

[35]In practice, both the windings are wound one over the other, usually primary over the secondary winding that is first wound next to the core. This is adopted to improve "magnetic coupling" between the windings.

Applications of Transformers

At Power Frequency

- step-up transformers are used at power stations to increase the voltage of generation to very high values, required for power transmission; for example, from 13 kV to 132/220/400 kV.
- step-down transformers are used
- at "receiving" end of a transmission line to bring down the voltage first to primary distribution level, say, 33 kV and later to "distribution" level of 11 kV; this is further stepped down to 400 V for domestic and commercial usage, in general. A typical "pole-mounted" distribution transformer used for the purpose is shown in Fig.4.43.

Fig.4.43 : A pole-mounted distribution transformer

See also Fig.3.4, Chapter III.

- in "rectifier" units in numerous audio and video devices to bring down the voltage from a supply voltage of 230 V to 12/6/4.5/1.5 V before rectification for use of DC supply/output variously in the device; typical 1-phase transformers for such an application are shown in Fig.4.44 .

Fig.4.44 : Examples of low-voltage rectifier transformer

The concept of transformer is utilised in developing a special transformer, shorn of conventional design aspects discussed above. This is described in Appendix A.

APPENDIX A

A Special Demonstration Transformer

This is a uniquely designed and fabricated electromagnetic device, similar to a single-phase transformer with an "open-ended" core meant to demonstrate a variety of aspects of electromagnetic induction following Faraday's laws.

A photographic view of the transformer and its basic constructional details are shown in Fig.4.45. The device consists of a soft iron core in strip or wire form, tightly packed inside a PVC tube of about 20 mm diameter. The tube/core is firmly inserted inside a coil wound around an insulated bobbin at the bottom, with the core projecting upward up to height of about 30 cm. The coil at the bottom is wound using 24 SWG copper enamelled wire with 2300 turns so that when connected across a 230 V supply, the turns per volt will be 10.

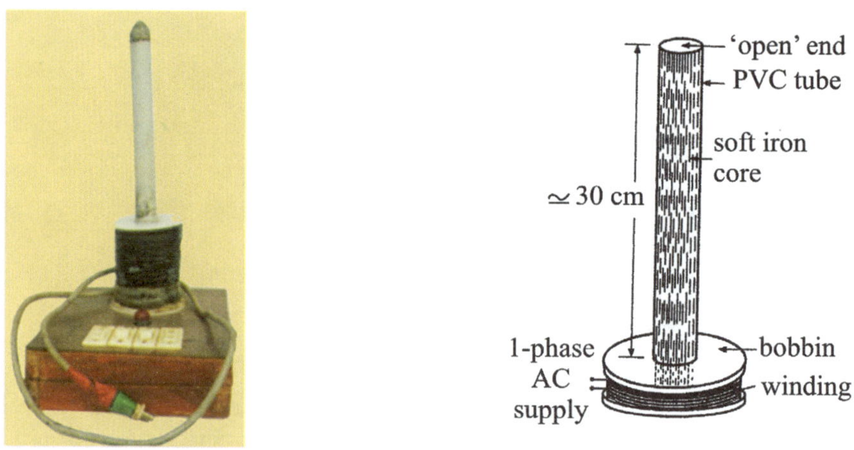

(a) actual device (b) constructional details

Fig.4.45 : The special "open core" transformer

A sketch of the device and its transformer equivalent is given in Fig.4.46.

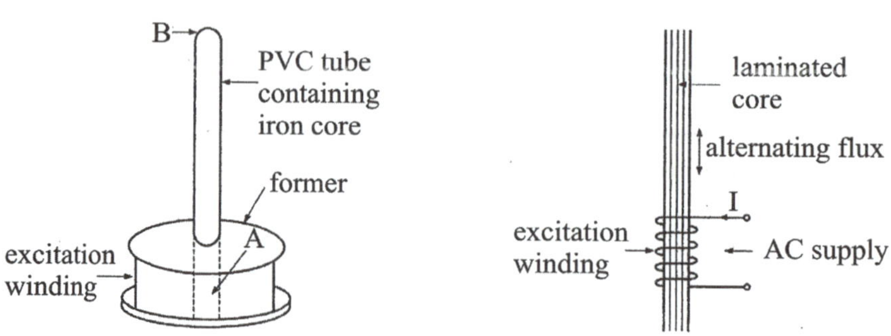

Fig.4.46 : Sketch of the transformer and its transformer equivalent

When excited by an alternating current, the core is magnetised with a time-varying field of the same frequency, the field being maximum at the bottom and gradually reducing towards the "open" end. A variety of experiments can be performed using the device.

A. To Show Transformer Principle

To demonstrate transformer action, a 'secondary' coil of a given, known number of turns is wound on a ring-shaped former that can slide along the core, the secondary terminals being connected to a low-voltage bulb, say of 3 V. The EMF induced in the coil can be related to the 'primary' voltage applied to the main coil. The EMF would be maximum when the coil is at the bottom, resting on the top coil surface, the value being given almost exactly by the turns ratio. On moving the coil upward, the EMF would gradually reduce owing to decreasing magnetic field in the core and decreasing flux linkage. This is reflected by the glow of the bulb connected across the (secondary) coil. See Fig.4.47.

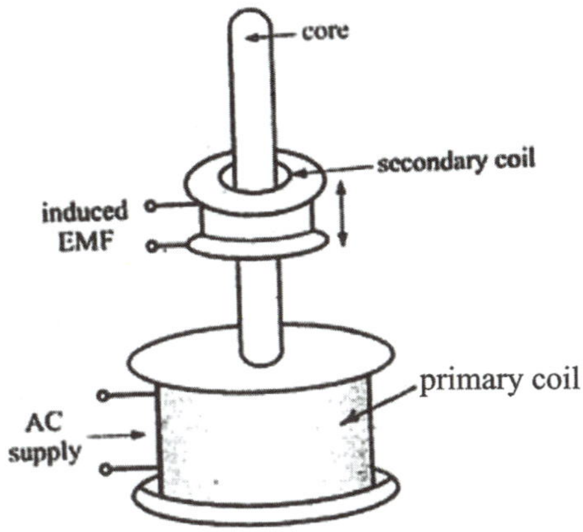

(a) primary and secondary winding

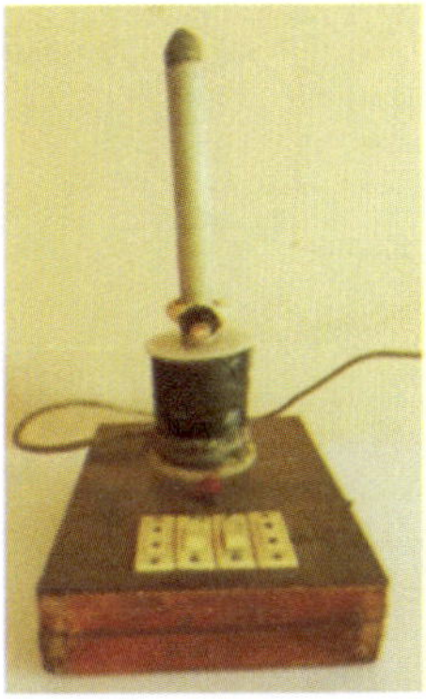

(b) maximum output

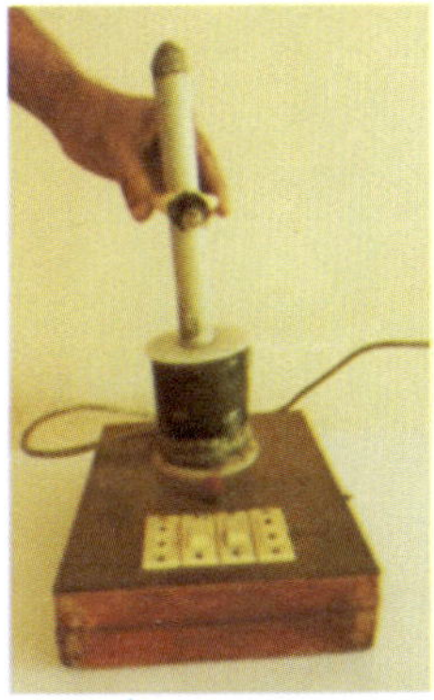

(c) 'small' output

(d) 'zero' output

Fig.4.47 : Experiment showing transformer action

B. Addition and Subtraction of Two EMFs

By fabricating two identical, movable coils of small axial length and placing them near the bottom of the core so that EMF induced in each coil is very nearly equal, the addition and subtraction of two (AC) EMFs can be demonstrated by connecting the coils first in series and then in series opposition. The resultant EMF in the first case will be 2E and zero in the second case, E being the induced EMF in each coil.

C. Magnetic Levitation

This is a special case of electromagnetic induction where an AC field from a source (for example the "main" coil here) interacting with the field from induced currents in a metallic body would result in levitation of the latter. This can be demonstrated in the present case by inserting an aluminium ring, about 10 cm in outside diameter and 3 mm thick in the core, so as to slide freely about the later, and exciting the main coil. It is seen that the field produced by the currents induced in the ring, acting as a short-circuited secondary of a transformer, the ring levitates or floats at a certain height depending on its weight, dropping to bottom of the core as the excitation is switched off. A photographic view of the phenomenon is illustrated in Fig.4.48(a); an schematic explanation being given in Fig.4.48(b).

(a) photographic view

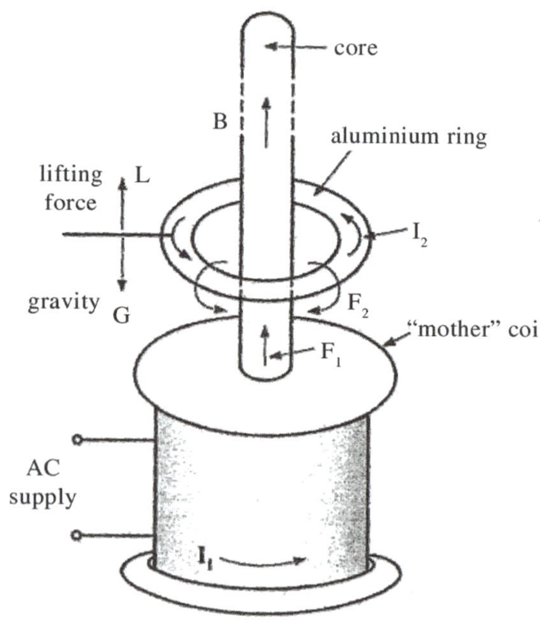

(b) schematic

Fig.4.48 : Magnetic levitation

5. Heat

Unit 5: Topics covered

Heat

(i) Calorimetry: meaning, specific heat capacity; principle of method of mixtures;- Numerical Problems on specific heat capacity using heat loss and gain and the method of mixtures.

Heat and its units (calorie, joule), temperature and its units (0C K); thermal (heat) capacity C' = Q/Δ T..... (SI unit of C): Specific heat Capacity C = Q/mΔ T (SI unit of C) Mutual relation between Heat Capacity and Specific Heat capacity, values of C for some common substances (ice, water and copper) Principle of method of mixtures including mathematical statement. Natural phenomenon involving specific heat. Consequences of high sp heat of water. (Simple numerical problems].

(ii) Latent heat; loss and gain of heat involving change of state for fusion only.

Change of phase (state); heating curve for water; latent heat; sp latent heat of fusion (SI unit) Simple numerical problems. Common physical phenomena involving latent heat of fusion.

I Calorimetry

Calorimetry: meaning, specific heat capacity; principle of method of mixtures; numerical problems on specific heat capacity using heat loss and gain and the method of mixtures. *Hear and its units (calorie, joule), temperature and its units (^oC, K); thermal (heat) capacity $C' = Q/\Delta T$ (SI unit of C): specific heat capacity $C = Q/m\Delta T$ (SI unit of C); mutual relation between heat capacity and specific heat capacity, values of C for some common substances (ice, water and copper); principle of method of mixtures including mathematical statement; natural phenomenon involving specific heat; consequences of high sp. Heat of water. [simple numerical problems].*

CALORIMETRY

Meaning

Calorimetry[1] is the science associated with determining the changes in energy of a system by measuring the heat exchanged with the surroundings for the purpose of deriving the heat transfer associated with changes.,

Typical changes may involve, for example

- heating of water to its boiling point and conversion into steam
- forming of ice from water or melting of ice into water
- heating of a metal to its molten state

The changes are thus associated with change of its *state*, at molecular level, from solid to liquid or from liquid to gas etc. All such changes require exchange of heat for conversion from one state to the other.

Thus, **calorimetry is the science to *measure* exchange of heat in the process.**

[1]The word "*calorimetry*" is derived from the Latin word *calor*, meaning heat and the Greek word*μέτρον* (metron), meaning measure. Scottish physician and scientist Joseph Black, who was the first to recognize the distinction between heat and temperature, is said to be the founder of the science of calorimetry.

Heat

In physics, **heat** is the energy that 'spontaneously' passes between a "system", comprising, for example, matter in the form of solid or liquid, and its surroundings in some way other than through work or the transfer of matter.
When a suitable physical 'pathway' exists, heat begins to flow from a hotter to a colder body.

Modes of Heat Flow or Heat Transfer

The flow of heat can take place in one or more of the following modes

- By conduction

In this mode the transfer of heat can take place only by *contact* between the source of heat to body receiving heat. For example, if someone touches a hot metal rod accidentally, the heat passes on to the person by conduction, making the rod to feel hot.

- By convection

In convection, usually occurring in liquids, the transfer of heat is due to *circulation* taking place within the liquid; sometimes in the form of "eddies". A common example is the process of boiling milk in a vessel.

- By radiation

Heat transfer by radiation takes place through a gaseous medium between remotely located hot bodies, in no way connected with each other. A classic example is the heat energy reaching Earth from the Sun, being millions of kilometre away. Another example is that of a room heater used for heating air inside the room.

Units of heat

Heat being a form of energy, it can be measured in appropriate units.

In CGS system, heat is measured in calories. The unit **calorie** is defined as
"the heat energy required to increase the temperature of 1gm of water through one degree Celsius" (for example, from 14.5 degrees Celsius to 15.5 degrees Celsius).

In SI system, heat energy is measured in joules (J), similar to any other form of energy, such that one calorie = 4.185 joules[2].

---

[2]In practice, this may be rounded to 1 calorie = 4.2 J.

A larger unit called kilocalorie is also in use. 1 kilocalorie = 1000 calories. One kilocalorie is defined as "the quantity of heat energy required to change the temperature of one kilogramme of water by one degree Celsius".

A unit of heat energy often used is known as BTU (British Thermal Unit), defined as "amount of heat required to raise the temperature of one pound of water from 63degrees Fahrenheit to 64 degrees Fahrenheit", one BTU being equal to 252 calorie (approx.).

Both definitions specify that the temperature changes are to be measured at a *constant pressure of one atmosphere*, because the amount of energy involved depend in part on pressure.

The calorie used in measuring the energy content of foods is the large calorie, or kilogramme-calorie, equal to 1,000 gramme-calories. For example, the energy value of 100 g of baked potato is equal to 100 kilocalories. This is usually interpreted as "calorific value" of the substance. More specifically, calorific value of a substance is the "amount of energy that is released when burning a given amount of the substance".

Calorific Values of Some Common Food Items

Food Item	Quantity	Calorific Value, kcal
peanuts	100 g	560
whole milk	100 g	70
skimmed milk	100 g	34
bread	4 'mini' slices	280
apple	1 (small)	60
tomato	1 (small)	20

Temperature

What is Temperature?

An exact definition of the concept "temperature" in physics is difficult to obtain. However, an *understanding* of the entity called **temperature** can be had in terms of

- The degree of 'hotness' or 'coldness' of a body or environment
- A measure of the warmth or coldness of an object or substance *with reference to some standard value*
- A measure of the average kinetic energy of the particles in a sample of matter, expressed in terms of units or degrees designated on a standard scale

- A measure of the ability of a substance, or more generally of any physical system, to transfer heat energy to another physical system
- Any of various standardised numerical measures of this ability, such as the Kelvin, Fahrenheit, and Celsius scale

The first two 'statements' point to one's ability to "measure" the relative coldness or hotness of an object or body (solid, liquid or gaseous) by suitable means to express (or define) temperature of the body under reference. Such a device is known as a "thermometer", available in a variety of forms.

Thus, in simple terms, temperature is what the thermometer reads. Whatever it is that temperature is a measure of, it is reflected by the reading on a thermometer.

Thermometer(s)[3]

A thermometer has two important elements

- a temperature *sensor* (for example the bulb of a mercury-in-glass thermometer) in which some physical change occurs with temperature

- some means of converting this physical change into a numerical value (for example the visible scale that is marked on a mercury-in-glass thermometer) or other such suitable means: for example, LCD display in a "digital" thermometer.

Principle of Operation of Thermometers[4]

There are various principles by which thermometers operate. The commonest is the thermal expansion of solids or liquids with temperature, the latter being employed in all mercury thermometers, typically used as a clinical thermometer to measure body temperature; for the latter, digital thermometers are now in use in which the sensor may be a thermistor – a resistance that changes its value with temperature of a body in contact - and using appropriate electronics and LCD display.

A typical clinical thermometer and its digital counterpart are shown in Fig.1.1.

---

[3]Observe that a thermometer, although encompassing the word meter, is not a meter in the usual sense; there is no deflection of a pointer on a graduated scale as in an ammeter or a meter, nor an indication as in an energy meter. Nevertheless, it may be identified as a 'means' to measure thermal status of a body.

[4]Note that all types of thermometer or temperature-measuring devices are not 'absolute' instruments and must be calibrated using appropriate means to make them "direct reading".

(a) glass-capillary type

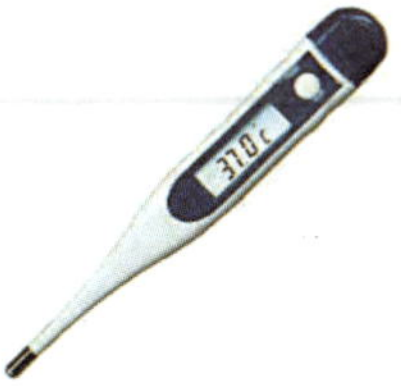

(b) digital type

Fig.1.1 : Clinical thermometer

A "wall type" thermometer used to measure room temperature, based on the expansion of ethanol in a glass tube is shown in Fig.1.2. The thermometer is hung vertically on the wall at a convenient height and ethanol is added with a red dye for easy visibility.

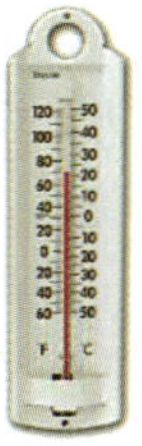

Fig.1.2 : A wall thermometer

Thermocouple

A common type of thermometer, doing away with any liquid that may pose problem of leak etc., makes use of an encapsulated "thermocouple" that can be used to measure a wide range of temperatures – from -250 °C to 1400°C. A thermocouple is a bimetallic device consisting of junctions of copper-constantan or iron-constantan thin wire pairs. When the two junctions are at two different temperatures, an EMF appears across the junction as shown schematically in Fig.1.3.

A thermometer based on the principle of a thermocouple and used numerously is shown in Fig.1.4. Observe the probe with long lead and digital visual display device indicated in the figure.

Principle of working

This is clear from Fig. 1.3.

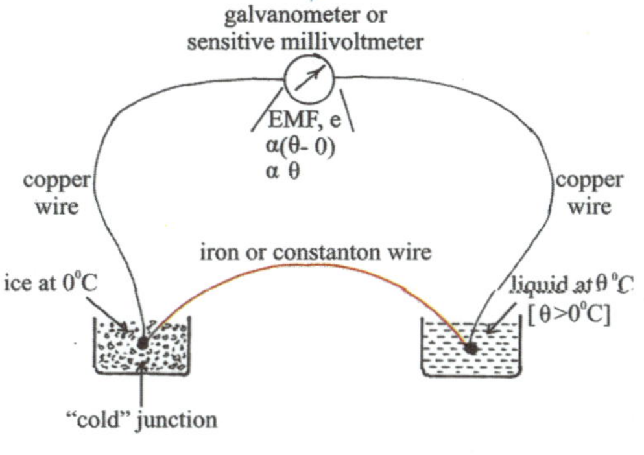

Fig.1.3 : Principle of working of a thermocouple

Fig.1.4 : A thermometer using thermocouple probe and digital display

Yet another type is the "infrared" thermometer, extensively used in industry for measuring very high temperatures of molten metal etc., is based on the principle of radiation of infrared energy emitted from a hot object or body. The thermometer has an important advantage that it allows measurement of temperature remotely without any contact with the object.

A typical such thermometer with digital display is shown in Fig.1.5.

Fig.1.5 : An infrared thermometer

Thermometer scales

There are *three* different ways or norms, called "scales", commonly used to express temperature of a body or object

A. Fahrenheit Temperature Scale

In this scale, the temperature is denoted by the unit "degrees Fahrenheit"[5], (symbol °F), named after Polish-Dutch physicist[6] who proposed the scale in 1724. The lower defining point in this scale, 0 °F, was established as the temperature of a solution of brine made from equal parts of ice and salt.
The unit is still popular and widely in use in the USA.

B. Celsius Temperature Scale

This scale is credited to its originator Celsius[7]. Also called "centigrade scale" owing to the thermometer based on this scale having one hundred divisions from zero degree to hundred degree. The zero degree mark in this scale refers to freezing point or temperature whilst the 100 degree mark corresponds to boiling point or temperature of "pure" water. The scale is

---

[5]The use of the word "degrees" would imply a relative measure of 'hotness' or 'coldness' of the body akin to that indicated in defining the term temperature of a body.
[6]Daniel Gabriel Fahrenheit: 1686-1736.
[7]Swedish astronomer Anders Celsius: 1701–1744.

now universally used throughout the world, with the temperature of bodies being expressed as "degrees Celsius or centigrade" (symbol °C)[8]. This is also the SI unit of temperature.

When compared to Fahrenheit scale, the freezing and boiling points or temperatures of water are noted to be 32 °F and 212 °F, respectively, and a Fahrenheit thermometer is graduated accordingly. From this comparison, a temperature in °C can be related to °F by the expression

$$^{\circ}C = \frac{^{\circ}F - 32}{9}$$

Similarly, degrees Fahrenheit in terms of degrees Celsius is given by

$$^{\circ}F = \frac{^{\circ}C}{5} \times 9 + 32$$

The relationship between the two scales for temperatures from -50 °C to +50 °C is illustrated in Fig.1.6 in a dial-type thermometer.

It is interesting that a value of - 40 degree is common to both the scales as indicated in the figure and also derivable from either of the above expressions.

Fig.1.6 : Temperatures in °C vs. °F

C. Kelvin Temperature Scale

Kelvin[9] temperature scale, named after Lord Kelvin, is the standard *metric* system of temperature measurement and the most widely used temperature scale among *scientists.* The scale is similar to the Celsius temperature scale such that there are still 100 equal degree increments between the normal freezing point and the normal boiling point of (pure) water. However, the zero-degree mark on the Kelvin temperature scale is 273.15 units cooler than it is on the Celsius scale. So a temperature of 0 Kelvin is equivalent to a temperature of -273.15 °C. *Observe that the degree symbol is not used with this system.* So a temperature of 300 units above 0 Kelvin is referred to as 300 Kelvin and *not* 300 degree Kelvin. Conversions between Celsius temperatures and Kelvin temperatures (and *vice versa*) can be performed using one of the following two equations

---

[8]The name "centigrade" of the unit was adopted from the Latin *centum* which means 100 and *gradus* which means steps.

[9]William Thomson, 1st Baron Kelvin: 26 June 1824 – 17 December 1907, was a Scotch-Irish mathematical physicist and engineer . At the University of Glasgow he did important work in the mathematical analysis of electricity and formulation of the first and second laws of thermodynamics.

$$°C = K - 273.15°$$
$$K = °C + 273.15$$

A ready-reckoner of temperatures in the three scales is provided in Fig.1.7.

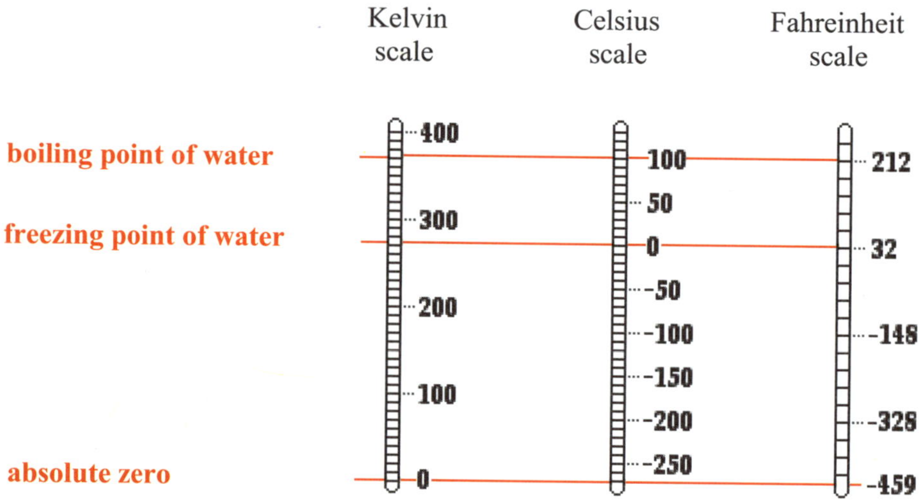

Fig.1.7 : Temperatures in three scales

Common temperature reference points

Case	Celsius(°C)	Kelvin	Fahrenheit (°F)
Surface of the Sun	5600	5900	19100
Boiling point of water	100	373	212
Body temperature	37	310.2	98.6
Sweltering day	40	313	104
'Hot' day	30	303	86
Room temperature	20	293	68
'Cold' day	19	283	50
Freezing point of water	0	273	32

Calibration of thermometers

Thermometers must be calibrated periodically for accurate measurement of temperatue. This can be accmoplished either by comparing them with other calibrated thermometers or by checking them against known *fixed points* on the temperature scale; for example that of "pure" water since it has 'convenient' (fixed) freezing and boiling points or temperatures at NTP[10].

The traditional way of marking a scale on a liquid-in-glass or liquid-in-metal thermometer can be carried out in three stages:

1. Immerse the sensing portion of the thermometer in a stirred mixture of pure ice and water at atmospheric pressure and mark the point indicated when it had come to thermal equilibrium
2. Next, immerse the sensing portion in a steam bath at *standard atmospheric pressure* and again mark the point indicated
3. Divide the distance between these 'limiting' marks into equal parts according to the temperature scale being used; for example 0 and 100 for the first and second point, respectively, for a centigrade thermometer.

Heat Energy for a Body

Heat (energy) is frequently used to 'heat up' a body or raise its temperature from, say, T_1 to T_2, the latter being higher, so the difference of temperature is $\Delta T = T_2 - T_1$ in a given unit. It is observed that the heat input to a body depends on, or proportional to, three parameters

- mass of the body
- difference of temperature (rise) such as ΔT above
- a constant that depends on the *material* of the body, called **specific heat**

Accordingly, the heat input, Q, required for raising temperature of a body of mass m by ΔT degree will be given by

$$Q = m\, c\, \Delta T$$

where c denotes the specific heat, discussed later.

When m is in kg and ΔT is in °C, the heat input is expressed in joule (J) and the unit of c is obtained as J/kg/°C[11]. Thus, specific heat of a substance can be defined as **"the amount of heat energy required to raise the temperature of unit mass of a substance by one Kelvin or 1 °C"**.

Thermal (heat) capacity

Thermal capacity, also referred to as **heat capacity**, is the "amount of heat required to change the temperature of an object by a certain degree".

---

[10]Note that boiling point of water varies with pressure, so it must be controlled or a correction may have to be applied.

[11]At times, specific heat is related to volume of the body (for its known density value) when *volumetric* specific heat is specified as $J/m^3/°C$.

Accordingly, thermal capacity *of a given material* is an "extensive" property and would vary with the *amount* of the material or substance. For example, for the same temperature rise 100 gm of iron will have twice the thermal capacity than that of 50 gm of iron.
The unit of thermal capacity is joule/kelvin (or J/K) or J/°C[12].

Thus, if a heat input of Q J is required to raise temperature of a material or substance by ΔT K, its thermal or heat capacity will be

$$C' = Q/\Delta T \quad J/K$$

Specific heat or specific heat capacity

Specific heat capacity, also referred to as the specific heat of a material, is **"the amount of heat needed to raise the temperature of an object *per unit mass* of that object"**. Using the example above, 50 gm of iron will have the same specific heat as that of 100 gm.
Accordingly, specific heat capacity or (simply) specific heat of a material or substance is **"the amount of heat energy required to raise the temperature of unit mass of the material through one degree Kelvin or one degree centigrade"**, denoted by the letter c.
Thus

$$\text{specific heat, } c = \frac{\text{heat input, } Q\,(J)}{\text{mass, } m(kg) \times \text{temperature rise, } \Delta T\ (K \text{ or } ^\circ C)}$$

It follows that the unit of specific heat is J/kg/K (or °C).
Note that specific heat capacity is an "intensive" property of a substance.

Relationship Between Heat Capacity and Specific Heat

Comparing the expressions for the two quantities, the relationship between heat capacity and specific heat can be expressed as

heat capacity, C' = mass, m x specific heat, c

Specific Heat of Some Substances

Substance	specific heat, at 25 °C in J/kg/°C
H_2 gas	14267
He gas	5300
water	4184
ice at 0 °C	2010

[12]Since SI unit of temperature in all modern scientific or engineering practice is Kelvin (or K), it is best to express thermal capacity in J/K.

steam at 100 °C	2010
air	1020
aluminium	900
brass	380
copper	385
gold	129
iron	440
lead	160
mercury	140
silver	240
zinc	390
glass	840

Measurement of Specific Heat

Method of mixture

Principle of the Method

If two or more bodies initially held at two different temperatures are brought into thermal contact, the net heat lost by the hot bodies is equal to net heat gained by the cold bodies *until they attain thermal equilibrium*, it being assumed that NO heat is lost by any other process to the surroundings or otherwise.

The relationship *at equilibrium*

Net heat lost (by hot bodies) = net heat gain (by cold bodies or bodies at lower temperature)

This is known as the **method of mixture;** also sometimes called the "principle of calorimetry".

Thus, let a (heat-conducting) body of mass m_1 having a specific heat c_1 and at a temperature of T_1 is brought in 'close' contact with a body of mass m_2, specific heat c_2 at (a lower) temperature T_2 such that after an elapse of some time the *two bodies together* attain a steady temperature of T

where $T_1 > T$ and $T_2 < T$

and all parameters are expressed in SI units.

Then heat lost by the hot body = $m_1 \times c_1 \times (T_1 - T)$

and heat gained by the body at lower temperature = $m_2 \times c_2 \times (T - T_2)$

Following the method of mixture

$$m_1 \times c_1 \times (T_1 - T) = m_2 \times c_2 \times (T - T_2)$$

Example

1. 50 g of a metal piece at 27°C requires 2,400 J of heat energy so as to attain a temperature of 327 °C. Calculate the specific heat capacity of the metal.

[ICSE, 2014]

Initial temperature of the metal, T_1 = 27 °C
Final temperature, T_2 = 327°C
The difference of temperature, $\Delta T = T_2 - T_1 = 327 - 27$ or 300 °C
The heat energy taken, Q = 2,400 J
From the expression $Q = m \times C \times \Delta T$
specific heat capacity of the metal, $C = Q/(m \times \Delta T)$
$= 2{,}400/(50 \times 300)$ or 0.16 J/g/°C

2. A calorimeter of mass 50 g and specific heat capacity 0.42 J/g/°C contains some mass of water at 20°C. A metal piece of mass 20 g at 100 °C is dropped into the calorimeter. After stirring, the final temperature of the mixture is found to be 22°C. Find the mass of water used in the calorimeter.
[specific heat capacity of the metal piece = 0.3 J/g/°C; specific heat capacity of water = 4.2 J/g/°C]

[ICSE, 2013]

Let the mass of water used in the calorimeter be m g
Heat lost by metal piece = mass of metal x specific heat capacity x temp. difference
$= 20 \times 0.3 \times (100 - 22)$ or 468 J
Likewise, heat gained by the calorimeter = $50 \times 0.42 \times (22 - 20)$ or 42 J
and heat gained by water = $m \times 4.2 \times (22 - 20)$ or 8.4 m J

Now by the principle of calorimetry or "method of mixture"
heat lost = heat gained

∴ by substitution

$$468 = 42 + 8.4\, m$$

whence

$$m = 50.71 \text{ g}$$

3. In order to find specific heat capacity of copper, a copper calorimeter of mass 0.16 kg and at 15 °C is used. Into the calorimeter is poured 0.22 kg of water at 70 °C. After stirring, the final temperature of water is found to be 45°C. If specific heat capacity of water is 4,200 J/kg/°C, calculate
a) amount of heat energy released by the water when it cools from 70 °C to 45 °C
b) the specific heat capacity of the calorimeter, assuming that all the energy released by water is absorbed by the calorimeter.

a) now heat released by the water $= m\, c\, \theta_w = 0.02 \times 4{,}200 \times (70 - 45)$
$= 2{,}100$ J

b) heat gained by the calorimeter = heat lost by water

Or $0.16 \times c_c \times (45 - 15) = 2{,}100$, c_c being the specific heat capacity of calorimeter

whence

$$c_c = 437.5 \text{ J/kg/}^\circ\text{C}$$

Natural phenomenon involving specific heat

Specific heat, or rather relative values of it, gives rise to natural phenomena in and around surroundings.

An example

Production of Sea Breeze

Along a sea shore, natural breeze is produced as a result of different specific heats of sea water and sand on the shore. The phenomenon occurs since

- land has lower heat capacity than sea water (almost one-fifth); therefore, in day time, the temperature of the land increases faster than the sea
- hot air (of lower density) above the land rises whilst cooler air from the sea flows towards land and hence produces sea breeze

A pictorial schematic of the phenomenon is shown in Fig.1.8.

As a result of the phenomenon, places adjoining sea and ocean may have rather moderate climate.

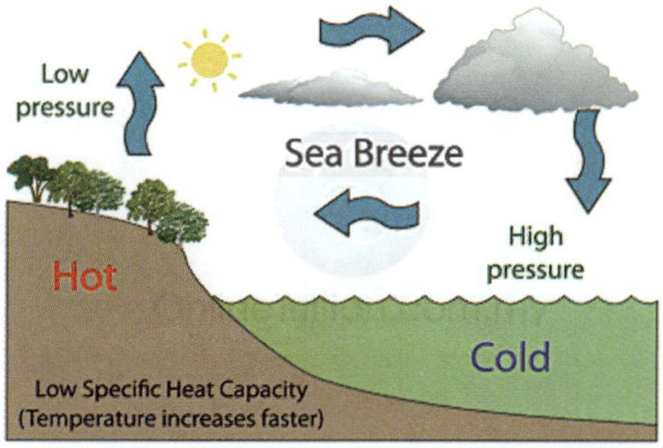

Fig.1.8 : Production of sea breeze

Consequences of high specific heat of water

Compared to most substances, water has a high specific heat, being nearly 4200 J/kg/K. This has numerous useful consequences in practice. Some of these are

- Fomentation

Hot water bottles are used for fomentation, as the water remains hot in the bottle for a long time in spite of giving off large quantities of heat.

- Car radiators

Water is used as coolant in car radiators. Due to its high specific heat capacity, it can absorb a large amount of heat energy from the engine of the car, but its temperature does not rise too high.

■ Bottled wines

In cold countries, wine bottles are kept under water. Due to the high specific heat capacity of water, it takes a long time for the temperature to fall to very low level, even 0 °C.

■ Water in a swimming pool

Due to its high specific heat, water in a swimming pool remains cool even in summer and people enjoy a lot staying inside the pool.

■ Protection of crops

Farmers generally fill their field with water during winter season at night to protect the crops from frost. In the absence of water, if on a cold night the temperature of the surroundings fall below 0^0C, then the veins of the plants shall freeze. But water sprinkled on the crops shall not allow the temperature of the veins to fall below 0^0C.

Applications involving high and low specific heats

As detailed in the table given above, substances differ appreciably in the values of specific heat. This can result in many applications in practical usage. Some of these, typically, are

- Utensils used for making tea or coffee, or cooking vegetables or rice are made of materials of *low* specific heat; for example copper, aluminium or stainless steel. In addition, they are polished at the bottom so that they get heated quicker.

- Materials of *high* specific heat can be used as "insulators". For example, wood has a high specific heat. This allows construction of wooden houses in countries having cold climate where wooden construction would keep the house warm in winter, not allowing the heat of hot air from within to conduct outside.

- The handles of utensils at home are made of materials which have *very high* specific heat. They act as thermal insulators. The insulating capability or conductivity is dependent on specific heat of the material.

- In an instrument like thermometer, the body may be made of a material of high specific heat (such as glass) and the tip or the sensor element is made of material of low specific heat (such as mercury).

- Steam has a *high* specific heat (more than water) and hence can carry a lot of heat energy at high pressures to run, for example, a steam turbine used as a prime mover for an AC generator.

Calorimeter

A device frequently used in calorimetry is called a **calorimeter**[13], used in the process of measuring the heat of chemical reactions or physical changes as well as heat capacity.

Construction

A typical calorimeter – also known as a "coffee-cup" calorimeter – comprises two vessels: one an outer vessel made of an insulating material and the other a thin inner vessel made of a conducting metal such as copper or aluminium, usually the former, and well polished. The air between the inner vessel and the outer vessel works as a heat insulator, so there is no or very little heat exchange between the contents of the inner vessel with the outside.

It also has provision for a thermometer and a stirrer. The thermometer is used to measure the temperature of the liquid (usually water) in the inner vessel. The stirrer is used to stir the liquid and distribute heat 'uniformly' in the vessel.

A fiber ring, made of an insulating material, is provided to hold the inner vessel hanging in the center of the outer vessel. A calorimeter also has an insulating cover or lid at the top with holes for the stirring rod and thermometer. The lid prevents any loss of heat from the top.

A photograph of an actual calorimeter is shown in Fig.1.9(a) whilst a schematic view is given in Fig.1.9(b).

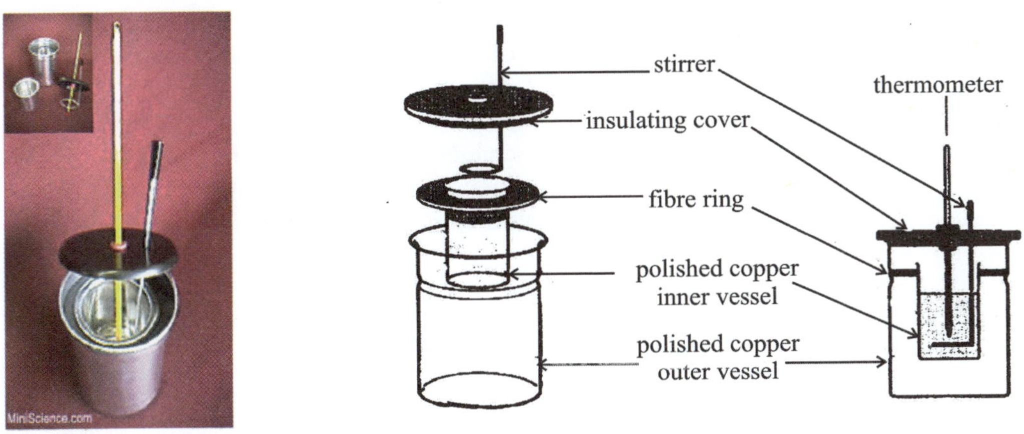

(a) an actual calorimeter

(b) schematic of a calorimeter

Fig.1.9 : Details of a calorimeter

[13]As in the case of a thermometer, a calorimeter does not comprise or include a meter in the usual sense. However, it does involve measurement of heat as described.

How to assemble a calorimeter?

To assemble the calorimeter, place the fiber ring in the larger vessel and then insert the small vessel in the center of the fiber ring.
Insert the thermometer in the center hole of the lid or insulating cover; position it so that when you close the cover, the bulb of the thermometer will be in the center of the lower vessel, about one inch from the bottom.
Insert the stirring rod and attach the handle of the insulating rod. An assembled calorimeter is shown in Fig.1.10.

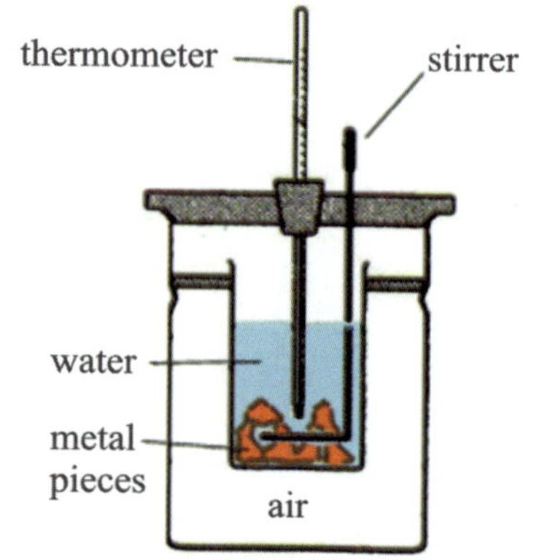

Fig.1.10 : Assembling a calorimeter

Applications of Calorimeter

Example: to measure specific heat of iron

■ Fill the inner vessel with 125 ml water, so that its mass will be 0.125 kg. Close the lid and leave it for about 30 minutes. At this time the bulb of the thermometer must be in the water. (If it is not, push the thermometer lower). After 30 minutes, read and record the water temperature. Let it be T_1 (°C).

• Separately weigh the inner vessel of the calorimeter. Let its mass be m_c kg.

▪ Fill up a test tube half way with iron nails. Weigh and record the weight of the nails. Let it be m_i(kg). Place the test tube with nails in hot-boiling water for about 10 minutes. This will allow the nails to acquire a steady temperature of 100 °C without getting wet.

■ Open the lid of the calorimeter and transfer the nails quickly to the inner vessel. Immediately close the lid. Move the stirrer up and down and read the temperature. Record the highest temperature. This will be the final temperature of the water *and* nails. Let it be T °C.

Then,

heat lost by iron nails $H_i = m_i \times c_i \times (100 - T)$ J

where c denotes the specific heat of iron, to be determined.

heat gained by water $= 0.125 \times c_w \times (T - T_1)$ J

where c_w is the specific heat of water, and known.

heat gained by calorimeter $= m_c \times c_c \times (T - T_1)$ J

where c_c is the specific heat of material of calorimeter inner vessel, usually copper and known.

Then applying the principle of "method of mixture"

$$\text{heat lost} = \text{heat gained}$$

Substituting

$$m_i \times c_i \times (100 - T) = 0.125 \times c_w \times (T - T_1) + m_c \times c_c \times (T - T_1)$$

whence

$$c_i = \frac{[0.125 \times c_w + m_c + c_c \times (T - T_1)]}{m_i \times (100 - T)} \text{ J/kg/°C}$$

Example

A metal of mass 0.2 kg at 100 °C (boiling point of water) is quickly dropped into 0.08 kg of water at 15 °C contained in a calorimeter of mass 0.12 kg and specific heat capacity 400 J/ kg/K. The final temperature reached is 35°C. Calculate specific heat capacity of the metal.
Assume specific heat capacity of water to be 4,200 J/kg/K.

Heat gained by water = 0.08 x 4,200 x (35-15) = 6,700 J
Heat gained by the calorimeter = 0.12 x 400 x (35-15) = 960 J
Heat gained by water + calorimeter = 6,700 + 960 or 7,680 J
Heat lost by metal = 0.2 x c x (100 – 35) J

∴ 0.2 x c x 65 = 7,680

whence

c = 590 J/kg/K

Measurement of specific heat of a solid by electrical method[14]

The specific heat capacity of a metal can be found by an electrical method. The method is illustrated in Fig.1.11.

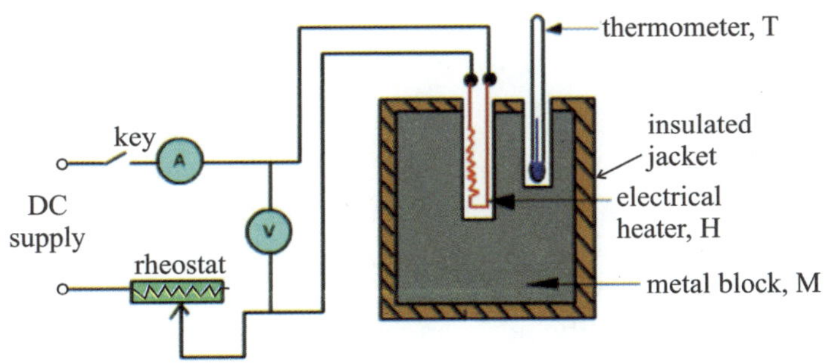

Fig.1.11 : Determination of specific heat by electrical method

[14]In practice, more ingenious methods can be devised to measure specific heat of a substance – solid or liquid – where the principle of measurement is based on the "method of mixtures", the known heat lost in the device being equated to heat gained by the substance of which the specific heat is to be obtained.

Referring to Fig.1.11, M is a thick, solid block of metal, say aluminium, of mass m (kg). A suitably insulated (electrical) heater element of relatively small size, H, is inserted completely in a hole bored in the block. Another deep hole is bored to fit a thermometer, T, as shown. It is essential that both H and T make good thermal contact with the metal. An insulating jacket is provided round the metal to prevent conduction of heat to the exterior.

In the experiment, suppose the voltmeter connected to an appropriate *DC* electric supply reads V volt whilst the ammeter reads a current I ampere. At the start, let the thermometer reads a temperature T_1°C. At the end of time t seconds, measured by means of a stop watch, let the temperature of the metal be T_2°C.
Then, the power supplied electrically = V x I W
and the energy supplied = V x I x t Ws or J
If c denotes the specific heat of aluminium, the heat gained by the block will be
= m x c x $(T_2 - T_1)$ J
whence the specific heat of the metal/aluminium is obtained as

$$c = \frac{V \times I \times t}{m \times (T_2 - T_1)} \ J/kg/°C$$

Note that thermal heat capacity of the calorimeter is given by $C_c' = c \times (T_2 - T_1)$ J/kg.

Water equivalent of calorimeter

Definition

In a reaction, chemical or otherwise, the quantity of heat that raises the temperature of some substance by some amount, the same quantity of heat can simultaneously raise the same temperature of a certain mass of water. The mass of water is then termed as **water equivalent (WE)**.

Specifically, the term is frequently used in calorimetry involving a common calorimeter. With an input of heat, H (J), however produced, is given out to, say, a (mixture) of water in a calorimeter, the given heat is used to raise the temperature of water *and* the calorimeter from an initial value of T_1 to the final value of T_2. Then the heat taken by the water and calorimeter would be, respectively,
H_{water} = mass of water x specific heat capacity of water x $(T_2 - T_1)$
and
$H_{calorimeter}$ = mass of calorimeter x specific heat capacity of calorimeter x $(T_2 - T_1)$
so that
$H = H_{water} + H_{calorimeter}$
If the $H_{calorimeter}$ can be expressed as
$H_{calorimeter} = WE \times (T_2 - T_1)$

Then WE is termed as "water equivalent of calorimeter".

The concept is better explained by the following example:

200 g of water at 100 °C is mixed with 140 g of water at 10 °C contained in a calorimeter. If the final temperature of the mixture is 40 °C, calculate the water equivalent of the calorimeter. [given specific heat capacity of water = 1 cal/g/°C].

Heat given out by the (boiling) water = 200 x 1 x (100 – 40)
= 12,000 cal

Heat taken by water in the calorimeter = 140 x 1 x (40 – 10)
= 4,200 cal

∴ heat taken by the calorimeter = 12,000 - 4,200
= 7,800 cal

Then water equivalent of the calorimeter = 7,800/(40 – 10) or 260 g

EXERCISES

1. A hot solid of mass 60 g at 100 °C is placed in 150 g of water at 20 °C. The final steady temperature recorded is 25°C. Calculate the specific heat capacity of the solid. [Specific heat capacity of water = 4,200 J/kg/°C].

[ICSE, 2012]
[0.7 J/g/K

2. 200 g of hot water at 80°C is added to 300 g of cold water at 10 °C. Calculate the final temperature of the mixture of water. Consider the heat taken by the container to be negligible. [specific heat capacity of water is 4,200 J/kg/°C].

[ICSE, 2011]
[38 °C

3. 40 g of water at 60°C is poured into a vessel containing 50 g of water at 20 °C. The final temperature recorded is 30°C. Calculate the thermal capacity of the vessel. (Take specific heat capacity of water as 4.2 J/g/°C).

[ICSE, 2009]
[294 J/°C

4. Some hot water was added to three times its mass of cold water at 10°C and the resulting temperature was found to be 20 °C. What was the temperature of the hot water?

[ICSE, 2007]
[50 °C

5. A solid of mass 60 g at 100 °C is placed in 100 g of water at 18 °C. The final temperature reached is 20°C. Calculate the specific heat capacity of the metal.

[0.17 J/g/°C

6. Water falls from a height of 50 m. Calculate the rise in temperature of water when it strikes the bottom. Specific heat capacity of water = 4,200 J/kg/°C; g = 10 m/s/s.

[0.119 °C

II Latent Heat

Latent heat: loss and gain of heat involving change of state for fusion only.
Change of phase (state); heating curve for water; latent heat; specific latent heat of fusion (SI units), simple numerical problems; common physical phenomena involving latent heat of fusion.

Change of Phase

Most matters can usually exist in THREE states: solid, liquid and gas[1]. The basis for the matter to be in any phase derives from the distance between the molecules or atoms of the matter. The factors responsible for the state are temperature and pressure. Under *constant* pressure when a matter is heated, its speed of motion increases and as a result the distance between the atoms or molecules becomes larger. If a solid is heated, its temperature increases up to a specific point and after this point temperature stays constant and it starts to change its phase from solid to liquid.

A classic example of a matter in various phases is that of water. It can exist in solid form, that is ice, which when heated melts into water or turns into liquid state. When water is heated it boils and changes into steam and on continued heating starts to evaporate, that is changes into gaseous state or phase.

The terms **melting** and **freezing** are usually associated with the processes of change of a substance from solid to liquid state and *vice versa*. The change of state from liquid to gas is termed as **boiling**.

In all, there are FIVE stages of change in a matter.

---

[1]A fourth state is sometimes encountered called "plasma", from Greek πλάσμα, "anything formed". Plasma is a cloud of protons, neutrons and electrons where all the electrons have come loose from their respective molecules and atoms, giving the plasma the ability to act as a whole rather than as a bunch of atoms. A plasma is more like a gas than any of the other states of matter because the atoms are not in constant contact with each other, but it behaves differently from a gas.

The five changes of phase

In general, the five stages of phase change can be described as

Description of phase change	Term for phase change	Heat movement during phase change	Temperature change during phase change
solid to liquid	melting	heat *into* the solid	none
liquid to solid as it freezes	freezing	heat *leaves* the liquid	none
liquid to gas	vapourisation (including boiling and evaporation)	heat *into* the liquid as it vaporises	none
gas to liquid as it condenses	condensation	heat *leaves* the gas	none
solid to gas[2] as it sublimates	sublimation	heat goes *into* the solid	none

The temperature at which a solid changes into liquid is called its "melting point"; likewise, the temperature at which a liquid changes into solid is called its "freezing point". Also, the temperature at which a liquid starts to boil is called its "boiling point".

Effects of Pressure and Impurity on Boiling Point

Boiling occurs only when the vapor pressure of liquid and pressure of outside equals each other. If the pressure of outside increases then the boiling point of the liquid also increases. On the contrary, if the pressure of the outside decreases, then boiling point of the liquid also decreases. For example, at the top of a mountain atmospheric pressure is lower than the atmospheric pressure at the sea level and therefore water would boil at lower than 100 °C on a mountain top.

[2]There are solids that turn directly into gas. As dry ice (frozen carbon dioxide gas) gets warmer, it turns into carbon dioxide gas. This process is called *sublimation.*

In addition to this, impurity of the liquid matter also affects the boiling point of that matter. For instance, if water is mixed with a pinch of salt or sugar, it would increase the boiling point of the water

So, how could there be a change in heat during a state change without a change in temperature?

During a change in state the heat energy is used to change the *bonding* between the molecules. In the case of melting, added energy is used to break the bonds between the molecules. In the case of freezing, energy is subtracted as the molecules bond to one another. These energy exchanges are *not* changes in kinetic energy. They are changes in bonding energy between the molecules.

Heating curves

To consider a substance that is heated to change its state or phase from solid to liquid and then from liquid to gas, the temperature changes with time. A curve or graphical relationship between temperature of the substance at a given time is termed its heat or heating curve.

Heating Curve for Water

The diagram below in Fig.2.1 depicts the heating curve for water.

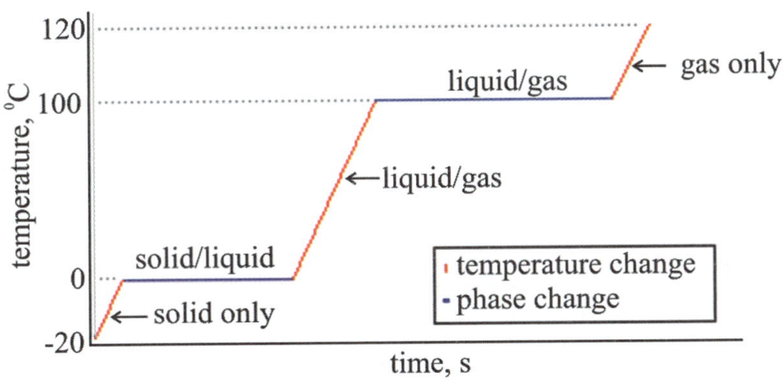

Fig.2.1 : Heating curve for water

Observe that there are three slopping sections and two horizontal sections on the temperature-time plot. The first slopping section corresponds to a change in temperature of the ice from -20°C to 0°C. The water in its solid state is warming up to the melting point - the temperature at which transition of water begins from solid to liquid state, that, is from ice to water. The heat transferred to the ice causes a temperature change. Once the transition temperature (melting point) of 0°C is reached, the heat added to the water causes the water to change from its solid state to its liquid state. This is referred to as melting. The melting occurs at a constant temperature.

The next section of the *heating curve* is another sloping section. The (liquid) water is increasing its temperature from 0°C to 100°C, the boiling point of water at which water transitions from the liquid state to the gaseous state. Once water reaches this temperature, boiling occurs. Large bubbles of gas would be observed forming throughout the bulk of the liquid.
The heat added to the liquid during this stage of the experiment causes a loosening of the attractions that hold the water particles in the liquid state. The temperature remains constant while the state of water changes. Once all the water transitions from the liquid to the gaseous state or steam, water (now in the gaseous state) begins to increase its temperature again.

In summary, the three slopping sections represent heat causing a temperature change in the substance that absorbs it. And the two plateau sections represent heat causing a change of state in the substance that absorbs it. The temperature changes are the result of the added energy causing the particles of water to move more vigorously. Either the particles of solid vibrate more vigorously about their fixed positions or the particles of liquid and gas move about their container more rapidly. Either way, the addition of heat is causing an increase in the average kinetic energy of the particles in water. The changes of state are the result of the added energy causing changes in the strength of the inter-particle attractions. The attractions that hold water in the solid or in the liquid state are being overcome. The energy is being used to loosen these attractions and change to a state of greater potential energy. A 'pictorial' view of change of phases is depicted in Fig.2.2.

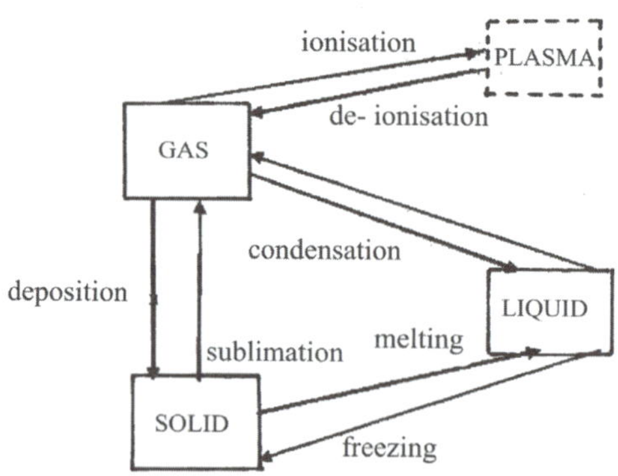

Fig.2.2 : Transition of phases

Example

1. Some ice is heated at a constant rate, its temperature being recorded after every few seconds till steam is formed at 100°C. Draw a temperature time graph to represent the change. Label the two phase changes in the graph.

[ICSE, 2016]

The graph is drawn below.

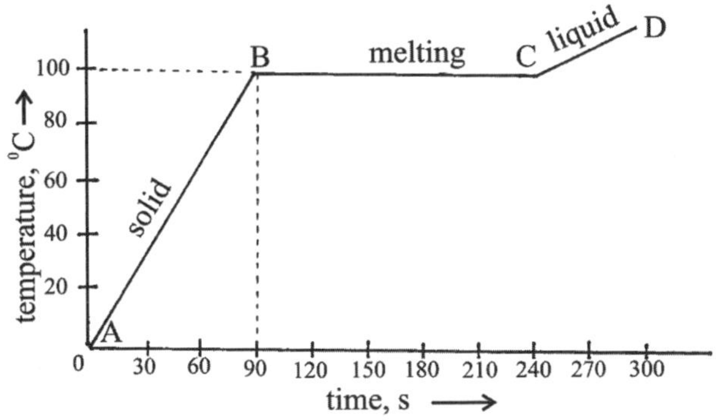

2. A piece of ice at 0 °C is heated at a constant rate and its temperature recorded at regular intervals till steam is formed at 100 °C. Draw a temperature-time graph to represent the change in phase. Label the different parts of the graph.

[ICSE 2012]

The graph is shown below.

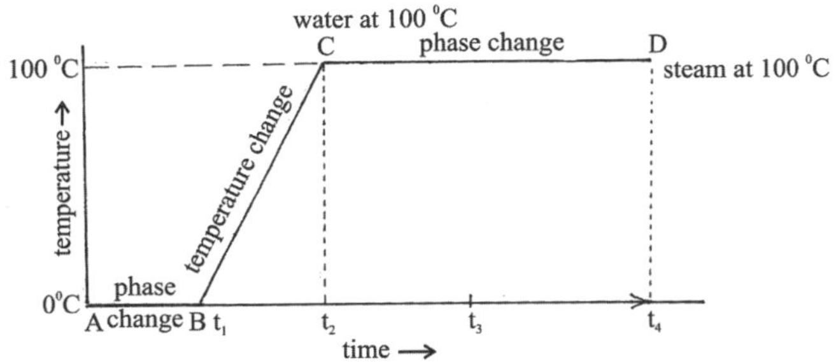

LATENT[3] HEAT

Heat absorbed or released as the result of a phase change is called **latent heat**[4], for example the heat required to convert a solid into a liquid or vapour, or a liquid into a vapour. *There is no temperature change during a phase change*, thus there is no change in

---

[3]Dictionary meaning: existing but not manifest; concealed; dormant.

[4]The term was introduced around 1762 by Scottish chemist Joseph Black, derived from the Latin *latere* (*to lie hidden*).

the kinetic energy of the particles in the material. The energy released comes from the potential energy stored in the bonds between the particles.
Accordingly, latent heat is involved in all the three cases of change of state.

The latent heat of fusion, Q_f.
This is the heat needed to make the change between the solid and liquid phases; for example when water turns to ice or ice turns to water.

The latent heat of vapourisation, Q_v
This is the heat needed to make the change between the liquid and gas phases, as when water boils or when steam condenses into water.

The latent heat of sublimation, Q_s.
This is the heat needed to make the change between the solid and gas phases, as when dry ice evaporates.

Specific Latent Heat

This is a constant for a given substance, defined as *latent heat per unit mass* of the substance. The SI unit of specific latent heat is joule per kilogram (J/kg).

Specific latent heat of fusion

Specific latent heat of fusion, L_f, of a substance is defined as the amount of heat required to change a unit mass of the substance from solid to liquid state, without any change in the temperature.

Specific latent heat of vapourisation

Specific latent heat of vapourisation L_v of a substance is defined as the amount of heat required to change unit mass of the substance from *liquid state to gas state* without a change of temperature.

Example

1. 10 g of ice at 0°C is added to 10 g of water at 80 °C such that the temperature of the mixture is 0 °C. Calculate specific latent heat of fusion of ice. Specific heat capacity of water = 4.2 J/g/°C.
Let latent heat of ice be L J/g
Heat gained by ice = 10 x L
Heat lost by water in melting from 80 °C to 0 °C = 10 x 4.2 x 80 = 3,360 J

Or $10\,L = 3360$
and $L = 336$ J/g

2. A metal ball of mass 0.5 kg at 900 °C is placed on a block of ice till it attains the temperature of ice. If the specific heat capacity of the metal ball is 850 J/kg/K, calculate the amount of ice that would melt. Specific latent heat of fusion of ice is 34×10^4 J/kg.

Heat lost by the metal ball = 0.5 x 850 x 900
= 3,82,500 J

Heat gained by the ice to melt to 0 °C = m x 3,40,000, m being the mass of the ice

Equating

3,40,000 m = 3,82,500

or m = 1.125 kg

3. a) 4 kg of water (at 25 °C) and 2 kg of ice (at 0 °C) are mixed inside an isolated container. What is the final temperature ? Is there any ice left?
b) What happens if there is only 1 kg of ice?

a) First, check how much heat is needed to melt 2 kg of ice. The ice is at 0 °C so that in this case it is only necessary to compute the latent heat. The specific latent heat of water is $L_f = 3.33 \times 10^5$ J/kg. Therefore heat required to melt the ice
$Q_1 = L_f\, m = 3.33 \times 10^5 \times 2 = 6.66 \times 10^5$ J

Next, check how much heat is needed to bring all the water to 0 °C. The specific heat of water is $c_{water} = 4186$ J/kg/ °C. With the difference of temperature ΔT being 25 °C, the heat required to bring all the water to 0°C will be
$Q_2 = c \times m \times \Delta t = 4186 \times 4 \times 25 = 418600$ J $= 4.186 \times 10^5$ J

Conclusion: There will be ice left, because Q_2 is lower than Q_1. The mixture will stay at 0 °C and no further heat transfers will occur (heat can only be transferred if there is a temperature difference).

b) When the ice is only 1 kg
$Q_1 = L_f \times m = 3.33 \times 10^5 \times 1 = 3.33 \times 10^5$ J
In this case since $Q_1 < Q_2$ all the ice will be melted.
What will be the final temperature of the mixture?

Let the final temperature of the mixture be T_f. It will be somewhere between 0 °C and 25 °C. Then making use of the method of mixtures,
$4186 \times 4 \times (25 - T_f) = Q_1 + 4186 \times 1 \times (T_f - 0)$
whence

$$T_f = 4.1\ ^\circ\text{C}$$

4. Calculate the mass of ice required to lower the temperature of 300 g of water at 40 °C to water at 0°C.
(specific latent heat of ice = 336 J/g, specific heat capacity of water = 4.2 J/g/°C)
[ICSE, 2016]

Mass of water = 300 g
Initial temperature of water = 40 °C; final temperature = 0 °C
Let the mass of ice required be m g

$\therefore$ heat to be removed to lower temperature of 300 g of water from 40 °C to 0 °C
= 300 x 4.2 x (40 – 0) or 50,400 J
Heat lost by ice in melting = m x 336 J

$\therefore$ 336 m = 50,400, giving m = 150 g

5. A copper vessel of mass 100 g contains 150 g of water at 50 °C. How much ice would be needed to cool it to 5°C? Given: specific heat capacity of copper = 0.4 J/g/°C; specific heat capacity of water = 4.2 J/g/°C; specific latent heat of fusion of ice = 336 J/g.
[ICSE, 2016]

Given
mass of ice copper vessel = 100 g
mass of water = 150 g
initial temperature, T_1 = 50 °C
final temperature, T_2 = 5 °C
Let the mass of ice required be m g
Now, heat lost by water for decrease in temperature from 50 to 5 °C
= 150 x 4.2 x (50 – 5) or 28,350 J
heat lost by the calorimeter = 100 x 0.4 x (50 – 5) or 1,800 J
heat gained by ice = m x 336 x (5 – 0) or 1680m J
heat lost by water + calorimeter = 28,350 + 1,800 or 30,150 J
By the principle of calorimetry
heat lost by water and calorimeter = heat gained by the ice

$\therefore$ 30,150 = 1,680 m or m = 17.94 g

6. A refrigerator converts 100 g of water at 20°C to ice at -10°C in 35 minutes. Calculate the average rate of heat extraction in terms of watts.
Given: specific heat capacity of ice = 2.1 J/g/°C
specific heat capacity of water = 4.2 J/g/°C
specific latent heat of fusion of ice = 336 J/g

[ICSE, 2015]

Heat extracted to convert 100 g water at 20°C to water at 0°C

= 100 x 4.2 x (20 – 0) or 8,400 J

Heat extracted to convert 100 g water at 0°C to ice at 0°C

= mass x sp. latent heat of fusion of ice

= 100 x 336 or 33,600 J

Heat extracted to convert 100 g ice at 0°C to ice at -10°C

= 100 x 2.1 x (0 – (-10)) or 2,100 J

∴ total heat extracted in 35 minutes = 8,400 + 33,600 + 2,100

= 44,100 J

and average rate of heat extraction in watts

= 44,100/(35x60) or 21 W

7. Heat energy is supplied at a constant rate to 100 g of ice at 0 °C. The ice is converted into water at 0°C in 2 minutes. How much time will be required to raise the temperature of water from 0 °C to 20°C?
[given: specific heat capacity of water = 4.2 J/g/°C; specific latent heat of ice = 336 J/g]
[ICSE, 2014]

Mass of ice = 100 g

Heat extracted from ice in melting at 0 °C = mass x sp. latent heat

= 100 x 336 J

Since it takes 2 minutes or 120 s to melt the ice, if P is the power applied (energy at constant rate)

P x 120 = 100 x 360

or P = 280 W

Now, heat required to raise the temperature of 100 g of water from 0 °C to 20 °C will be

100 x 4.2 x 20 J

If this heat (energy) is supplied from the same source of power and the time taken is T s

then 100 x 4.2 x 20 = 280 x T

whence T = 30 s

8. 250 g of water at 30 °C is present in a copper vessel of mass 50 g. Calculate the mass of ice required to bring down the temperature of the vessel and its contents to 5 °C.
Specific latent heat of fusion of ice = $336 x 10^3$ J/kg
Specific heat capacity of copper vessel = 400 J/kg/°C
Specific heat capacity of water = 4,200 J/kg/°C.
[ICSE, 2011]

By the principle of calorimetry or method of mixtures

heat gained during the process = heat lost

Let the mass of ice required be m g

Then heat gained by ice in acquiring 5 °C temperature is

m x 336 + m x 4.2 x 5 J

and heat lost by water and copper vessel is

$$250 \times 4.2 \times (30 - 5) + 50 \times 0.4 \times (30 - 5) \text{ J}$$

$\therefore$ $$m \times 336 + m \times 4.2 \times 5 = 250 \times 4.2 \times (30 - 5) + 50 \times 0.4 \times (30 - 5)$$

$$357 \text{ m} = 26,259 + 500$$

or $$\text{m} = 74.9 \text{ g}$$

9. Calculate the amount of ice which is required to cool 150 g of water contained in a vessel of mass 100 g at 30 °C, such that the final temperature of the mixture is 5 °C. (Take specific heat capacity of material of the vessel as 0.4 J/g/°C; specific latent heat of fusion of ice = 336 J/g; specific heat capacity of water = 4.2 J/g/°C).

[ICSE, 2009]

Let the mass of ice required be m g
Now the heat gained by ice to melt and come to final temperature =
heat lost by water and vessel

$\therefore$ $$m \times 336 + m \times 4.2 \times 5 = 150 \times 4.2 \times (30\text{-}5) + 100 \times 0.4 \times (30\text{-}5)$$

Or $$357 \text{ m} = 16,750 \quad \text{or} \quad \text{m} = 46.91 \text{ g}$$

10. A substance is in the form of a solid at 0 °C. The amount of heat added to this substance and the temperature of the substance are plotted on the following graph:

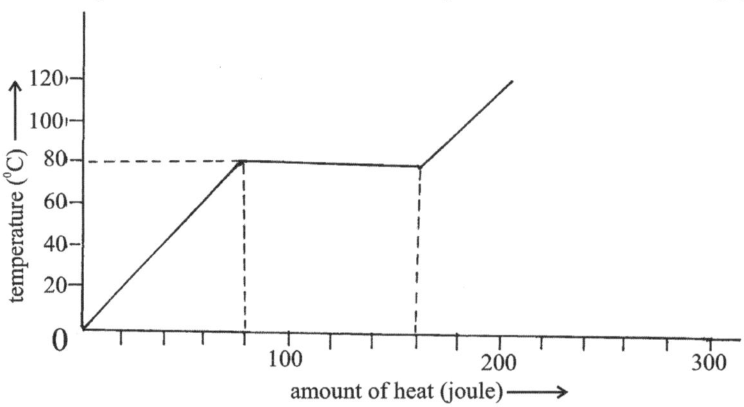

If the specific heat capacity of the solid substance is 500 J/kg/°C, find from the graph:
(i) the mass of the substance
(ii) the specific latent heat of fusion of the substance in the liquid state

[ICSE, 2007]

(i) From the graph, the substance is heated from 0 °C to 80 °C, using heat energy of 800 J.

$$\text{Now specific heat capacity of the substance} = \frac{\text{heat supplied}}{\text{rise in temp.} \times \text{mass}}$$

Substituting

$$500 = 800/(80 \times \text{mass})$$

From which mass of substance

$$= 800/(80 \times 500)$$
$$= 0.02 \text{ kg} \quad \text{or} \quad 20 \text{ g}$$

(ii) Again, from the graph, the heat absorbed by the substance during fusion at a (constant) temperature of 80 °C is seen to be 800 J

∴ specific latent heat of fusion of the substance $= 800/0.02$
$$= 4.0 \times 10^4 \text{ J/kg}$$

11. A 30 W immersion heater just keeps 600 g of molten metal at its melting point. The heater is switched off and temperature stars falling after 6 min. Calculate specific latent heat of metal.

Time for which temperature of metal does not fall = time in which metal changes its state
Heat given out (or lost) by metal at its melting point = 600 x L, L being its latent heat
Heat gained by the metal from immersion heater = power x time (in s)
= 30 x 360 or 10,800 J

From the principle of calorimetry

heat lost = heat gained

Or

600 L = 10,800
L = 18 J/g

Table of Specific Latent Heat

The values of specific latent heat of some selected substances are listed below*.

Substance	Specific latent heat, J/kg	
	fusion, L_f	vaporisation, L_v
water	$3.34\text{x}10^5$	$2.26\text{x}10^6$
sulphur	$3.81\text{x}10^4$	$3.26\text{x}10^5$
lead	$2.45\text{x}10^4$	$8.75\text{x}10^5$
aluminium	$3.97\text{x}10^5$	$1.14\text{x}10^7$
silver	$8.82\text{x}10^4$	$2.33\text{x}10^6$
gold	$6.44\text{x}10^4$	$1.58\text{x}10^6$
copper	$1.34\text{x}10^5$	$5.06\text{x}10^6$
mercury	$1.14\text{x}10^4$	$2.6\text{x}10^3$

*values may vary slightly

Experimental determination of specific latent heat

This is described for fusion of ice using "method of mixture" and electrical method and for vapourisation of a liquid.

A. Specific Latent Heat of Fusion of Ice, L_f[5]

■ Method of mixture

Specific latent heat of fusion of ice, L_f

- Place some warm water at a temperature of θ_1°C, a few degrees above the room temperature inside a calorimeter, having weighed the water. Let this be m_1 kg
- Add small lumps of ice, having been 'dried' on a blotting paper, to the water until the mixture of water and melting ice reaches a temperature θ_2, as much below room temperature as θ_1 was above it
- Weigh the mixture to obtain the mass m of the ice
- The specific latent heat of fusion of ice can then be derived as follows

$$\begin{pmatrix}\text{heat given by calorimeter} \\ \text{and water in cooling}\end{pmatrix} = \begin{pmatrix}\text{heat used up in} \\ \text{melting ice}\end{pmatrix} + \begin{pmatrix}\text{heat used up in warming} \\ \text{melted ice from } 0^\circ\text{C to } \theta_2\end{pmatrix}$$

Or

$$(m_1 x c_w + C) \text{ x } (\theta_1 - \theta_2) = m \text{ x } L_f + m \text{ x } c_w (\theta_2 - 0)$$

where c_w is the specific heat capacity of water, C = heat capacity of calorimeter whence

$$L_f = \frac{(m_1 \times c_w + C) \times (\theta_1 - \theta_2)}{m} - c_w \theta_2$$

Note that ice of mass m requires heat for fusion and also for raising the melted water from 0°C to θ_2.

---

[5]Even though described for ice, the method would equally apply to any solid with appropriate changes/means for supplying heat required for the purpose.

■ Electrical method

The experimental set-up for this method is given in Fig.2.3.

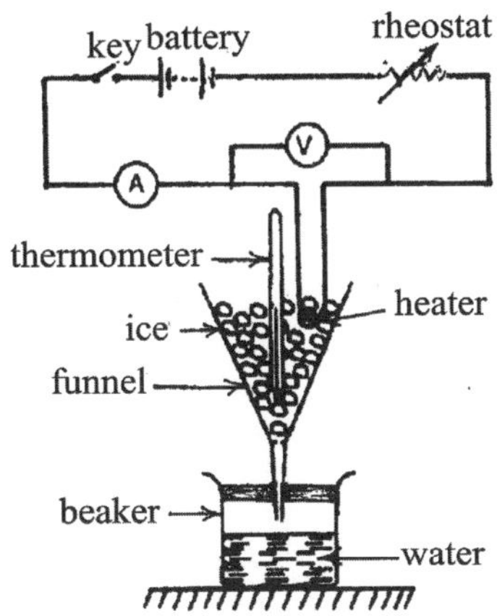

Fig.2.3 : Electrical method to determine latent specific heat of fusion

Procedure

- Determine the mass of the empty beaker. Let this be m_o kg.
- Fill the filter funnel with *crushed* ice
- Place an immersion heater completely covered with ice in the funnel as shown, the heater being connected to a DC supply as shown
- Adjust the rheostat to a suitable value and close the switch, to allow a steady current to flow, and the clock at the same instant
- Note the readings of the ammeter and voltmeter
- After a reasonable amount of water is collected in the beaker, switch off the heater and note the time elapsed
- Remove the beaker and weigh it with the condensed water in it. Let this be m_1 kg
- Then the mass of condensed ice is given by $m_1 - m_o$ kg

If the specific latent heat of fusion of ice is L, the heat 'gained' by ice = J
whilst heat supplied electrically = V x I x t
where V and I are voltage and current in volts and ampere, respectively, and t the time recorded by the clock in *seconds*.

Equating

$$(m_1 - m_o) \times L = V \times I \times t$$

whence

$$L = \frac{V \times I \times t}{(m_1 - m_0)} \text{ J/kg}$$

B. Specific Latent Heat of Vapourisation of a Liquid

The specific latent heat of vapourisation of a liquid may be a method described below. The apparatus is set up as shown in Fig.2.4.

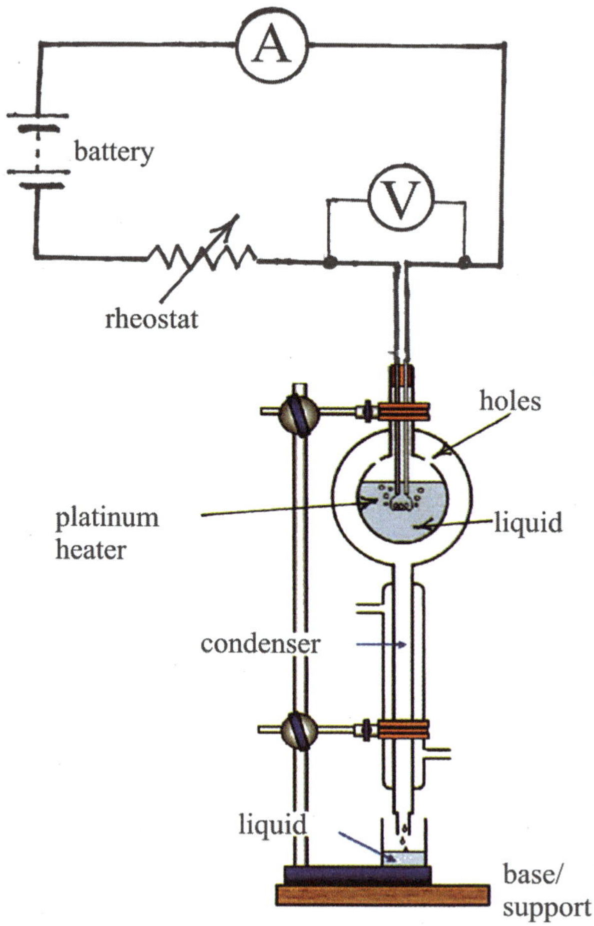

Fig.2.4 : Determination of specific latent heat of vapourisation

Procedure

The double-walled, round (spherical) glass vessel is fitted to a condenser and mounted vertically using a rigid stand or support. The inner section of the vessel contains the liquid of which the specific latent heat of vapourisation is to be determined. A suitably designed, insulated heater/heating element is placed inside the inner vessel as indicated.

Switch on the electric supply to the heater, adjusting the rheostat for a steady current to flow. Note the voltmeter and ammeter readings. When the liquid starts to boil, the vapour passes through small holes into the outer vessel and then down into the condenser where it condenses, runs down and is collected in the beaker placed below the condenser outlet[6].

When a steadystate has been reached, that is, when liquid drips into the beaker at a constant rate, a clean beaker is placed under the condenser and the mass of liquid, m, condensing, and hence being evaporated, in a measured time t is obtained.

The specific latent heat of vaporisation L of the liquid can then be found from the equation

$$L = (V \times I \times t)/m \quad \text{J/kg}$$

where

V is the applied voltage, V
I the measured current, A
t the time measured, s, corresponding to condensation of liquid in the beaker
m mass of the liquid in kg

Example

An electric kettle with a 2 kW heating element has a heat capacity of 400 J/K. It is filled with 1 kg of water at room temperature of 20 °C. The kettle is switched on and it is found that after 13 minutes, the mass of water in it is 0.5 kg. Disregarding stray heat losses, calculate the specific latent heat of vapourisation of water. Assume specific heat capacity of water as 4200 J/kg/K.

Total heat supplied, by the heater of the kettle $= 2000 \times 13 \times 60 \text{ J} = 1.56 \times 10^6 \text{ J}$

Heat used for the kettle $= C\,\theta = 400 \times (100 - 20)$

$= 32000 \text{ J}$ or $0.032 \times 10^6 \text{ J}$

Heat used to raise temperature of 1 kg of water from 20°C to 100 °C $= m\,C\,\theta$

$= 1.0 \times 4200 \times (100 - 20)$

$= 0.336 \times 10^6 \text{ J}$

So total heat used to change water at 100 °C to steam at 100 °C

$= 1.56 \times 10^6 - (0.032+0.336) \times 10^6 \text{ J}$

$= 1.192 \times 10^6 \text{ J}$

This must equal the latent heat of water converted to steam

Or $m \times L = 1.192 \times 10^6 \text{ J}$

---

[6]It is essential that evaporation is rapid, the power input to the heater being adjusted accordingly, for then the vapour in the outer vessel acts as a heat shield and eliminates heat losses from the inner vessel.

where m = mass of water/steam, equal to 0.5 kg here, and L its specific latent heat of vapourisation

whence $L = 1.192 \times 10^6 / 0.5$ or 2.38×10^6 J/kg

Some Common Physical Phenomena Involving Latent Heat of Fusion

The latent heat of fusion, esp. that of ice, manifests itself in numerous phenomena of great natural importance. Some of these are

- Snow on the mountain does not melt all at once.
  Each kilogram of ice would require about 336 kJ of energy to melt; this energy being supplied from the atmosphere or the sun.

- Cold drinks are best cooled by adding 'soft' (not entirely crushed) pieces of ice and not cold water at 0 °C.
  As each gram of ice absorbs 336 J of energy as it melts, it results in keeping the drink cold for longer.

In cold countries/climate

- Water in lakes and ponds does not freeze all at once as water has to absorb heat energy from the atmosphere. The water starts to freeze first from the surface.
  This allows water below the surface to be rather warm and helps fish and sea life to sustain themselves.

- The weather becomes very cold as soon as snow starts melting as the heat is taken from the atmosphere.

- The weather becomes pleasant when freezing of water starts as every kg of water releases 336 kJ of heat as it turns into ice.
  Thus a very large amount of heat is slowly released in the atmosphere.

- The temperature becomes very low after a hail storm.
  This is because every kilogram of ice absorbs 336 kJ of heat to melt, bringing down the atmospheric temperature.

- The global warming may lead to climate changes as a result of heat unbalance owing to melting of ice in glaciers and polar regions.

EXERCISES

1. Calculate the amount of heat released when 5 g of water at 20 °C is changed into ice at 0 °C. Specific latent heat of fusion of ice = 336 J/g; specific heat capacity of water = 4.2 J/g/°C.

[2,100 J

2. How much heat energy is released when 5 g of water at 20 °C changes to ice at 0 °C? [Specific heat capacity of water = 4.2 J/g/°C; specific latent heat of fusion of ice = 336 J/g/°C]

[ICSE, 2013]
[2,100 J

3. 40 g of ice at 0 °C is used to bring down the temperature of a certain mass of water at 60 °C to 10 °C. Find the mass of water used.
[Specific heat capacity of water = 4,200 J/kg/°C]
[Specific latent heat of fusion of ice = 336x103 J/kg]

[ICSE, 2012]
[72 g

4. 50 g of ice at 0 °C is added to 300 g of a liquid at 30 °C. What will be the final temperature of the mixture when all the ice has melted? The specific heat capacity of the liquid is 2.65 J/g/°C while that of water is 4.2 J/g/°C. Specific latent heat of fusion of ice = 336 J/g.

[ICSE, 2010]
[7 °C

5. A certain quantity of ice at -10 °C is heated till it changes into steam at 100 °C. Draw a time-temperature heating curve to represent it. Label the two phase changes in the graph.
[ICSE, 2008]

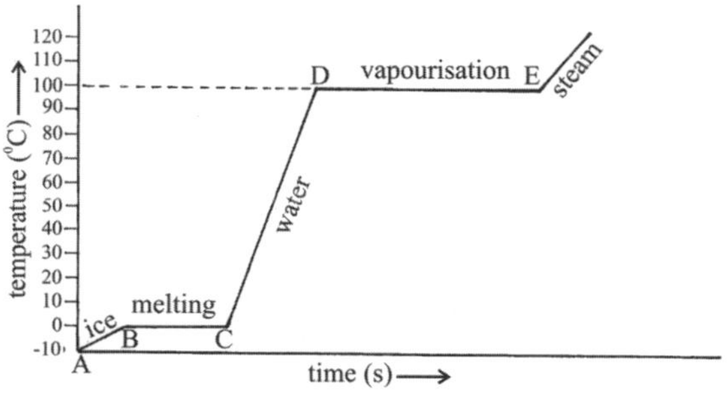

6. A piece of ice of mass 40 g is dropped into 200 g of water at 50 °C. Calculate the final temperature of water after all the ice has melted. [ICSE, 2008]
[28.33 °C

7. A vessel of mass 80 g (specific heat capacity = 0.8 J/g/°C) contains 250 g of water at 35 °C. Calculate the amount of ice at 0°C that must be added to it so that the final temperature of the mixture would be 5 °C. Assume specific latent heat of fusion of ice to be 340 J/g.

[92.57 g

8. Calculate the temperature of a furnace when 400 g of copper ball, taken out from it, melts 400 g of ice to form water at 0 °C. Specific heat capacity of copper = 0.4 J/g/°C; specific latent heat of fusion of ice = 336 J/g.

[840 °C

9. A metal ball of 0.20 kg and at a temperature of 200 °C when placed on an ice block melts 100 g of ice, when its temperature stops falling. If specific latent heat of fusion of ice is 30 J/g, calculate the specific heat capacity of metal ball.

[0.85 J/g/°C

6. Modern Physics

Unit 6: Topics covered

Modern Physics

(i) Radioactivity and changes in the nucleus; background radiation and safety precautions.

Brief introduction (qualitative only) of the nucleus, nuclear structure , atomic number (Z), mass number (A). radioactivity as spontaneous disintegration. α, β *and* γ*- their nature and properties; changes within the nucleus. One example each of* α *and* β *decay with equations showing changes in Z and A. Uses of radioactivity – radio isotopes. Harmful effects. Safety precautions. Background radiation.*

Radiation: X-ray; radioactive fallout from nuclear plants and other sources.
Nuclear Energy: working on safe disposal of waste. Safety measures to be strictly reinforced.

(ii) Nuclear fission and fusion; basic introduction and equations.

I Radioactivity

Radioactivity and changes in the nucleus; background radiation and safety precautions.
Brief introduction (qualitative only) of the nucleus, nuclear structure, atomic number (Z), mass number (A).Radioactivity as spontaneous disintegration.α, β and γ - their nature and properties; changes within the nucleus. One example each of α and β decay with equations showing changes in Z and A. Uses of radioactivity – radio isotopes. Harmful effects. Safety precautions. Background radiation.
Radiation: X-rays; radioactive fallout from nuclear plants and other sources.
Nuclear energy: working on safe disposal of waste. Safety measures to be strictly reinforced.

Thermionic Emission

This is a precursor to the discovery of alpha, beta and gamma particles leading to concepts of radioactivity and evolvement of modern physics.

Thermionic emission: definition

The process by which free electrons are emitted from the surface of a metal when external heat energy is applied is called thermionic emission.

In other words

Thermionic emission, also known as thermal electron emission or Edison effect[1], is the thermally induced flow of charge carriers from a surface or over a potential-energy barrier.

The classical example of thermionic emission is the emission of electrons from a "hot cathode" into a vacuum in a vacuum tube. The hot cathode can be a metal filament, a coated metal filament or a separate structure of metal. Vacuum emission from metals tends to become significant only for temperatures over 1,000 K (730 °C; 1,340 °F)[2]. It follows that

[1]After American inventor Thomas Alva Edison: 1847-1931.
[2]The term "thermionic emission" is now also used to refer to any thermally-excited charge emission process. This process is crucially important in the operation of a variety of electronic devices such as TVs, CROs and X-ray tubes, and can even be used for electricity generation.

the key to thermionic emission is the application of (high) heat energy to the metal for escape of free electrons from its surface. This is shown conceptually in Fig.1.1.

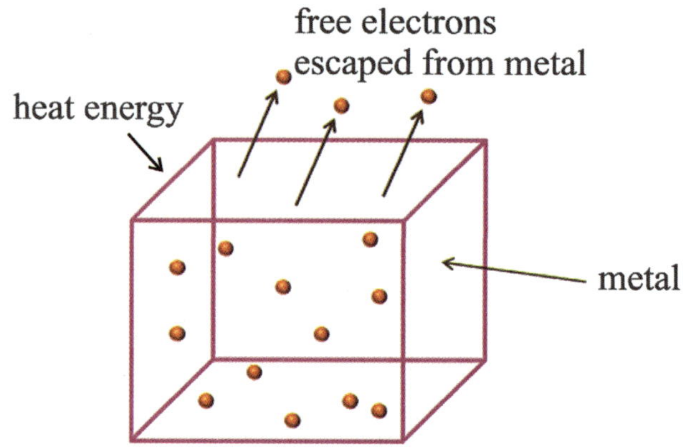

Fig.1.1 : Emission of electrons from a metal surface

Metal(s) Under the Influence of Electric Field

When a strong *positive* electric field is applied to the metal, the negatively charged free electrons in the metal experience an attractive force. In many cases, these electrons gain extra energy from the external electric field and move to the source of positive charge. This phenomenon is depicted in Fig.1.2.

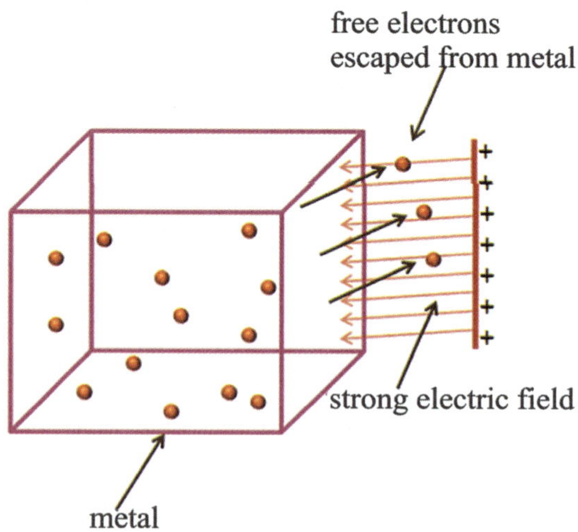

Fig.1.2 : Free electrons from a metal accelerated by a positive electric field

Cathode Ray Tube

The phenomenon of thermionic emission forms the basic requirement in various applications that incorporate a device called *cathode ray tube* or CRT.

The design features of a typical CRT are illustrated in Fig.1.3(a) and (b)[3].

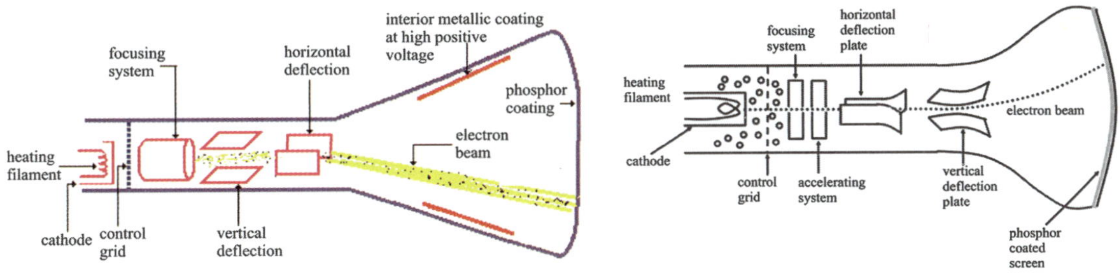

(a) (b)

Fig.1.3 : Design details of a cathode ray tube

An image of an actual CRT used in a 'conventional' TV is depicted in Fig.1.4.

A typical cathode ray tube can be divided into THREE sections:

- electron gun
- deflecting system
- screen

enclosed appropriately in a specially designed and highly evacuated glass tube. See the schematic in Fig.1.3(b).

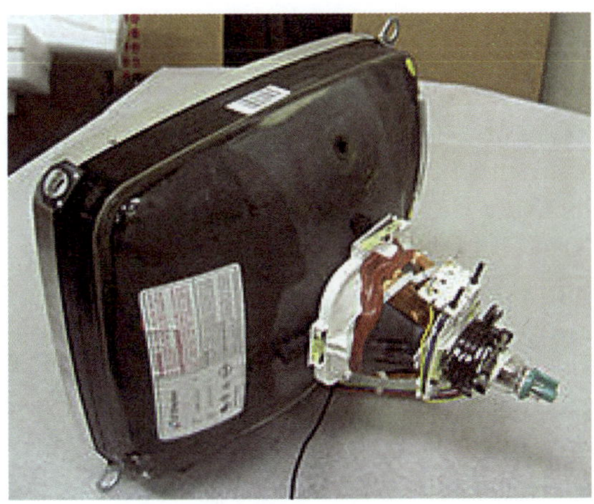

Fig.1.4 : Constructional details of a CRT used in a TV

Electron Gun

A well-designed electron gun constitutes the heart of a CRT. As shown in Fig.1.3(b), the basic component is the cathode in a tubular form, heated by a heater filament to raise the temperature of the cathode to facilitate emission of free electrons. The emitted electrons pass through a cylindrical control grid containing a diaphragm having a negative potential with respect to the cathode. Varying its

[3]See also
S.C.Bhargava: Electrical Measuring Instruments and Measurements (book), Ch. 4, B S Publishers, India, 2013.

potential varies the cathode emission current and hence the brilliance or intensity of the electron beam or spot striking the screen. The ring-shaped electrode with a hole forms the accelerating electrode to impart a high velocity to the electrons along the tube axis. There is also a focusing system or electrode in a cylindrical form whose potential can be varied to focus the electron beam to a sharp spot on the screen. This is achieved in association with the high-voltage anode system located following the focusing system which also imparts further acceleration to the electrons.

Deflecting System

After leaving the anode the focused beam of electrons passes in turn between *two* pairs of deflecting plates or electrodes as shown in Fig.1.3(b). The former causes the beam movement in the *horizontal* direction whilst the later leads the deflection of the beam in *vertical* direction when voltages are applied across the plates.
The resulting trace on the screen would thus be a function of the *relative* values of the voltage across these plates *and their relative movement with time*. For example, if a time-varying voltage is applied only across the vertical plates and no voltage across the horizontal plates, the spot of beam would simply appear to move up and down on the screen.

Screen

This is the front end of the glass tube containing all other parts as above for viewing of the electron beam movement externally. Since the electrons are invisible in themselves, the visibility of the beam is achieved by providing a very thin fluorescent coating on the *inside* surface of the screen. When the beam strikes this coating, a visible trace movement appears on the screen. The preferred colour of the trace is usually green, blue or light grey with the beam spot itself being 'suitably' brilliant.

Cathode Ray Oscilloscope or CRO

This represents a very potential application of a CRT to form a versatile and most effective visual apparatus frequently used in labs or elsewhere to observe and analyse the various unknown voltage inputs or "signals" applied across the vertical deflecting plates or electrodes. A common example is to study the waveform of alternating-current PD of a certain amplitude and frequency, or a waveform of audio output from a microphone corresponding to a given sound[4]. See Fig.1.1, Chapter I, Unit 3.

A commonly used CRO is shown in Fig.1.5.
An image showing comparison of frequencies of two alternating voltage waveforms when

---

[4]For example, if a tuning fork be placed in front of a microphone producing, an electric signal corresponding to the frequency and amplitude of vibrations of the tuning fork, the same could be seen *visually* on a CRO screen when the microphone output is fed to the vertical plates (in actual usage to the vertical amplifier) of the CRO.

connected *simultaneously* to horizontal and vertical plates (or amplifiers) of a CRO is depicted in Fig.1.6. Called a Lissajous figure[5], it shows a harmonic relationship of ONE horizontal oscillation cycle to THREE vertical oscillation cycles.

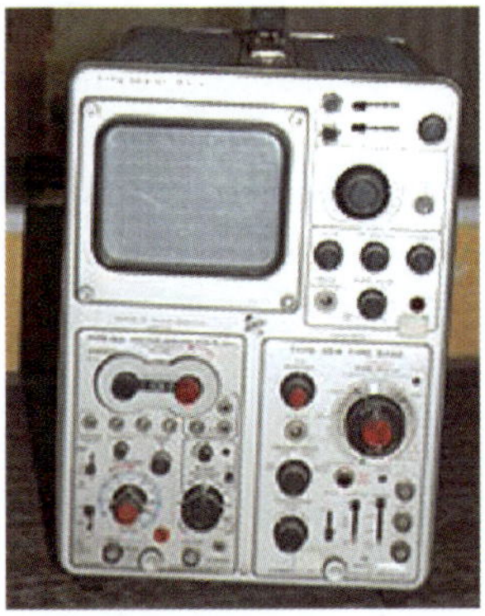

Fig.1.5 : A commonly used CRO showing various controls

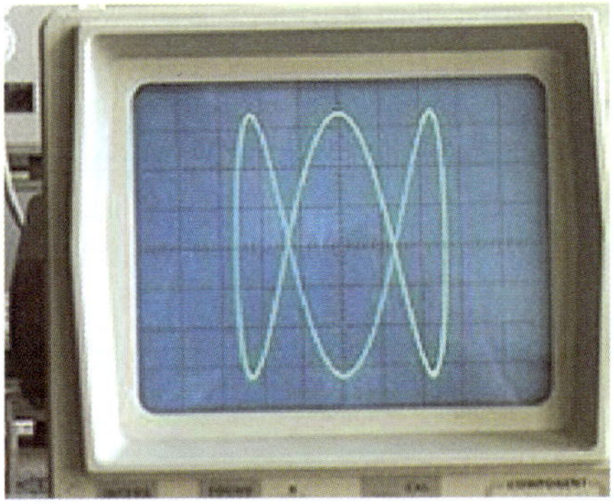

Fig.1.6 : Comparison of frequencies of two AC signals on a CRO

Atomic Structure

All matter is primarily composed of 'elements' called **atoms**. An atom is a complex arrangement of negatively charged **electrons**[6] arranged in well-defined shells (or orbits) around a centrally placed, positively charged **nucleus**. The nucleus is composed of **protons** and **neutrons**[7] and contains most of the atom's mass. All atoms are roughly the same size. A convenient unit of length for measuring atomic sizes is the angstrom (Å), which is defined as 1×10^{-10} m. The diameter of an atom is approximately 2-3 Å.

The mass of an electron is known to be 9.1×10^{-31} kg. Most properties of atoms of different elements of matter in the universe are based on the number and arrangement of their (negatively-charged) electrons in various shells around the nucleus. The number of electrons in a given shell is given by the 'index' $2n^2$ where $n = 1, 2, 3, 4, 5, 6, \ldots$ and the shells are identified as K, L, M, N, O, P, Q etc. Accordingly, the number of electrons in various shells follows the "magic numbers"

2, 8, 32, 50, 72, 98

and so on.

Nucleus Structure

The **nucleus** is the small, 'dense' region consisting of protons and neutrons at the 'centre' of an atom. After the discovery of the neutron in 1932, models for a nucleus composed of protons and neutrons were quickly developed.

---

[5]First studied by French mathematician Jules-Antoine Lissajous, *circa* 1857.

[6]Discovered by J. J. Thompson in 1897.

[7]Except for common hydrogen that has only one proton.

Almost all of the mass of an atom is located in the nucleus, with a very small contribution from the electron 'cloud' or the electrons in various shells. Protons and neutrons are bound together to form a nucleus by the nuclear force. A proton has a mass of 1.67×10^{-27} kg.

The diameter of the nucleus is in the range of 1.75×10^{-15} m for hydrogen (the diameter of a single proton) to about $15\text{x}10^{-15}$ m for the heaviest atoms, such as uranium, being much smaller than the diameter of the atom itself. The proton in the nucleus carries a *positive* charge of about $1.6\text{x}10^{-19}$ C (+e) whilst an electron possesses an opposite or *negative* charge equal to $-1.6\text{x}10^{-19}$ C (-e). The neutrons are electrically neutral, carrying no charge; the mass of each neutron being equal to that of a proton[8].

Atomic and mass number of an atom

Atomic Number, Z

The number of protons in the nucleus of an atom determines the **atomic number** of an element. In other words, each element has a unique number that identifies how many protons are in one atom of that element.

Examples

A hydrogen atom, and only hydrogen atom, contains only one proton and hence has an atomic number of 1. All carbon atoms, and only carbon atoms, contain six protons and hence have an atomic number of 6. Oxygen atoms contain 8 protons and have an atomic number of 8.
The atomic number of an element never changes, meaning that the number of protons in the nucleus of every atom in an element is always the same. The atomic number is denoted by the letter Z.

Mass Number, A

As stated above, the main constituents of a nucleus are protons and neutrons, often called nucleons. The total number of nucleons, or sum total of number of protons and neutrons of a atom, is called its **mass number** and is denoted by the letter A. The concept of mass number of an atom is represented pictorially in Fig.1.7.

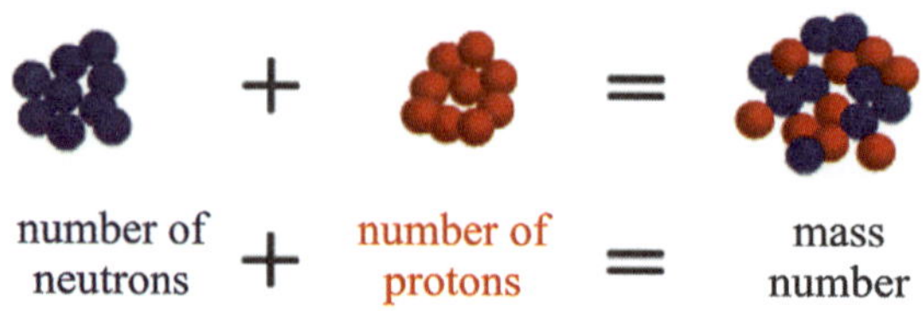

Fig.1.7 : The mass number of an atom

[8]Note that an electron has negligible mass. Hence, mass of an *atom* is the same as that of its nucleus, that is, sum of the mass of protons and neutrons.

Typical "models" of a representative atomic symbol, element X, and those of helium and carbon are depicted in Fig.1.8.

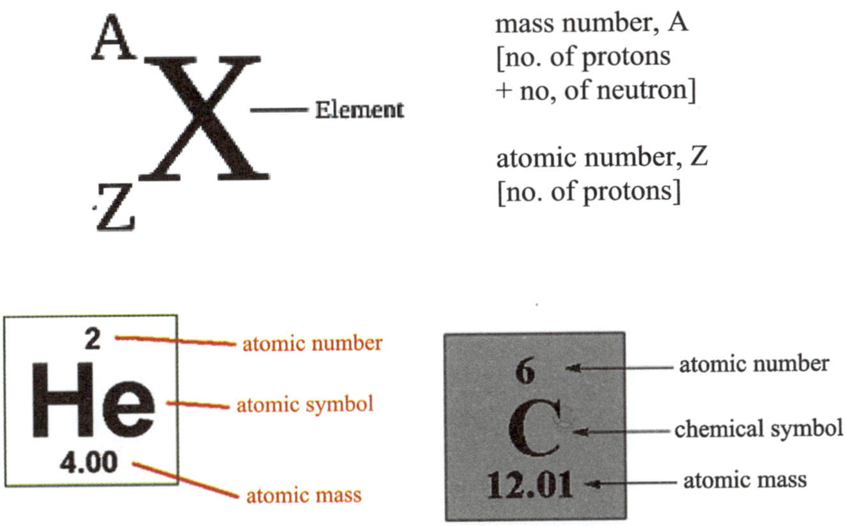

Fig.1.8 : Atomic symbols of an element "X" and that of helium and carbon

Radioactivity[9]

Radioactive decay (also known as **nuclear decay** or simply **radioactivity**) is the process by which the nucleus of an *unstable* atom loses energy by emitting radiation. A material that spontaneously emits such radiation is considered **radioactive**.

In 1896 Henri Becquerel was using naturally fluorescent minerals to study the properties of X-rays which had been discovered in 1895 by Wilhelm Roentgen[10]. He exposed potassium sulphate to sunlight and then placed it on photographic plates wrapped in black paper, believing that the uranium absorbed the sun's energy and then emitted it as X-rays. This hypothesis was disproved on the 26th-27th of February, when his experiment "failed" because it was overcast in Paris. For some reason, Becquerel decided to develop his photographic plates anyway. To his surprise, the images were strong and clear, proving that the uranium emitted radiation without an external source of energy such as the Sun.
Becquerel had discovered ***radioactivity.***
The term radioactivity was actually coined by Marie Curie[11], who together with her husband Pierre, began investigating the phenomenon discovered by Becquerel.

---

[9]Discovered by Henry Becquerel in 1896. Antoine Henri Becquerel (15 December 1852 – 25 August 1908) was a French physicist and a Nobel laureate.
[10]Wilhelm Rontgen, German mathematical engineer: 1845-1923.
[11]Marie Curie, Polish, but naturalized-French physicist:1867-1934.

Note that radioactive decay occurs only in *unstable* atomic nuclei, that is, ones that do not have enough binding energy to hold the nucleus together *due to an excess of either protons or neutrons*[12].

The decay results in the emission of three main forms identified as

- **alpha (α)**
- **beta (β)**
- **gamma (γ)**

radiations[13].

Of the three, alpha and beta are only *particles* whereas gamma is a kind of *radiation*.

Alpha Decay

An **alpha** particle is identical to a helium nucleus, being made up of two protons and two neutrons bound together.
It initially escapes from the nucleus of its parent atom and is repelled from it by electromagnetism as both the alpha particle and the nucleus are positively charged.
The process changes the original atom from which the alpha particle is emitted into a different element.
Its mass number decreases by four and its atomic number by two; for example, uranium-238 will decay to thorium-234.

Beta Decay

Beta decay itself comes in two forms: β^+ and β^-.
β^- emission occurs by the transformation of one of the nucleus's neutrons into a proton, an electron and an antineutrino. Byproducts of fission from nuclear reactors often undergo β^- decay as they are likely to have an excess of neutrons.
β^+ decays is a similar process, but involves a proton changing into a neutron, a positron and a neutrino.

Gamma Decay

After a nucleus undergoes alpha or beta decay, it is often left in an excited state with excess energy.
Just as an electron can move to a lower energy state by emitting a photon somewhere in the ultraviolet to infrared range, an atomic nucleus loses energy by emitting a gamma ray.

---

[12]Ernest Rutherford, who did many experiments studying the properties of radioactive decay, named these alpha, beta, and gamma particles, and classified them by their ability to penetrate matter.
[13]Named after the three letters of Greek alphabet.

Gamma radiation is the most penetrating of the three, and will travel through, say a plate of lead, several cm thick.

Beta particles can be absorbed by an aluminium sheet, of thickness of a few mm whilst alpha particles will be stopped in their tracks by a column of air of a few mm or a sheet of paper. This type of radiation does the most damage to materials it hits.

The property of absorption of the three particles is illustrated pictorially in Fig.1.9.

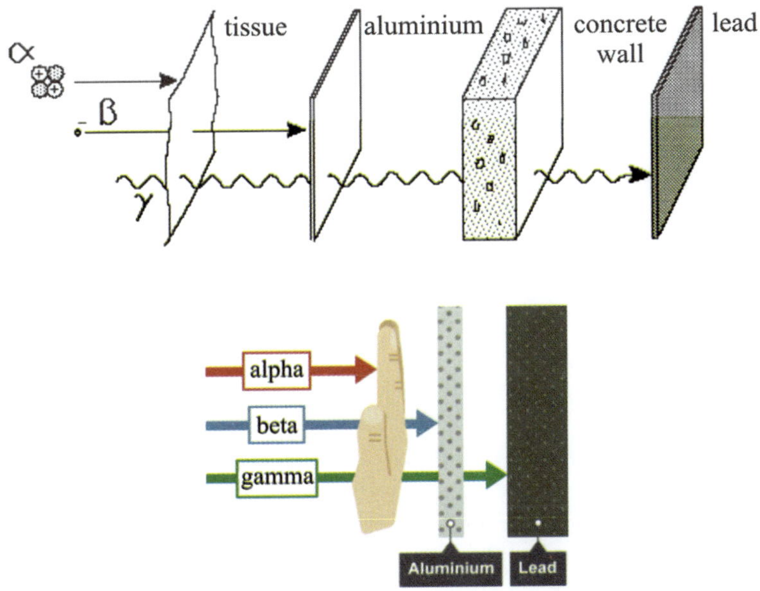

Fig.1.9 : Absorption of alpha, beta and gamma particles

Deflection of Particles

The three particles possess an important property of getting deflected in magnetic or electric field as indicated in Fig.1.10.

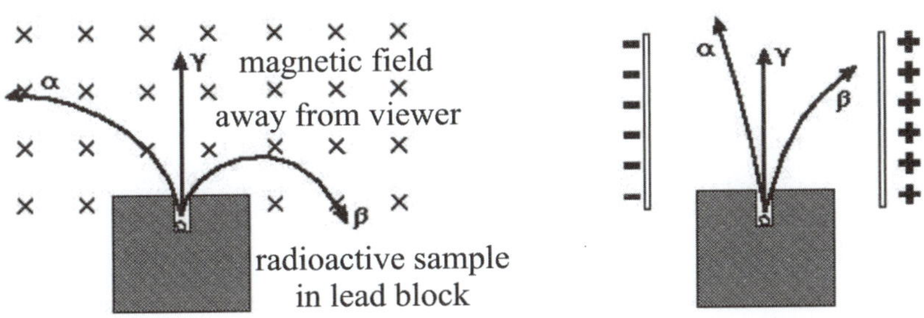

(a) deflection in magnetic field

(b) deflection in electric field

Fig.1.10 : Deflection of alpha, beta and gamma particles

Properties of alpha, beta and gamma particles

Alpha Particles

- Alpha rays or alpha particles are the positively charged particles and hence can be deflected in a magnetic or electric field as shown in Fig.1.10.
- An alpha particle comprises two protons and two neutrons, such as a highly energetic helium nucleus.
- Alpha particles have the least penetration power, but the greatest ionisation power.
- The mass of an alpha particle is nearly four times that of a proton and its charge twice that of the proton.
- The speed of motion of alpha particles is typically 10^{-7} m/s; different for different radioactive substances.
- The particles can cause fluorescence on striking a fluorescent material; likewise they can affect a photographic plate.
- They possess large kinetic energy and momentum owing to the finite mass, but lose the energy rapidly as they move away from the source, see Fig.1.9.

Beta Particles

- Beta particles are highly-energetic *electrons* as released from inside of a nucleus.
- They are negatively charged, having negligible mass, being about $9.1\text{x}10^{-31}$ kg.
- Being charged particles, these are also deflected in a magnetic or electric field, the deflection being 'steeper' than alpha particles owing to their near-zero mass and in an opposite direction. See Fig.1.10(a).
- The particles have greater penetration power than alpha particles.
- Beta particles can also ionise a gas through which they pass; however, their ionising power is only about $1/100^{th}$ of that of alpha particles. These particles affect a photographic plate and cause fluorescence.

Gamma Radiation

- Gamma radiation or r*ays* belong to high-frequency end of "electromagnetic spectrum", having a wavelength of about 10^{-13} m.
- Being a form of electromagnetic rays, they possess velocity of light for propagation.
- They have the greatest power of penetration as compared to alpha or beta particles as shown in Fig.1.9.
- They are the least ionising in comparison to alpha or beta particles, but can penetrate most substances including human skin.
- Unlike alpha and beta particles, gamma rays are not deflected by electric or magnetic fields since they are not charged particles. However, they also can affect photographic plates or films and can cause fluorescence similar to alpha and beta particles.

Characteristic properties of the three radiations

Radiation → Property ↓	alpha	beta	gamma
Nature	positively charged	negatively charged	no charge
[electromagnetic radiation]			
Charge	+ 2e	-e	0
Mass	6.6466×10^{-27} kg	9.1×10^{-31} kg	0
Range	approx. 10 cm in air;	up to a few m in air;	several m in air;
Can be stopped by	1 mm of Al	1 cm of Al	1 cm of Pb
Effect of electric or magnetic field	(less) deflected	(more) deflected	not deflected
Ionising power	most	less	very feeble

Changes Within the Nucleus

■ Alpha decay

Alpha decay or radiation is a type of radioactive decay in which an atomic nucleus emits an alpha particle (helium nucleus) and thereby transforms or 'decays' into an atom with a mass number that is reduced by four and an atomic number that is reduced by two. This can be represented as follows

$$ {}^{A}_{Z}X \rightarrow {}^{A-4}_{Z-2}Y + {}^{4}_{2}He \text{ [alpha particle]} $$

where X is the "parent" nucleus of atomic number Z and mass number A undergoing decay, Y the resulting nucleus called "daughter" nucleus and He the "alpha" particle that is identical to the nucleus of a "helium-4" atom, consisting of two protons and two neutrons. A common example to illustrate the phenomenon is the emission from radioactive uranium nucleus that emits an alpha particle and a new, "daughter" nucleus identified as thorium, represented by the reaction

$$ {}^{238}_{92}U \rightarrow {}^{234}_{90}Th + {}^{4}_{2}He $$

Observe that thorium nucleus has an atomic number of 234 (238-4), 90 (92-2) protons and 144 neutrons.

■ Beta decay

In nuclear physics, beta decay (*β*-decay) is a type of radioactive decay in which a beta ray, and a respective neutrino are emitted from an atomic nucleus. By beta decay emission, a neutron is transformed into a proton by the emission of an electron, or conversely a proton is converted into a neutron by emission of a positron (positron emission).
The beta particle and its associated neutrino do not exist within the nucleus prior to beta decay, but are created in the process. The process can be represented as follows

$$^{1}_{0}N \rightarrow \; ^{1}_{1}p + \; ^{0}_{-1}e$$

[neutron] [proton] [β particle]

A continuous stream of beta particles is called a *beta ray.*
Thus, a radioactive nucleus X with a mass number A and atomic number Z when undergoing decay emits a beta particle to form a "daughter" nucleus Y with a mass number A, but atomic number Z+1, the change being expressed by

$$^{A}_{Z}X \rightarrow \; ^{A}_{Z+1}Y + \; ^{0}_{-1}e$$

[parent nucleus] [daughter nucleus] [beta particle]

Consider the examples

- $$^{234}_{90}Th \rightarrow \; ^{234}_{91}Pa + \; ^{0}_{-1}e$$

  [thorium] [protactinium] [alpha particle]

or

- $$^{14}_{6}C \rightarrow \; ^{14}_{7}N + \; ^{0}_{-1}e$$

  [radioactive carbon nucleus] [nitrogen] [alpha particle]

■ Gamma decay

After a nucleus undergoes alpha or beta decay, it is often left in an excited state with excess energy. This extra or excess energy is released in the form of an electromagnetic radiation known as gamma radiation or simply gamma rays.
Gamma rays assume no mass and no electric charge from the parent nucleus, that is, no neutrons or protons are lost or gained, meaning that the nucleus does not decay or change into a different nucleus. This means there is no change in the mass or atomic number of the nucleus following gamma radiation. The phenomenon can be represented by

$$\underset{\text{[excited parent nucleus]}}{{}^{A}_{Z}X} \rightarrow \underset{\text{[daughter nucleus]}}{{}^{A}_{Z}X} + \underset{\text{[gamma radiation]}}{\gamma}$$

The essential features of the various reactions are shown in Fig.1.11.

Decay Type	Radiation Emitted	Generic Equation	Model
Alpha decay	${}^{4}_{2}\alpha$	${}^{A}_{Z}X \longrightarrow {}^{A-4}_{Z-2}X' + {}^{4}_{2}\alpha$	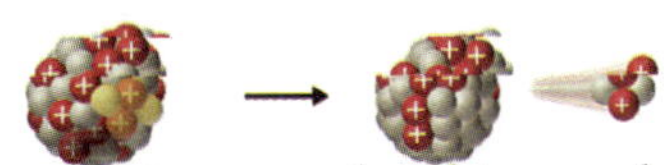 parent → daughter alpha particle
Beta decay	${}^{0}_{-1}\beta$	${}^{A}_{Z}X \longrightarrow {}^{A}_{Z+1}X' + {}^{0}_{-1}\beta$	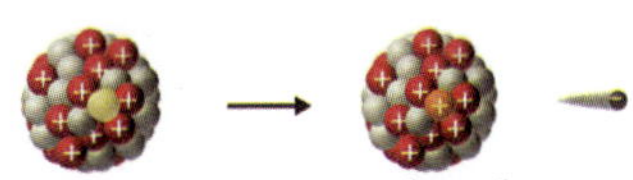 parent → daughter beta particle
Gamma emission	${}^{0}_{0}\gamma$	${}^{A}_{Z}X^{*} \xrightarrow{\text{Relaxation}} {}^{A}_{Z}X' + {}^{0}_{0}\gamma$	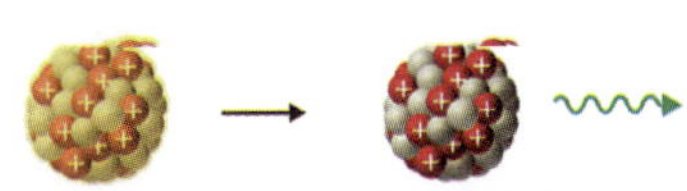 parent (excited nuclear state) → daughter gamma ray

Fig.1.11 : Alpha, beta and gamma reactions

SI Unit of Radiation

The *derived* SI unit of ionising radiation used to measure or express its dose(s) in practice is **Sievert** (symbol Sv). It is a measure of the health effect of low levels of ionizing radiation on the human body. Multiples or sub-multiples of the unit may be used to express very high or low doses of radiation.

Uses of Radioactivity

When used carefully in a directed and modulated form, radioactivity or radioactive phenomena can be utilised in numerous ways to serve the mankind. Some of these are

■ Medicine

Diagnostic

Hospitals use radiation in a wide range of ways. X-Ray, CT, and PET machines use X-ray (X-ray and CT) and Gamma radiation (PET) to produce detailed images of the human body, which provide valuable diagnostic information for doctors and their patients. Radioactive tracers and dyes are also used to be able to accurately map a specific area or system, such as in a cardiac stress test.

Radiography

Essentially the process of using high-powered X-ray machines very commonly used as cost-effective means of detecting fractures in bodies and to examine at hard-to-reach places.

Treatment(s)

Radiation therapy is extremely common and popular form of means for treating countless illnesses and diseases such as

- to stop cancer cells growth with less treatment
- for focused radiation to arrest spread of cancer
- to shrink tumors (if removing the cancer may not be possible)
- to effect low dose of particular radiation may provide improved immunity and lower stress effects
- for use of radio isotopes for radiation therapy and to keep track of activities of some organs[14]

■ Food safety

Radioactive sources are used to "sterilise" food stuff, the process being called "food irradiation". The radiation works by killing bacteria and viruses, or eliminating their ability to reproduce by severely damaging their DNA or RNA. Since neutron radiation is not used, the remaining food does not become radioactive by itself, leaving it safe to eat. This method is also used to sterilize food packaging.
The irradiation process is carried out by using radioactive isotopes. Besides preventing

---

[14]Radioactive isotopes, discussed later, have an unstable nucleus that decays or emits excess energy or radiation. Apart from being used in medicine, the isotopes find extensive applications in fields of food safety and industry etc.

illness, irradiating food can help it stay fresher longer because it gets rid of microorganisms and bacteria that cause early spoilage. The irradiated food does not come into direct contact with the radioactive isotope, so the food will not be radioactive.

To mark irradiation of a food product, an internationally recognised symbol is usually embossed on the food package(s) as depicted in Fig.1.12.

Fig.1.12 : The international symbol denoting that a food product has been irradiated

■ **Industry**

There are many applications of radioactivity or radiation in industry. Some of these are

- Weld Check

Appropriately designed X-ray machines are frequently used to examine welds for defects or irregularities, or examining other materials to locate structural anomalies or internal components. One such set-up is illustrated in Fig.1.13.

Fig.1.13 : An industrial radiography camera used to inspect a weld for defects

Industrial radiography is also very useful for secure, non-invasive scanning at security check points, such as airports, where X-ray baggage scanners are in routine use. Larger versions of the same machines are often used to examine shipping containers all over the world.

- Smoke Detectors

Alpha particles are most commonly used in smoke alarms. These alarms contain a tiny amount of a decaying radioactive element between two sheets of metal. The element emits alpha radiation. A small electric current is then passed through the sheets of metal.

When the field of alpha radiation is blocked by smoke, the alarm goes off. This alpha radiation is not harmful because it is very localised and any radiation that might escape would be stopped quickly in the air and would be extremely difficult to get into one's body.

Use of beta radiation

- Paper Industry

Beta radiation is mainly used in industrial processes such as paper mills and aluminium foil production. A beta radiation source is placed above the sheets coming out of the machines while a radiation reader is placed underneath. The purpose of this is to test the thickness of the sheets. Because the beta radiation can only partially penetrate aluminium foil, if the readings on the reader are too low, it means that the aluminium foil is too thick and the presses are adjusted to make the sheets thinner. Likewise, if the reading on the reader is too high, the presses are adjusted to make the sheets thicker.

Use of gamma radiation

- Pipe Leaks

Gamma rays are used to detect leaking pipes. In this application, a gamma ray source is placed into the substance flowing through the pipe. Then, the radiation given off is measured. The leak will be identified wherever the count on the reader spikes, indicating a large presence of gamma radiation coming out of the pipes.

Radio Isotopes[15]

Isotopes are variants of a particular chemical element which differ in neutron number. All isotopes of a given element have the same number of protons in each atom. The concept can be explained by reference to Fig.1.14.

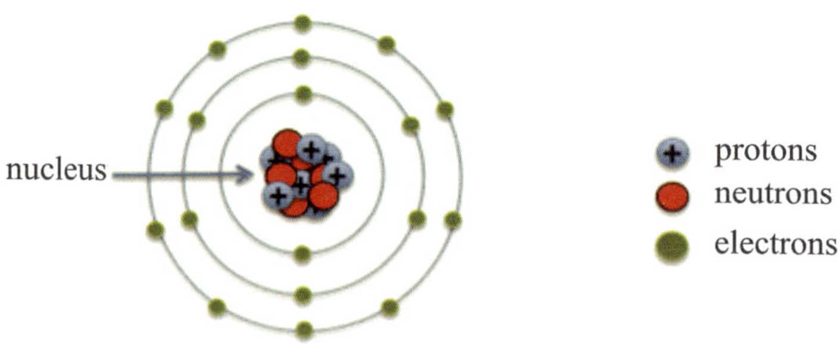

Fig.1.14 : An atom with protons, neutrons and electrons

The figure shows picture of an atom. It has a nucleus made up of protons and neutrons, with electrons around the nucleus. All the atoms of an element have the same number of protons, but the number of their neutrons can differ. If this happens, such atoms are called isotopes of

[15]The term isotope is formed from the Greek isos (σος "equal") and topos (τόπος "place"), meaning "the same place"; thus, the meaning behind the name is that different isotopes of a single element occupy the same position on the periodic table.

that element. Isotopes have the same number of protons, but they differ in the number of neutrons.

As an example, isotopes of hydrogen are shown in Fig.1.15 and how they differ on account of number of neutrons. Here, the number of protons and electrons are the same, but for the three isotopes of hydrogen, the number of neutrons differ.

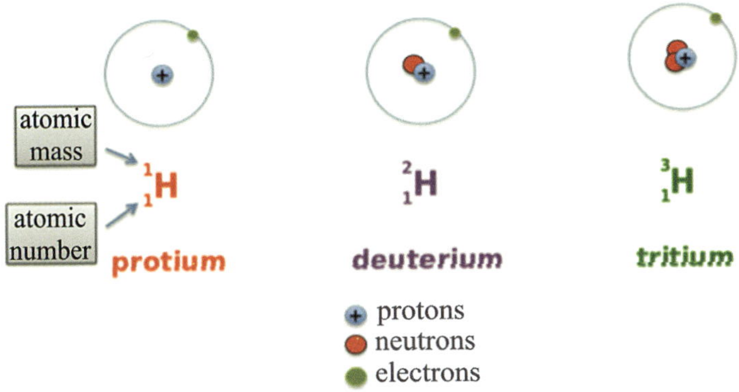

Fig.1.15 : Isotopes of hydrogen

Radioactive isotopes

If an isotope has an *unstable* nucleus that breaks down to emit radiation, then these are called **radioactive isotopes.** Thus, a radioactive isotope has an unstable nucleus that decays or emits excess energy or radiation *until the nucleus becomes stable*. Radioactive isotopes are extensively used in medical treatments. Some of the commonly used isotopes used for the purpose are listed below.

Radioactive Isotope	Applications in Medicine
Cobalt-60	Radiation therapy to prevent cancer
Iodine-131	Locate brain tumors, monitor cardiac, liver and thyroid activity
Carbon-14	Study metabolism changes for patients with diabetes, gout and anemia
Carbon-11	Tagged onto glucose to monitor organs during a PET scan
Sodium-24	Study blood circulation
Thallium-201	Determine damage in heart tissue, detection of tumors
Technetium-99m	Locate brain tumors and damaged heart cells, radiotracer in medical diagnostics (imaging of organs and blood flow studies)

Harmful Effects of Radiation

If not handled with (great) care, nuclear radiation can lead to many severe harmful effects all around. In this regard, it is important to understand the controlling factors in relation to radiation.

Controlling factors

There are *three* factors that control the amount, or dose, of radiation received from a source and resulting effects.

► Time

Reducing the time of an exposure reduces the effective dose proportionally.

► Distance

Increasing distance reduces dose following inverse square law. Controlling the distance can be as simple as handling a source with forceps rather than fingers.

► Shielding

Providing appropriately designed (biological) shields between the source of radiation and object can considerably reduce the harmful effects. The term 'biological shield' refers to a mass of absorbing material placed around a radioactive source to reduce the radiation to a safe level. Almost any material can act as a shield from gamma or X-rays if used in sufficient measure.

Harmful effects

Plant Growth

In practice, natural radiation can be beneficial to plant growth; for example, for many plants that receive some form of non-ionising radiation can help in photosynthesis to occur.

However, ionised radiation that occurs from nuclear material(s) variously may result in weakening of seeds and consequent ill effects. For instance, following Chernobyl nuclear plant incident in Russia that leaked in 1986 caused excessive amounts of radiation pollution in that region. A huge cloud of radiation was formed which resulted in a massive amount of destroyed plant life; particularly pine trees in that area. High doses of radiation can be devastating to the environment.

Humans and Animals

High doses of radiation can be devastating to the environment. The effect of radiation in the environment can be dangerous and even fatal to humans and animals. The damage it causes depends on the level of radiation and the resiliency of the organism. Radiation causes molecules to lose electrons thus destroying it. Killing certain enzymes in the body can simply make one sick. However, once radiation damages DNA the body may not be able to repair itself. This can increase the chances of both humans and animals developing cancer.

After two nuclear explosions in Hiroshima and Nagasaki, survivors experienced higher cases of cancer and child deformities. The nuclear explosions are examples of high levels of radiation. A victim of radiation fallout in Hiroshima is pictured in Fig.1.16.

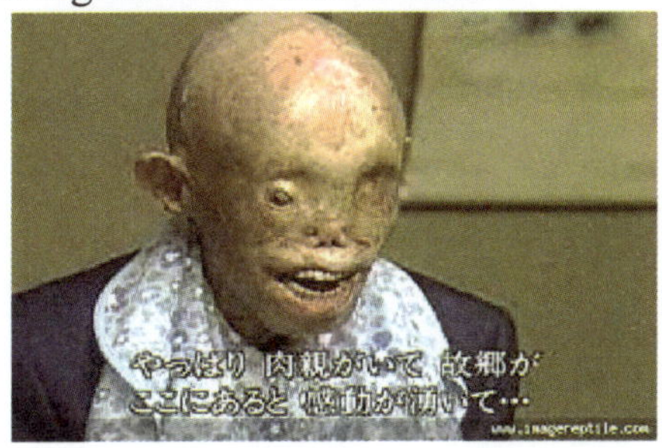

Fig.1.16 : A victim of Hiroshima atomic bomb explosion fallout

The various ways in which radiation can affect humans and animals are listed below.

Radiation in Food and Water

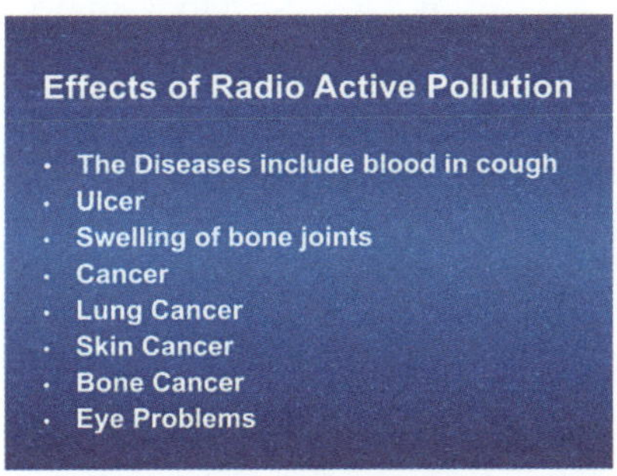

A source of concern is the prospect of contamination of their food or water caused by radioactivity. This can occur from radioactive contaminants being ingested by animals that are then used for food, such as Strontium-90 found in cows' milk after Chernobyl, or absorbed by plants through their root systems. This is how Brazil nuts become radioactive, due to their root system taking up radium from the soil.

As the cows graze in pastures exposed to radioactivity they become victim of radiation that can manifest in their milk produce (and hence in all milk products). See Fig.1.17.

Fig.1.17 : Cows grazing on grass contaminated by radioactivity

In situations where surface contamination of food is a concern, detection with handheld survey equipment can be useful. However, in cases such as contaminated meat or seafood, or contamination of water or other liquids such as milk, it generally requires the examination of samples using laboratory equipment to determine the presence and quantity of any contaminants.

Safety Precautions

It is a MUST to be *extremely careful and vigilant* whilst handling or inadvertently being exposed to radioactive substances or environment. Note that nuclear radiation is *essentially invisible*, without any colour or odor etc.

It is customary to have a warning sign displaced in the vicinity of major sources of nuclear activities or process industries, esp. near nuclear plants etc. Such an internationally approved warning sign is depicted in Fig.1.18.

Fig.1.18 : Sign to warn off against radiation

In general, the safety precautions listed below must be observed without fail when faced with possibility of nuclear radiation.

- *No radioactive substance should be handled with bare hands*. Alpha and beta emitters can be handled using thick gloves. Gamma ray emitters must be handled only by remote control, that is, by mechanical means. Whilst alpha and beta particles/radiation can cause severe burns on the skin, gamma rays are the most dangerous and over exposure can lead to serious biological damage mainly cancer etc.

- Radioactive materials **must** be stored in *thick lead containers only*.

- Reactor and laboratories dealing with, and/or conducting experiments with radioactive metals, must be surrounded with thick concrete walls lined with lead.

- People working with radioactive isotopes must wear protective clothing which should be *left in the laboratory*, and not taken away from there. The workers must be checked regularly with dosimeters (instruments to measure radiation), and appropriate measures should be taken in cases of any overdose.

- Radioactive waste must be sealed and buried deep in the ground[16].

- In nuclear power plants, there should be a back-up cooling system for cooling of reactor core to warrant against failure of the primary cooling system.

---

[16]Although this may not be the foolproof or most appropriate measure.

- The reactor building itself must be of foolproof design to withstand natural calamity such as earthquakes.

Background Radiation

Background radiation is the ionizing radiation present in the environment. Background radiation originates from a variety of sources, both natural and artificial. These sources include cosmic radiation, naturally occurring radioactive materials and fallout from nuclear weapons testing and nuclear accidents.

► Natural Background Radiation

Radioactive material is found throughout the nature. Detectable amounts occur naturally in soil, rocks, water, air and vegetation from which it is inhaled and ingested into the body. In addition to this *internal exposure*, humans also receive *external exposure* from radioactive materials that remain outside the body and from cosmic radiation from space.

- Air

The biggest source of natural background radiation is airborne radon, a radioactive gas that emanates from the ground. It can be inhaled into the lungs, along with its decay products, where they will reside for a period of time after exposure.

- Cosmic radiation

The Earth and all living things on it are constantly bombarded by radiation from outer space. This radiation primarily consists of positively charged ions, in turn interacting with atoms in the atmosphere to create an air shower of secondary radiation, including X-rays, alpha particles, electrons, and neutrons.

► Food and Water

Some of the essential elements that make up the human body, mainly potassium and carbon, have radioactive isotopes that add significantly to our background radiation dose. These may internally affect various body function in the long term.

► Artificial Background Radiation

- Medical

The global human exposure to artificial radiation comes primarily from medical imaging: X-rays for routine investigation to confirm fractures etc. or for screening of lungs and even for dental treatment, but appreciably from special diagnostic procedures such as using CAT scans.

Radiation treatment for various diseases also accounts for some dose, both in individuals and in those around them.

- Atmospheric nuclear testing

Frequent above-ground nuclear explosions, mainly carried out during 1940s and 1960s, but continue even at present, scatter a substantial amount of radioactive contamination. Some of this contamination is local, rendering the immediate surroundings highly radioactive, while some of it is carried longer distances as nuclear fallout; some of this material is dispersed worldwide.

- Nuclear accidents

Under normal circumstances, nuclear reactors release small amounts of radioactive gases which cause small radiation exposures to the public and may not be too harmful. However, in the events of accidents that may occur, for example at Chernobyl and Fukushima, enormous radioactive radiation ensue, spreading far and wide, severely contaminating the atmosphere, and affecting millions of people.

► Others

Other human contributors include smoking, air travel, radioactive building materials, unguarded and ill-planned activities in laboratories and research stations etc.

Half Life of Radioactive Materials

A very important aspect of ALL nuclear activities involving radioactive materials is what is called **"half life"** of the particular material.

The rate at which a radioactive isotope decays is measured in half-life. The term half-life is defined as **"the time it takes for *one-half* of the atoms of a radioactive material to disintegrate"**. Half-lives for various radioisotopes can range from a few microseconds to *billions* of years.

This feature of radioactive materials acquires great significance when it comes to *safe* disposal of *processed* materials.

As a comparison, half-life duration of a few representative materials is given below.

Radioactive material	Half-life
Polonium-215	0.0018 seconds
Bismuth-212	60.5 seconds

Sodium-24	15 hours
Iodine-131	8.07 days
Cobalt-60	5.26 years
Radium-226	1600 years
Uranium-238	4.5 billion years

Radiation: X-rays[17]

X-radiation comprising **X-rays** is a form of electromagnetic radiation, having a wavelength ranging from 0.01 to 10 nm, corresponding to frequencies in the range of 30 peta hertz to 30 hexa hertz (3×10^{16} Hz to 3×10^{19} Hz). The rays are invisible to eyes being much above the "light" spectrum. However, they possess the extraordinary property of penetrating a variety of matter such as wood and can affect a photographic film or plate.

Uses of X-rays in Practice

Medical Use

■ Routine X-rays

The most beneficial medical use of X-rays is in detection of fracture or misalignment of bones in body following a grave injury. This follows from the fact that whilst X-ray can pass through skin they are obstructed by bones. The process of X-raying an affected limb or body part thus consists of directing X-ray from a source to the body part and receiving the 'shadow' of bones on a photographic plate which when developed in the usual way confirms the defect of the bone(s).

A commonly used arrangement of X-raying bones of an arm is illustrated in Fi.1.19.

[17]X rays were discovered in 1895 by W. C. Roentgen who called them "X rays" because their nature was at first unknown; they are sometimes also called Roentgen, or Röntgen, rays.

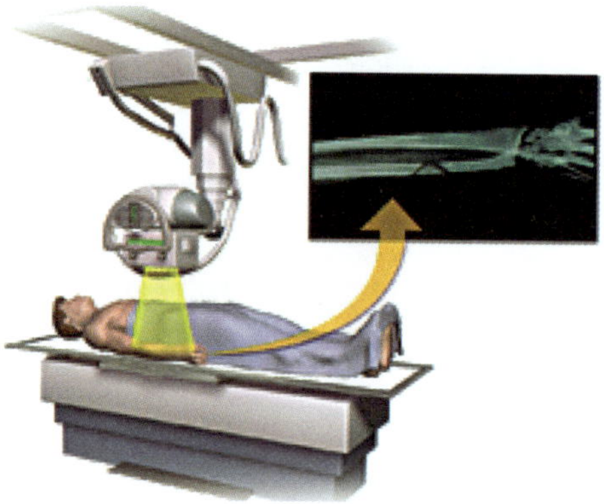

Fig.1.19 : Arrangement for X-raying a broken arm

■ Computerised Axial Tomography or CAT

In modern times, X rays are also used with computers in CAT scans to produce much improved cross-sectional images of the inside of the body parts or organs compared to routine X-rays.
A CAT scan of head showing clearly the inside of the brain is shown in Fig.1.20.

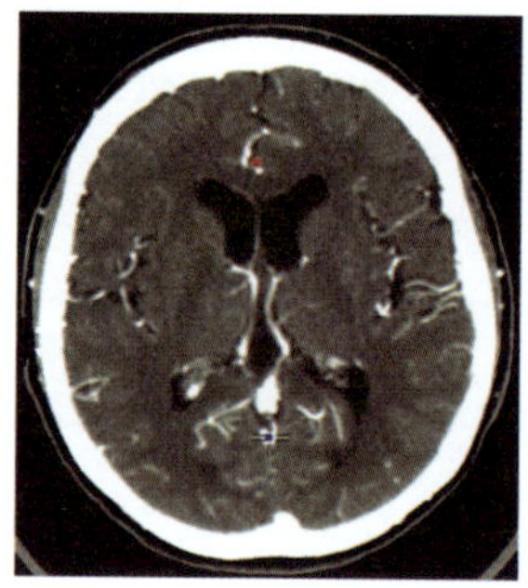

Fig.1.20 : CAT scan of brain

■ Radiography

The process of using X-rays to obtain photographic images of affected body parts is called radiography and the image or photograph itself is called a "radiograph".

Radiography can be used to identify lung diseases such as pneumonia, lung cancer, or pulmonary edema, and the abdominal X-ray which can detect bowel (or intestinal) obstruction.

Dental radiography is commonly used in the diagnoses of common oral problems, such as cavities.

■ Radiotherapy

The use of X-rays as a treatment is known as radiation or radio therapy and is largely used for the management of cancer; however, it requires higher radiation doses than those received for imaging alone.

Industrial Applications

■ Inspection of welds etc.

X-rays are used to inspect the quality of weld, for example of pipes etc., in industry, similar to fracture of bones in a body.

■ Food inspection

X-ray inspection in the food production is ever becoming an important factor in the competition among manufacturing companies.

■ Wheels and tyre inspection

Inspection of tyres of a vehicle is a common practice with radiography for the determination of defects during production processes.

Radioactive Fallout from Nuclear Plants and Other Sources

Radioactive or nuclear fallout, or simply termed as fallout, is the radioactive material or particles *propelled* into the upper atmosphere as a result of activity in a nuclear power plant or, more significantly, following a nuclear blast or a nuclear reaction *conducted in an unshielded facility*. It is so called because it "falls out" of the sky after the event.

Nuclear fallout can be typically classified in two categories:

- Global Fallout

It relates to fallout from the plant or any nuclear source/event that reaches upper atmosphere from where it may spread "globally" in course of time.

- Local Fallout

As the name implies, this type of fallout would relate to that surrounding the nuclear source, not necessarily spreading to surroundings or long distances.

Nuclear Energy: Working on Safe Disposal of Waste

Nuclear power has at least three waste streams that may impact the environment

1. Spent nuclear fuel at the reactor site (including fission products and plutonium waste)
2. Waste rock at uranium mining mills, during reactor operation
3. Releases of large quantities of dangerous radioactive materials during accidents, if ever

Most commercial nuclear power plants release gaseous and liquid radiological effluents into the environment as a by-product of the reaction.

Safe disposal[18]

All kinds of nuclear waste need to be disposed of, deep underground in 'engineered' facilities built in stable geological formations. A view of casks containing nuclear waste, buried underground is shown in Fig.1.21. Note the warning sign(s).

Fig.1.21 : Casks containing nuclear waste

In particular

- Solid waste must be sealed in a *red* plastic bag and placed in a cardboard radioactive waste box.
- Waste must be segregated by "waste type" unless otherwise permitted.
- All sharp items such as needles and broken glass etc. must be enclosed in a hard container (metal tin or plastic container) before being placed into waste boxes.
- Liquid radioactive waste may be disposed of via the sewer system. Such waste may only be disposed of via flushing sinks connected to approved radioactive drains provided for this purpose.

Safety Measures

Whilst dealing with or handling nuclear material(s) for *any* activity, particularly experiments in (research) labs, the following general precautions must be strictly enforced:

- **Try to avoid** any experiments that are traditionally performed with the help of radioactivity that can now be done using alternatives.
- **Be informed** while planning an experiment that involves the use of radioactivity about chemical form of the radioactivity, its radioactive concentration , total amount, and its chemical concentration. Order and use only as little as is really needed.
- **Familiarise yourself** with the designated working area. Perform a mental and practical 'dry' run (replacing radioactivity with a coloured solution) to make sure that all that the

[18]In India, currently about seven near-surface radioactive waste disposal facilities are operational in various parts of the country for disposal of nuclear waste that are routinely subjected to monitoring and safety performance assessment.

equipment would need is available and to get used to working behind a shield. Handle your samples as if sterility would be required to avoid contamination.

- **Always wear appropriate gloves,** lab coat, and safety goggles when handling radioactive material.
- **Check the work area** for contamination before, during and after your experiment (including your lab coat, hands, and shoes).
- **Localise your radioactivity.** Avoid formation of aerosols or contamination of large volumes of buffers.
- **Dispose of radioactive waste** only into designated, shielded containers.

II Nuclear Fission and Fusion

Nuclear fission and fusion; basic introduction and equations

NUCLEAR FISSION

The Basic Phenomenon

The word fission can be interpreted as disintegration or splitting in to fragments. Accordingly, nuclear fission would amount to splitting of an atom, or more specifically its nucleus, into various 'particles'[1].

Nuclear fission takes place when an atom's nucleus splits into two or more smaller nuclei, called" fission products", including particles such as neutrons, photons and alpha particles.

The process of nuclear fission is depicted schematically in Fig.2.1.

It is seen that the immediate action of fission of the given nucleus results into production of two sub-nuclei and *two neutrons*.

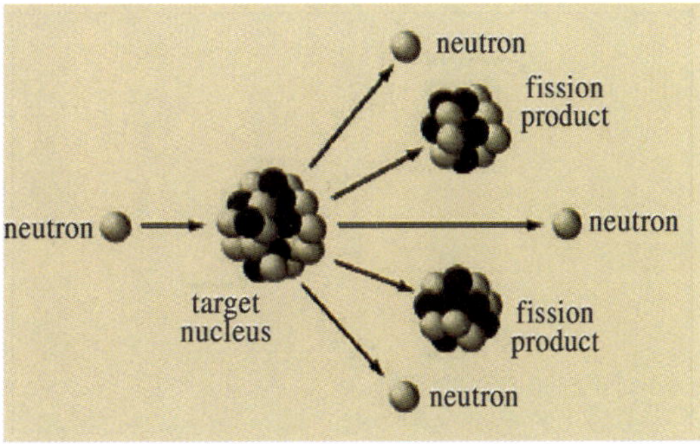

Fig.2.1 : Nuclear fission

As demonstrated by Fermi, the particle used to "bombard" the nucleus to set the phenomenon of fission in motion is a neutron. These particles are generally most effective for the purpose, that is, to initiate fission, because they have no charge and are therefore better able to penetrate more deeply into the positively charged nucleus and make it split.

[1]It was the Italian physicist Enrico Fermi [1901-1954] who in 1934 began using neutrons to lead to nucleus disintegration. Fermi is credited to have built the first nuclear reactor and is known as the "architect of the nuclear age – or even the atomic bomb"!

Fission Elements

Several heavy elements, such as uranium, thorium and plutonium can undergo fission. However, historically it is the element uranium-235 that has been found to be most suitable for nuclear fission[2]. Following its fission, the resulting 'picture' can be expressed in the form of the following equation

$$^{235}_{92}U + ^{1}_{0}n \longrightarrow ^{145}_{57}La + ^{88}_{35}Br + 3^{1}_{0}n$$

Uranium, when 'bombarded' by a neutron, splits into lanthanum[La] and bromine[Br] nuclei releasing *three* more neutrons

Example

1. An element $^{A}_{Z}S$ decays to $^{222}_{85}R$ after emitting 2 α particles and 1 β particle. Find the atomic number and atomic mass of the element S.

[ICSE, 2016]

The decay will follow the following sequence, considering the properties of α and β particles

$$^{A}_{Z}S \xrightarrow{\alpha} {}^{A-4}_{Z-2}P \xrightarrow{\alpha} {}^{A-8}_{Z-4}Q \xrightarrow{\beta} {}^{A-8}_{Z-3}R$$

Therefore, it follows that

$$Z - 3 = 85 \quad \text{or} \quad Z = 85 + 3 = 88$$

and

$$A - 8 = 222 \quad \text{or} \quad A = 222 + 8 = 230$$

Hence, the element S is expressed as

$$^{230}_{88}S$$

2. An atomic nucleus A is composed of 84 protons and 128 neutrons.

(i) The nucleus A emits an alpha particle and is transformed into nucleus B. What is the composition of nucleus B?

(ii) The nucleus B emits a beta particle and is transformed into a nucleus C. What is the composition of nucleus C?

[ICSE, 2015]

---

[2]It was in 1939 that physicists Frisch and Meitner showed that a uranium nucleus disintegrated into two relatively heavy nuclei and two neutrons.

According to the properties of alpha and beta particles
(i) the composition of nucleus B is: 82 protons and 124 neutrons
(ii) the composition of nucleus C is: 83 protons and 124 neutrons

3. A nucleus ${}^{24}_{11}\mathrm{Na}$ emits a beta particle to change into magnesium (Mg).

(i) Write the symbolic equation for the process
(ii) What are numbers 24 and 11 called

[ICSE, 2014]

(i) The equation of the process is

$${}^{24}_{11}\mathrm{Na} \longrightarrow {}^{24}_{12}\mathrm{Mg} + {}^{0}_{-1}e \text{ [beta particle]}$$

(ii) In the equation, 24 represents mass number and 11 the atomic number.

4. A radioactive nucleus undergoes a series of decays according to the sequence

$$X \xrightarrow{\beta} X_1 \xrightarrow{\alpha} X_2 \xrightarrow{\alpha} X_3$$

If the mass number and atomic number of X_3 are 172 and 69, respectively, what is the mass number and atomic number of X?

[ICSE, 2013]

Making use of the properties of α and β particles
if X_3 has atomic number of 69 and mass number 172, then working backward

$${}^{172}_{69}X_3 + {}^{4}_{2}\alpha \longrightarrow {}^{176}_{71}X_2$$

$${}^{76}_{71}X_2 + {}^{4}_{2}\alpha \longrightarrow {}^{180}_{73}X_1$$

$${}^{180}_{73}X_1 + {}^{0}_{-1}\beta \longrightarrow {}^{180}_{72}X$$

Hence X has a mass number of 180 and atomic number of 72

5. A nucleus $_{Z}^{A}X$ emits an alpha particle followed by γ emission; thereafter it emits two β particles to form X_3.

Complete the values of A and Z for X_3

$$_{Z}^{A}X \xrightarrow{-\alpha} X_1 \xrightarrow{-\gamma} X_2 \xrightarrow{-2\beta} \ldots\ldots X_3$$

[ICSE, 2009]

Applying the properties of α, β and γ particles

$$_{Z}^{A}X \xrightarrow{-\alpha} {}_{Z-2}^{A-4}X_1 \xrightarrow{-\gamma} {}_{Z-2}^{A-4}X_2 \xrightarrow{-2\beta} {}_{Z}^{A-4}X_3$$

This shows that X_3 has mass number of A-4 and atomic number of Z; that is it is an isotope of X.

6. Complete the following nuclear equations by filling the correct values in blanks

(i) $$_{92}^{238}P \xrightarrow{-\alpha} P_1 \xrightarrow{-\beta} P_2 \xrightarrow{-\beta} P_3$$

(ii) $$_{Z}^{A}X \xrightarrow{-\alpha} X_1 \xrightarrow{-2\beta} X_2 \xrightarrow{\gamma} X_3$$

(iii) $$_{80}^{290}X \xrightarrow[-1\beta]{-1\alpha} Y \xrightarrow{-2\alpha} T \xrightarrow{-2\beta} L$$

Following the properties of particles

(i) $${}^{238}_{92}P \xrightarrow{-\alpha} {}^{234}_{90}P_1 \xrightarrow{-\beta} {}^{234}_{91}P_2 \xrightarrow{-\beta} {}^{234}_{92}P_3$$

(ii) $${}^{A}_{Z}X \xrightarrow{-\alpha} {}^{A-4}_{Z-2}X_1 \xrightarrow{-2\beta} {}^{A-4}_{(Z-2)+1}X_2 \xrightarrow{\gamma} {}^{A-4}_{(Z-2)+1}X_3$$

(iii) $${}^{290}_{80}X \xrightarrow[-1\beta]{-1\alpha} {}^{286}_{79}Y \xrightarrow{-2\alpha} {}^{278}_{75}T \xrightarrow{-2\beta} {}^{278}_{76}L$$

7. The fission of ${}^{235}_{92}U$ nuclei takes place according to the reaction

$${}^{235}_{92}U + {}^{1}_{0}n \longrightarrow {}^{144}_{x}Ba + {}^{y}_{36}Kr + 3\,{}^{1}_{0}n + \text{energy}$$

What are the values of x and y?

The balancing equations are

a) $235 + 1 = 144 + y + 3x1$, giving $y = 89$
b) $92 + 0 = x + 36 + 0$, giving $x = 56$

Energy released in fission

When a nucleus fissions, it splits into several smaller fragments and two or three neutrons. The sum of the masses of these fragments is *less* than the mass of the original nucleus. The difference of the mass or 'missing' mass, that may be about 0.1 percent of the original mass, is equivalent to energy (given out by the fission) according to the equation by Einstein[3]

$$\boxed{\text{energy} = \text{mass x } c^2}$$

where c denotes the velocity of light in vacuum. If the mass is expressed in kg and velocity of light in m/s, the energy would be obtained in J.

[3]Albert Einstein, Germany-born theoretical physicist: 1879-1955.

Energy Released from Uranium Fission

Unit of energy

In nuclear phenomena, as in electronics, the unit of energy called *electron-volt*, denoted by eV, is often used. This is defined as the energy gained by charge of an electron, moving through a PD of one volt. Thus, with the charge of an electron being $1.6\text{x}10^{-19}$ C,

$$1 \text{ eV} = 1.6\text{x}10^{-19} \text{ x } 1 \text{ J} = 1.6\text{x}10^{-19} \text{ J}$$

An appropriate unit in practice is *mega-electron volt* defined as 1 million eV (or 1 MeV) so that

$$1 \text{ MeV} = 1.6 \text{ x}10^{-13} \text{ J}$$

Energy and unified atomic mass unit

This is yet another unit that may be related to mass (of an atom) since mass and energy are interchangeable as defined by Einstein. The unit is called the *unified atomic mass unit*, denoted by u, and is defined as $1/12^{th}$ of the mass of the carbon atom. It is known that $6.02\text{x}10^{23}$ atoms of carbon weigh 12 g.

$$\therefore \quad \text{mass of one atom of carbon} = \frac{12}{6.02\times10^{23}} \text{ g} = 12 \text{ u}$$

$$\therefore \quad 1 \text{ u} = 1/12\text{x } 6.02\text{x}10^{23} \text{ g}$$

or

$$= 1.66 \text{ x } 10^{-27} \text{ kg (very approx.)}$$

Note that a carbon atom is represented as $^{12}_{6}C$, with an atomic number of 12. It follows that for any other atom its unified atomic mass would be expressed by its atomic number; for example, an atom of uranium-235 has an atomic mass number of 235 u.

Unified atomic mass and energy

According to Einstein's energy equation, a *'change'* of 1 kg of "matter" would theoretically produce $1 \text{ x } (3\text{x}10^8)^2$ or $9 \text{ x } 10^{16}$ J of energy, with velocity of light, c, being $3\text{x}10^8$ m/s[4].

---

[4]As 1 kWh of energy is 1000x3600 or $3.6\text{x}10^6$ J, $9\text{x}10^{13}$ J is $2.5\text{x}10^7$ or 25 million kWh. This shows that a change in mass of just 1 g (equivalent to $3.6\text{x}10^{13}$ J) would be sufficient to keep the electric lamps in a million homes burning for about a week in winter on the basis of about seven hours' use per day!
Also, just about 0.6 grams of Uranium-235 were consumed by the atomic bomb which devastated Hiroshima in 1945.

Now since 1 u = 1.66×10^{-27} kg and 1 MeV = 1.6×10^{-13} J,

$$1\text{u} = \frac{1.66 \times 10^{-27} \times 9 \times 10^{16}}{1.6 \times 10^{-13}} \text{MeV}$$

Observe that an electron having a mass of 9.1×10^{-31} kg corresponds to about 0.5 MeV.

Natural uranium consists of uranium atoms $^{235}_{92}U$ and 140 parts by mass of uranium $^{238}_{92}U$. In a nuclear reaction with natural uranium and slow neutrons, it is usually the nucleus $^{235}_{92}U$ which is fissioned. If the resulting nuclei are lanthanum and bromine as given in the above equation, together with three neutrons, then the equation of reaction when uranium is bombarded by one neutron would be

$$^{235}_{92}U + ^{1}_{0}n \longrightarrow ^{148}_{57}La + ^{85}_{35}Br + 3\ ^{1}_{0}n$$

Now $^{235}_{92}U$ and $^{1}_{0}n$ together have a mass of (235.1 + 1.009) or 236.1 u (units).

The lanthanum, bromine and neutrons produced from the fission together have a mass

$$\begin{aligned} &= (48.9 + 84.9 + 3 \times 1.009) \\ &= 235.9 \text{ u} \end{aligned}$$

Therefore energy released

$$\begin{aligned} &= \text{mass difference} \\ &= 0.2 \text{ u} = 0.2 \times 931 \text{ MeV} \\ &= 298 \times 10^{-13} \text{ J (approx.)} \end{aligned}$$

This is the energy released *per atom* of uranium fissioned. In one kg of uranium there are about

$$\frac{1000}{235} \times 6 \times 10^{23} \text{ or } 26 \times 10^{23} \text{ atoms}$$

since the number of atoms in a mole of any element is 6.02×10^{23}. Hence, if all the atoms in one kg of uranium were fissioned, total energy released would be

$$= 26 \times 10^{23} \times 298 \times 10^{-13}$$
$$= 8 \times 10^{13} \text{ J (approx.)}$$
$$= 2 \times 10^{7} \text{ kWh}$$

which is the amount of energy given out by burning nearly 3 million tonnes of coal. Just for comparison, the equivalence of various fuels for production of energy is depicted in Fig.2.2.

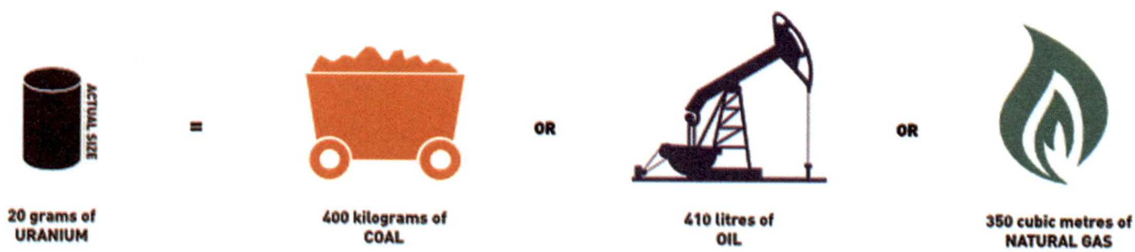

Fig.2.2 : Comparison of uranium with other kinds of fuel

Chain Reaction

So far the *theory* of fission has been confined to action of one uranium-235 atom. However, in practice it is important to consider what actually happens following fission of just one atom.

Fission in uranium-235

Uranium-235 is quite a stable atom[5].However, it was discovered that if an atom of uranium-235 is struck by a neutron, the latter initially sticks to the atom to make uranium-236. This is a very unstable isotope and decays rapidly by splitting into lighter atoms and particles. This is called *induced* fission, splitting into barium, krypton and three neutrons. This is shown schematically in Fig.2.3. Firstly, the neutron striking the uranium-235 atom to make a uranium-236 atom. Secondly, the new uranium-236 atom rapidly decays into an atom of Ba-141 (barium), an atom of Kr-92 (krypton) and three neutrons.

[5]If left by itself it has a half-life of 7.1×10^{8} years.

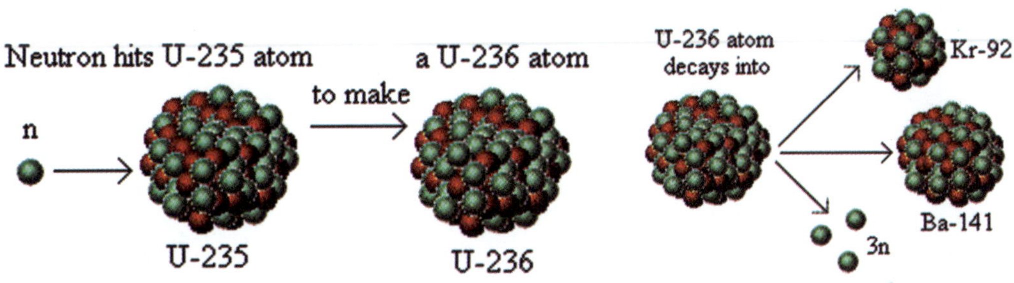

(a) uranium-235 resulting in uranium-236

(b) uranium-236 splitting into krypton, barium and neutrons

Fig.2.3 : Fission of uranium-235 and uranium-236 atom

Following the fission, all the resulting particles have kinetic energy deriving from conversion of a little of the mass of the original atom, being about 200 MeV.

The three neutron released in the first instance and having acquired the said energy hit the other uranium-235 atoms of the mass/matter contained appropriately. This results in fission of three uranium atoms, producing 3x3 or 9 more neutrons, apart from more barium and krypton elements, which again bombard nine more uranium atoms and the process continues. This is called the **chain reaction** and constitutes tremendous significance in *all* nuclear phenomena involving nuclear fission. The phenomenon of chain reaction is depicted schematically in Fig.2.4.

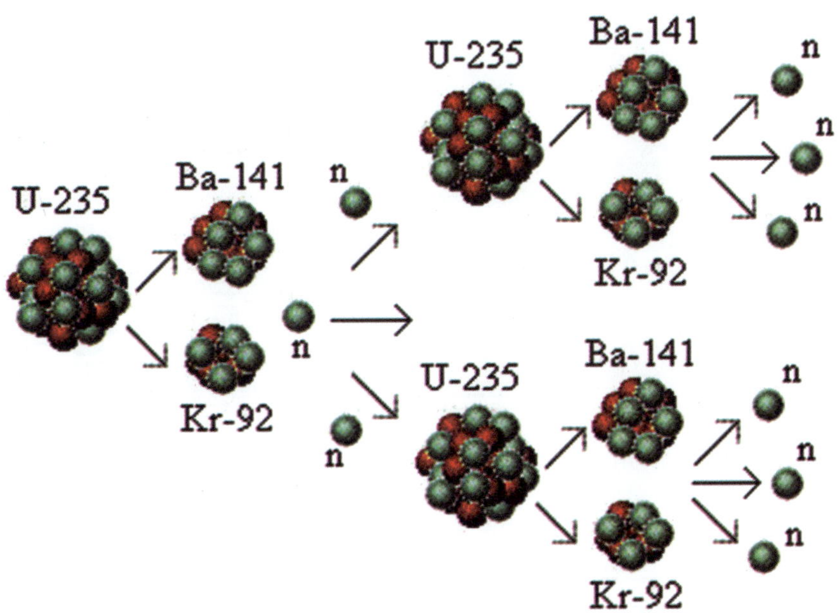

Fig. 2.4 : Chain reaction in uranium-235

Electricity Generation from Nuclear Energy

Nuclear power stations

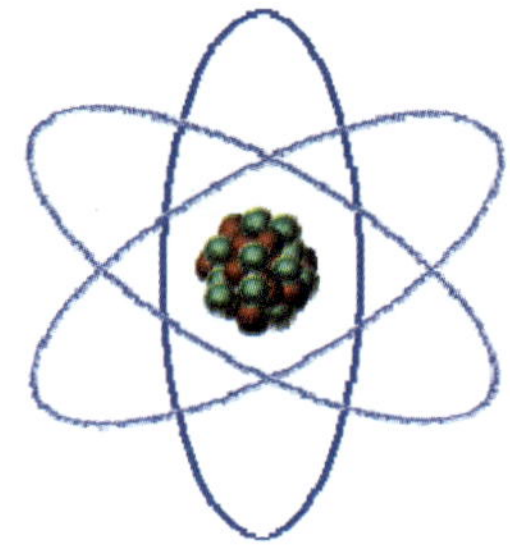

Even as the nuclear fission evolved with theoretical and experimental work using uranium-235, modern nuclear power stations generally use plutonium as the basic source of nuclear energy. This element is slightly heavier than uranium and it doesn't occur naturally in sufficient quantities. So it is made by bombarding uranium-238 with neutrons. In this way the vast bulk (99.3%) of mined uranium can still be used to provide energy. The process of turning uranium into plutonium is sometimes referred to as "breeding", and hence the modern nuclear reactors are known as "fast breeder reactors".

Principle of Nuclear Power Generation

A nuclear power station works in much the same way as any other thermal power station. The rods of fissionable material, usually plutonium, are pushed towards each other until a *controlled* amount of heat is produced. This heat is used to produce steam, which is used to drive steam turbines, as in a thermal power plant, that are coupled to generators producing electricity[6]. The basic principle of using a nuclear reactor for production of steam, and electricity, is illustrated in Fig.2.5.

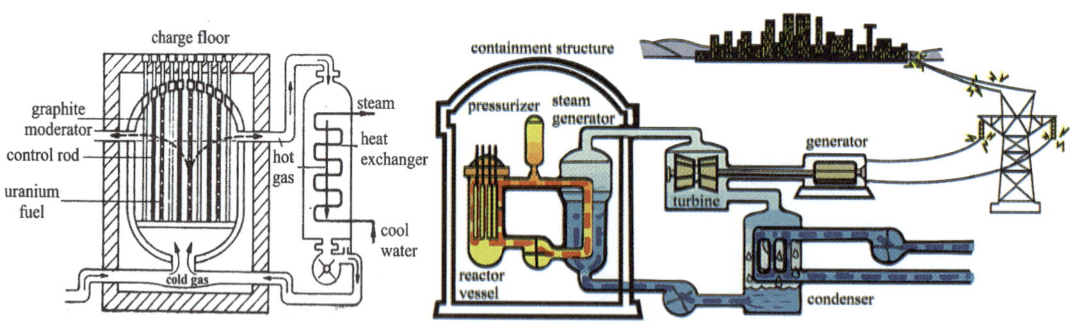

Fig.2.5 : Principle of operation of a nuclear reactor for electricity generation

The schematic of a historic nuclear power station using uranium as fuel is given in Fig.2.6.

[6]It is claimed that electricity produced using nuclear fuel would be the cheapest; in fact, during 1950s – the infancy of nuclear power generation – it was presumed that electricity so generated could be supplied free to the users! However, for various direct and indirect reasons it is far from that.

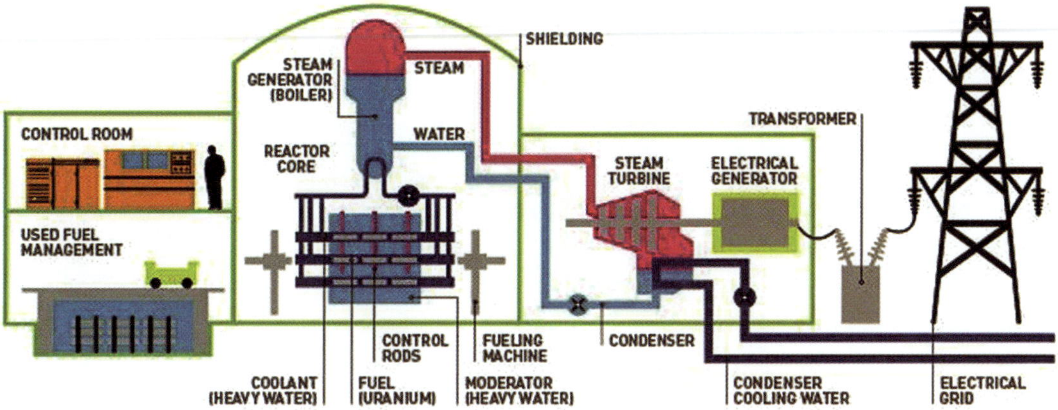

Fig.2.6 : Schematic of a 'historic' reactor and electricity generation

See also Fig.4.13, Chapter IV, Unit 1 for the latest atomic power generation in Tamil Nadu, India.

Atomic Bomb[7]

In contrast to the useful application of nuclear energy for electricity generation, a devastating use has been, and continues to be, in the production of atomic bombs.

The key to the 'working' of an atomic bomb is still nuclear chain reaction of any fissionable material. However, unlike in a reactor designed for electric power generation, the chain reaction in the bomb is ***uncontrolled*** or leading to a "meltdown". A meltdown is a condition when there is a runaway fission chain reaction. More and more neutrons are released hitting more and more U-235 atoms, producing so much heat that everything around the uranium melts. An atomic bomb can be thought of a *very* rapid melt down. The idea behind an atomic bomb is really very simple: bring lots of uranium-235 together very quickly so that it undergoes very rapid fission. This can be done in a number of ways; for example, by causing a conventional explosion around a lump of high-grade (i.e. high density) U-235. This causes an "implosion", crushing the U-235 together to the point at which fission very rapidly takes over and a huge explosion results as all the particles fly apart, together with a lot of heat and light that is also released during the fission process. It is sobering to realise that the enormous amount of explosive energy released when the first atomic bomb was dropped was produced by about the amount of uranium-235 that could be held in a coffee mug.

The second bomb, called the "fat man", that was dropped on Nagasaki in Japan that brought the Second World War to an end in 1945 is shown in Fig.2.7.

[7]It is incorrect to call it "atom bomb".

Fig.2.7 : The atomic bomb that ended the World War II

NUCLEAR FUSION

Nuclear fusion is a process in which two or more atomic nuclei are *fused* together to form heavier nuclei and subatomic particles such as neutrons and/or protons. The difference in mass between the products and reactants is manifested in the form of release of energy of gigantic magnitude. As in nuclear fission this follows from Einstein's energy equation. An important feature of nuclear fusion is the requirement of extremely high temperatures, of the order of 1.5×10^{7}°C, that can fuse the nuclei together, that is to initiate the fusion reaction[8].

Fusion reactions constitute the fundamental energy source of stars, including the Sun. The evolution of stars can be viewed as a passage through various stages of thermonuclear reactions, that is involving extreme temperatures to initiate the fusion, a requirement easily met with in stars. A main source of nearly all stars, hydrogen (H) "burning" initiates the fusion energy source of stars resulting in the formation of helium (He). Generation of fusion energy for practical use also relies on fusion reactions between the lightest elements that burn to form helium. In fact, the heavy isotopes of hydrogen, deuterium (D) and tritium (T), commonly identified as 'suitable' elements for fusion, react more efficiently with each other and, when they do undergo fusion, they yield more energy per reaction than do two hydrogen nuclei. The phenomenon of fusion is shown schematically in Fig.2.8.

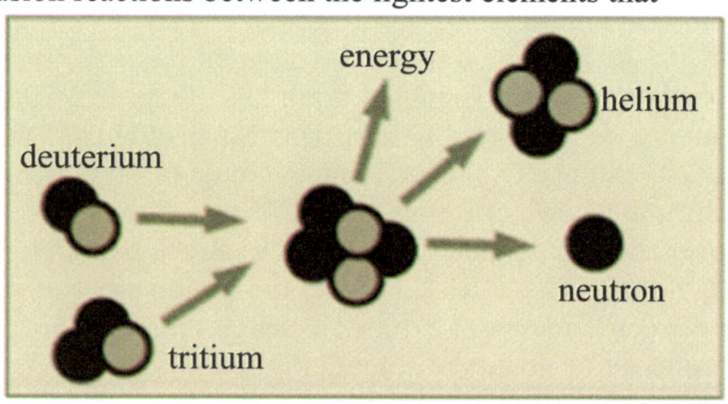

Fig.2.8 : Schematic of fusion of deuterium and tritium

[8]Till the modern times, the greatest deterrent to the useful, constructive application of nuclear fusion has been the impossibility of achieving such high temperatures in practice following established or special means.

An image of the Sun releasing tremendous energy as a result of nuclear fusion is shown in Fig.2.9. In its core, the Sun fuses 650 *million metric tonnes* of hydrogen in one second.

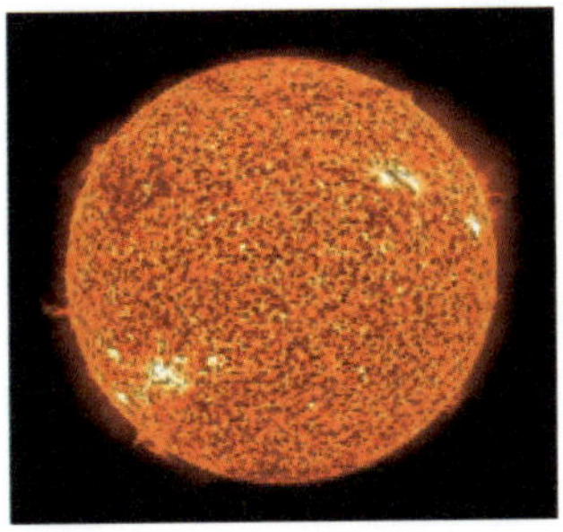

Fig.2.9 : The Sun releasing the energy for life on earth

The Equation Defining Nuclear Fusion

The equation defining nuclear fusion involving deuterium and tritium, the most commonly usable reaction on earth, can be expressed as shown in Fig.2.10.

$$ {}_1^2H + {}_1^3H \rightarrow {}_2^4He + {}_0^1n + 17.59 MeV $$

Deuterium-tritium
Fusion

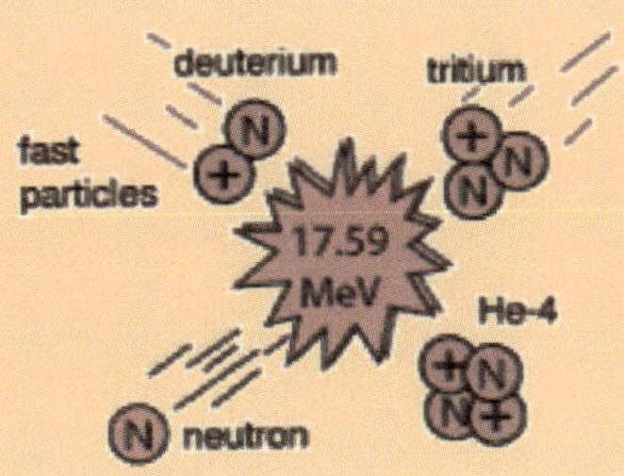

Fig.2.10 : Equation representing a nuclear fusion reaction

This reaction is more promising than the proton-proton fusion of the stars for energy sources. Note the tremendous amount of energy released due to fusion.

The Hydrogen Bomb

The atomic bomb that works on the principle of fission was found to be not 'very efficient'. It did not take scientists long to wonder if the opposite nuclear process -- fusion -- might work better. At extremely high temperatures, the nuclei of hydrogen isotopes deuterium and tritium can readily fuse, releasing enormous amounts of energy in the process. Weapons that take advantage of this process are known as **fusion bombs, thermo-nuclear bombs** or **hydrogen bombs**. Fusion bombs have higher kiloton yields and greater efficiencies than fission bombs.

How a hydrogen bomb works

The first stage of a hydrogen bomb involves a fission explosion, producing extreme heat and pressure. The explosion following fission, in turn, leads to a second stage — fusion.
A hydrogen bomb can be *700 times* more powerful than the explosion from Little Boy – the atomic bomb dropped on Hiroshima in August, 1945.

Comparison of Fission and Fusion Nuclear Reaction

	Nuclear Fission	**Nuclear Fusion**
Definition	Fission is the splitting of a large atom into two or more smaller ones. Can be controlled.	Fusion is the fusing of two or more lighter atoms into a larger one. Is un-controllable.
Natural occurrence of the process	Fission reaction does not normally occur in nature.	Fusion occurs in stars, such as the Sun.
Byproducts of the reaction	Fission produces many highly radioactive particles.	Few radioactive particles are produced by fusion reaction.
Conditions	Critical mass of the substance and high-speed neutrons are required. Can occur at room temperature.	High density, high temperature environment is required. Requires a very high temperature equal to 4×10^6 °C.
Energy requirement	Takes little energy to split two atoms in a fission reaction.	Extremely high energy is required to bring two or more protons close enough.
Energy released	The energy released by fission is a million times greater than that released in chemical reactions.	The energy released by fusion is three to four times greater than the energy released by fission.
Nuclear weapon	One class of nuclear weapon is a fission bomb, also known as an atomic bomb.	One class of nuclear weapon is the hydrogen bomb which uses a fission reaction to "trigger" a fusion reaction.
Energy production	Fission is used in nuclear power plants.	Fusion is an experimental technology for producing power.
Fuel	Uranium, and now plutonium, is the primary fuel used in power plants.	Hydrogen isotopes (Deuterium and Tritium) are the primary fuel used in experimental fusion power plants.

EXERCISES

1. Represent the change in the nucleus of a radioactive element when a β is emitted.

[ICSE, 2016]

$$\left[{}^{A}_{Z}P \longrightarrow {}^{A}_{Z+1}D + {}^{0}_{-1}e\right.$$

2. A certain nucleus X has a mass number 14 and atomic number 6. The nucleus X changes to ${}^{14}_{7}Y$ after the loss of a particle. Represent this change in the form of an equation.

[ICSE, 2012]

$$\left[{}^{14}_{6}X \longrightarrow {}^{17}_{7}Y + {}^{0}_{-1}e\right.$$

3. The nucleus ${}^{202}_{84}X$ emits an alpha particle and forms the nucleus Y. Represent this change in the form of an equation.

[ICSE, 2011]

$$\left[{}^{202}_{84}X \xrightarrow{\alpha} {}^{198}_{82}Y + {}^{4}_{2}He\right.$$

4. Complete the following nuclear changes.

(i) $${}^{24}_{11}Na \longrightarrow {}^{\ldots\ldots}_{\ldots\ldots}Mg + {}^{0}_{-1}e$$

(ii) $${}^{238}_{92}U \longrightarrow {}^{234}_{90}Th + \ldots\ldots\ldots + \text{energy}$$

[ICSE, 2010]

[(i) $^{24}_{11}Na \longrightarrow \, ^{24}_{12}Mg + \, ^{0}_{-1}e$

[(ii) $^{238}_{92}U \longrightarrow \, ^{234}_{90}Th + \, ^{4}_{2}He + \text{energy}$

5. A thorium isotope $^{232}_{90}Th$ undergoes an α decay and changes to radium. What is the atomic number and mass number of radium?

[atomic number: 88; mass number: 228

6. If the radium as above undergoes further disintegration and emits two β particles, represent the reaction in the form of an equation.

[$^{228}_{88}Ra \xrightarrow{2\beta} \, ^{228}_{90}Th$

7. In the nuclear reaction $^{27}_{12}Mg \xrightarrow{\beta} Al \xrightarrow{\gamma}$

(i) Mg emits a β particle and is transformed to aluminium. What is atomic number and mass number of aluminium?
(ii) If Al emits a γ ray, what is the resulting nucleus?

[atomic number: 13, mass number: 27; no change

8. A radioactive nucleus undergoes a series of decays according to the reaction sequence

$$X \xrightarrow{\beta} X_1 \xrightarrow{\alpha} X_2 \xrightarrow{\alpha} X_3$$

If the mass number and atomic number of X_3 are 172 and 69, respectively, what is the atomic number and mass number of X?

[mass number: 180, atomic number: 72

9. The isotope of uranium $^{238}_{92}U$ decays by alpha emission to an isotope of thorium. The thorium decays by a beta emission to protactinium. Write down the equation to represent these two nuclear changes.

$$\left[{}^{238}_{92}U \xrightarrow{\alpha} {}^{234}_{90}Th \xrightarrow{\beta} {}^{234}_{91}Pa \right.$$

Books of Reference

1. H.Cotton: APPLIED ELECTRICITY: Cleaver-Hume Press Ltd., Kensington, London WS
2. M.Nelkon and P.Parker: ADVANCED LEVEL PHYSICS: Arnold Publishers, Heinemann
   Educational Books, 22, Bedford Square, London WC1B 3HH
3. S.C.Bhargava: ELECTRICAL MEASURING INSTRUMENTS AND MEASUREMENTS
   B S Publications, Hyderabad, India

Index

C

D

M

N

O

S

S

U

V

W

X

Z

Authors / Scientists Index

NOTES